Kudankulam

Kudankulam

The Story of an Indo-Russian Nuclear Power Plant

RAMINDER KAUR

OXFORD
UNIVERSITY PRESS

Oxford University Press is a department of the University of Oxford.
It furthers the University's objective of excellence in research, scholarship,
and education by publishing worldwide. Oxford is a registered trademark of
Oxford University Press in the UK and in certain other countries.

Published in India by
Oxford University Press
22 Workspace, 2nd Floor, 1/22 Asaf Ali Road, New Delhi 110 002, India

ISBN-13 (print edition): 978-0-19-949871-0
ISBN-10 (print edition): 0-19-949871-7

ISBN-13 (eBook): 978-0-19-909997-9
ISBN-10 (eBook): 0-19-909997-9

Typeset in ScalaPro 10/13
by The Graphics Solution, New Delhi 110 092
Printed in India by Rakmo Press, New Delhi 110 020

Contents

Figures

Preface

There is something about headlands and precipices that hold the imagination captive. The encounter with raw nature at its rough edges, the prospect of falling over, being hauled into a great expanse is another sublime moment of composite awe and fear in the Kantian vein. It is true that when I first visited Kanyakumari Town in 2006 on the tip of India, I was bowled away by both the beauty and ferocity of waves crashing in from the Laccadive Sea. At the confluence of a unique *sangam* or convergence where the Arabian Sea flows into the Indian Ocean and the Bay of Bengal, it is known as Cape Comorin—a majestic spot with unobstructed sunrises arcing with sunsets over the choppy sea. I did not at that point see two nuclear reactors under construction about 30 km to the east of the tip. But they became larger in my vision the more I scrutinized and investigated, and the more the construction rose and was resisted (Figure P. 1).

This tip is also the place where many Indians bring the cremated ashes of their beloved. Carbon residues of well-known figures from the selfless to the remorseless—the non-violent freedom fighter, Mohandas Karamchand Gandhi, to the chauvinistic Shiv Sena supremo, Bal Thackeray—once wallowed in these waters. What residents worry about now is that this region and these waters could become the site for the scattering of radionuclides, isotopes, and low-level liquid waste whose dangerous half-life spans days to decades.

Figure P.1   Women in Idinthakarai Next to the Kudankulam Nuclear Power Plant, 2012.
Source: Amirtharaj Stephen/Pep Collective

Their constant emission from the plant means that they would linger well after one's own lifespan.

This book is dedicated to those who live with the prospect of elemental half-lives in their midst. The historical and ethnographic material in the book draws from an extenuated period of fieldwork mainly in Kanyakumari District in Tamil Nadu. My base was in the town of Nagercoil over a period of a year in 2006 with five return visits up to a month long, between 2012 and 2018. During these visits, I stayed in other nearby towns and villages for a few days at a time. This was on top of ongoing online communication and analyses complemented by literature and newspaper archival work on the region from the 1980s.

I am deeply grateful to all my interlocutors who made this research and book possible. Unless their statements appear in a public outlet or forum, I have anonymized their identities or used pseudonyms so as they are not identifiable. In view of the highly charged and sensitive nature of the subject, on occasion I have also occluded other details about them where it may compromise an anonymized person's identity.

Still, I would be delighted if some of them read this book. For this reason, I have written it for a scholarly as well as a more general audience. With a life-time of academic training, it is of course difficult not to cite disciplinary godmothers and godfathers and their offspring. If an obstacle presents itself, skipping the paragraph(s) is always an option. For those more theoretically inclined, further thoughts on the issues raised are in references in the endnotes. For ease of reading, any vernacular terms cited in the text are transliterated in the anglicized version rather than presented with diacritics.

Papers that informed this book on Kudankulam (also spelled Koodankulam) have been presented at various seminars, colloquia, workshops, and conferences including at the Universities of Sussex, Manchester, Heidelberg, Mumbai, Göttingen, Yale, Sydney, King's College, Jawaharlal Nehru University, and New York University. My thanks to all those who have supported and/or commented on the writing of the research in its various forms including the anonymous book reviewers. The book is the 18-year outcome of earlier work supported by a Simon-Marks Fellowship at the University of Manchester, and funded by a British Academy pilot project grant (2002), an Economic and Social Research Council research grant (RES-000-23-1312, 2006–2008), and an Arts and Humanities Research Council research leave (AH/HOO/3304/1, 2010–2011) at the University of Sussex.

Last but not least, I would like to thank from the depths of my heart my dear family and friends. Chief among them are two little cherubs who might one day read this book, realize what I get up to on the laptop, and carry the light forward.

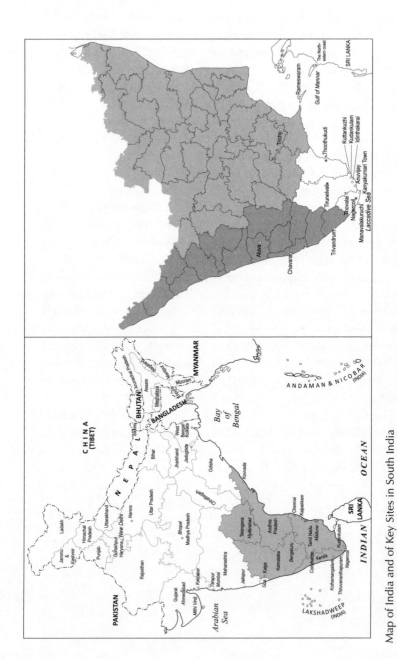

Map of India and of Key Sites in South India

Note: This map does not represent the authentic national and international boundaries of India. This map is not to scale and is provided for illustrative purposes only.

Abbreviations

AAP   Aam Aadmi Party
AERB   Atomic Energy Regulatory Board
AIADMK   All India Anna Dravida Munnetra Kazhagam
BARC   Bhabha Atomic Research Centre
BHAVINI   Bhartiya Nabhikiya Vidyut Nigam Limited
BJP   Bharatiya Janata Party
BRICS   Brazil, Russia, India, China, South Africa
CACC   Campaign against Criminalising Communities
CASA   Church's Auxiliary for Social Action
CFL   compact fluorescent lamps
CIA   Central Intelligence Agency
CID   Criminal Investigation Department
CP   Cerebral Palsy
DAE   Department of Atomic Energy
DMK   Dravida Munnetra Kazhagam
EDF   Électricité de France
EIA   Environmental Impact Assessment
EMI   Electro-Magnetic Interference
FCRA   Foreign Contributions Regulation Act
FDI   Foreign Direct Investment
FIR   First Information Report
IAEA   International Atomic Energy Agency

IB	Intelligence Bureau
ICRP	International Commission for Radiological Protection
IGCAR	Indira Gandhi Centre for Atomic Research
INTACH	Indian National Trust for Art and Cultural Heritage
IREL	Indian Rare Earths Limited
JOAR	Jharkhandi Organization Against Radiation
KKNPP	Koodankulam Nuclear Power Plant
KNPBP	Kalankulam Nuclear Power and Bomb Project
LTTE	Liberation Tigers of Tamil Eelam
MAPS	Madras Atomic Power Station
MDMK	Marumalarchi Dravida Munnetra Kazhagam
MISA	Maintenance of Internal Security Act
MR	mental retardation
NAAM	National Alliance of Anti-nuclear Movements
NAPM	National Alliance of People's Movements
NBA	Narmada Bachao Andolan
NEERI	National Environmental Engineering Research Institute
NGO	Non-Governmental Organization
NIMBY	Not in My Backyard
NIMHANS	National Institute of Mental Health and Neuro Sciences
NPCIL	Nuclear Power Corporation of India Limited
NSG	Nuclear Suppliers Group
PMANE	People's Movement Against Nuclear Energy
POSCO	Pohang Iron and Steel Company
RCC	Regional Cancer Centre
RSS	Rashtriya Swayamsevak Sangh
SACCER	South Asian Community Centre for Education and Research
SARCOP	Safety Review Committee for Operating Plants
SER	Site Evaluation Report
SWAN	South West Against Nuclear
TLD	Thermo-luminescent Dosimeter
UNEP	United Nations Environment Programme
UNSCEAR	United Nations Scientific Committee on the Effect of Atomic Radiation

1

Radiation Burdens

In the searing hot month of September 2012, women from a south Indian fishing village, Idinthakarai, prepared to advance to the compound walls of a nuclear power plant, a prohibited zone located about a kilometre away from the nearest home. They announced in a digital notice:

> We are women who live very close to the sea in the village of Idinthakarai.... Tomorrow after many years, we are going to lock and bolt all the doors and gate of our house and get out ... we will move en masse to the KKNPP [Kudankulam Nuclear Power Plant]. We will carry our children including babies. Our pregnant daughters will walk in the hot sun with us.... Our plea and demands have been side-lined.... We understand that we, our bodies, the food we eat, our animals and plants, the water we drink will soon be contaminated. Can you say it will not be? We know enough to say that you do not need a radiation Chernobyl/Fukushima style to carry radiation burdens.[1]

The radiation burdens they worried about were essentially of five overlapping kinds. First, and most obviously, the burden was *biomedical* in terms of the long-term somatic effects of living near nuclear reactors in the form of bone marrow cancer, infertility, and genetic mutations of their progeny among other health risks that would escalate should there be a nuclear accident. Second, it was *ecological* in terms of worries about radionuclide emissions into the environment and the release of hot water into the sea that would affect its biodiversity.[2] Third, it was *economic* in terms of the plant's effects on their livelihoods where fishing grounds were cordoned off and sea life would be affected by higher temperatures and perceived and actual contamination.[3] Fourth, it was *socio-psychological* in terms of people's opinions, anxieties, and social shame circulating around those living close to the nuclear plant that could mar their current status and imagined futures especially with regards to marriage prospects of the young. Fifth, it was what could be demarcated as *political* in terms of the authoritative mechanisms and severe policing and surveillance that they were subjected to, located as they were next to a high-security nuclear establishment.[4]

Inspired by the civil disobedience movement of the Indian freedom fighter, Mohandas Karamchand Gandhi, and after a series of other campaigns, they resolved to confront the concrete source of their burdens. The Madras High Court had dismissed a public interest petition for environmental clearance of the two reactors, and the Atomic Energy Regulatory Board had given a highly questionable clearance to the loading of fuel rods into one of the units.[5] In a show of defiance, more than five thousand people including inhabitants of neighbouring villages ended up camping outside the eastern walls of the Kudankulam Nuclear Power Plant on 9 September 2012.[6] The majority were women and children who came together to defy a prohibitory order to assemble under Section 144 of the Code of Criminal Procedure (1973) that had been imposed on the area six months earlier. Incredibly apprehensive as to how the authorities might react to their assembly at a time when it had become outlawed, they gathered with a collective resolve and commitment to camp outside the plant to register their peaceful dissent.

The following day, after a ten-minute warning from district and police officials, about two thousand police and paramilitary descended upon them with batons (*lathi*) and tear gas, pushing them away from the walls and into the sea. Several were harassed and injured in the onslaught. Children and pregnant women were not left unscathed, and even a woman physically challenged by polio was chased and molested by a Rapid Action Force officer, a specialized wing of the Indian Central Reserve Police Force. Around two hundred people of all ages from a 15-year-old boy to a semi-blind septuagenarian were arrested under charges to do with 'sedition' and 'war against the state' (Figure 1.1).

Figure 1.1   Paramilitary Descend upon Poeple Engaged in Non-violent Civil Disobedience Outside the Compound Walls of the Kudankulam Nuclear Power Plant, 2012.
Source: Amirtharaj Stephen/Pep Collective.

This onslaught was to add to the earlier paramilitary blockade in March 2012 against essential supplies of water, milk, food, and fuel to Idinthakarai village where many of the campaigners lived.[7] This was swiftly followed by the arrests and death of people involved in the struggle against the nuclear plant. Anthony John was shot by the police in a nearby port the day after the March siege, when police fired on a solidarity protest in Thoothukudi. The others were killed through state neglect when they were arrested and imprisoned, and through intimidating actions that led to accidental deaths.[8] Sahayam Francis was one of them: a father of three young children who fell from a height after an Indian Coast Guard surveillance plane dived within a few meters above peoples' heads (Figure 1.2). He was watching those engaged in civil disobedience who were standing in a chain among the waves and on the beach for a *jal satyagraha,* Literally meaning 'water-based truth-force', thousands of people engaged in an oceanic demonstration. It was adapted from Gandhian tactics of non-violence against a colonial state to evoke the greater truth embodied in people and the environment.

Figure 1.2   Indian Coast Guard Plane Flies over *Jal Satyagraha* (Water-based 'Truth-force' Struggle),Idinthakarai,2012.
Source: Amirtharaj Stephen/Pep Collective.

While this is not the first time a movement has emerged against the nuclear industries in India, it is one that sporadically grew in magnitude such that it drew widespread debate on an issue that is ordinarily shunted away under the blanket of national nuclear security. Concerns about the environment and democratic and human rights abuses lifted the anti-Kudankulam Nuclear Power Plant struggle to national and global significance as solidarity was expressed across the subcontinent and as far afield as Japan, Taiwan, Indonesia, Britain, France, Germany, Canada, and Australia aided by social media. With a touch of irony, the US-based Nuclear Information and Resource Service stated in 2013 that 'the largest, most important anti-nuclear protest you don't know about is in southern India'.[9] This book is written with a view to rise to that challenge.

People Power vs Nuclear Power

The case of the Indo–Russian collaboration for the Kudankulam Nuclear Power Plant has become a significant test case for a struggle over national development and the 'Indian people'—a term that is specific as it is loose, and empty as much as it is politically potent with its potential to be operationalized in a variety of ways.[10] While state functionaries and their allies promote the nuclear development as one *for the people* of the nation, they are wont to dismiss any resistance to it as anti-people, anti-national, anti-development, and as a front for foreign funders. Those who can see through such mechanisms of dismissal, identify the struggle as one *by the people*, and thus with a greater claim to decide their futures. This is particularly the case for those project-affected people who have hard-line experiences and grievances on how development and democracy is pursued that contrasts markedly with their ideals of inclusion, transparency, accountability, popular sovereignty, and 'good governance'. Discursively produced and consolidated through a collective voice against monolithic operations, people of the sea and soil stood up against a state–corporate–industrial–military complex abbreviated here as the *nuclear state*.[11] The authority of demos against the control of its elected representatives and allied forces, and more specifically, the power of people's numbers levellled against the power of nuclear industries: people power against nuclear power.[12]

As with other terms in the debate, development is an acutely contested and ambivalent practice. On the face of it, a metanarrative of modernity that promises progress and economic growth has, since at least the 1980s, been viewed with a more critical lens: one that highlights the darker side of development with its disempowering effects on people that it is purported to aid.[13] Among a wide array of studies and approaches, others have highlighted the way power is concealed in the instrumental practices of development posing as anti- or non-political.[14] Nor is development simply about an institutional or Western regime of control over other regions of the world. It might be internalized as a desired practice and become an essential part of the 'postcolonial predicament'.[15]

With respect to postcolonial state formation, development has an intricate relationship with narratives of democracy.[16] The Independence of India in 1947 has come with two main compulsions: emancipation and equality for all after a history of slavery to the 'yoke of colonialism'; and to make advances on the techno-scientific, infrastructural, and industrial fronts comparable to countries in the global north.[17] Crystallized in terms such as the post-independent developmental state, democracy is deeply entangled in development discourse as integral expressions of a denied and incomplete project of modernity. Chief among this, as the political theorist Itty Abraham has shown, are nuclear investments and imaginaries about a future India that combine development, sovereignty, and security in one.[18] To disturb this paradigm is to shake the very foundations of modern India with punitive comeback as the people around the Kudankulam Nuclear Power Plant found to their detriment. Democratizing and modernizing impulses have come with a chequered record against the populace despite claims to the contrary as 'the world's largest democracy'.[19]

Democratic Development?

This book provides another lens on assumptions that democratization processes necessarily lead to the widening of public engagement and consultation on matters that affect people's lives.[20] Instead, while civic avenues for redress might be possible, a host of mechanisms are deployed to ensure that these routes are shut down such that any dissenting views and opinions are paid short shrift. With the intensification of neo-liberal transactions and the paramilitarization of law enforcement agencies, we could even assert that the velvet language of democracy becomes the smooth coating for the iron fist of development. This velvet-iron Janus means that global and national powers can continue with the invisible hand of business-as-usual while turning a blind eye to the impact of their decisions on local lives.[21]

Based on her overview, anthropologist Judith Paley suggests 'that democracy is not a single condition that countries do or do not have, but rather a set of processes unevenly enacted over time'.[22] Yet the understanding of democracy in the subcontinent, founded as it is on basic principles such as freedom, justice, the free participation of citizens, human rights, and the denied goods of modernization, is resolutely tied with a narrative of compensation for colonialism.[23] Paley goes on to point out how 'democracy has been an aspiration for many who have lived within oppressive regimes'.[24] This is true of India, then as it is now—formally a representative democracy, yet substantively and effectively, far from it.[25] As we shall see in Chapter 3, the test for many social movements is to marry actuality with formality in a constant replenishment of the socio-political order with their demands to follow due process and justice.

Of course, all democracies have shortcomings with respect to the ideals enshrined in metanarratives of the Enlightenment era. Instead, democratizing principles inform permanent projects for consolidation and quality.[26] Another large democracy often cited as among the oldest is the USA, but it reached its apogee with the Cold War spread of capitalism and the support of counterinsurgency in the global South—free speech as the expression of the free market in response to its opposites epitomized by the communist bloc and socialist solidarity and, nowadays, terrorism.[27] As examined by the anthropologist Joseph Masco, this democratic ideal might be unravelled domestically too for its part in social and economic control of another order. He contends that the 'counterterror state' has become 'undemocratic and unrelenting' as it 'prioritizes speculative practices over facts ... in an unprecedented effort to anticipate and eliminate terror threats—real, imagined, and emergent'.[28] The dark vagaries of national security make for a blind spot in democratic norms. In his philosophical work, Ferit Güven goes on to propose that democracy is in fact an impossibility.[29] Controlling and regulatory mechanisms are inherent to the paradoxes of democracy—namely to uphold its sanctity, it must be fundamentally undemocratic against anyone or anything that might threaten its preserve.

Ambiguities between democratic license and control are no strangers to the subcontinent. Nevertheless, on a comparative note, an idealized democracy continues to act as a 'symbolic operator', especially when it came to large-scale developments where people could imagine themselves engaging with more rigorous regulations and avenues of redress comparable to the global North.[30] Democracy becomes a salient theme when it is either denied and therefore the basis of a struggle for a more perfect goal; or under threat and therefore the basis of a more defensive good. By the same token, democracies can lead to the rise of exclusivist identity-based movements as is clear in Thomas Blom Hansen's ethnography on the rise of the Hindu right while reviving liberal ideals to counter such surges relevant for the Indian landscape.[31]

While we reflect on the history of the struggle around the Kudankulam Nuclear Power Plant, we can begin to appreciate local meanings and contestations over what 'democratic development' means and how it ought to be pursued.[32] On the one hand, democratic proceduralism enshrined in laws and regulations promise project-affected people a fair hearing and compensation. They informed appeals for people's rights to consultation with reference to the Environment (Protection) Act (1986) and Environmental Impact Assessment (EIA) Notification (1994); and later, information in the public interest with reference to the Right to Information Act (2005), an act that replaced the erstwhile Freedom of Information Act (2002).[33] These were in addition to appeals to pursue mandatory protocols in the construction of a nuclear power plant as endorsed by the International Atomic Energy Agency and adapted by India's Atomic Energy Regulatory Board. Such recourse was framed by India's constitution, privileges supported by international treaties that the Indian government had signed up to, notably the International Covenant on Civil and Political Rights in 1979. Drafted by the Dalit jurist and social reformer, Dr Bhimrao Ramji Ambedkar, and coming into force in 1950, the Constitution of India is often referred to as the supreme law of the land, enshrining fundamental principles of a democracy such as the right to peaceful dissent and the right to live in a healthy environment.

On the other hand, there is, as the anthropologist Cris Shore might state, a 'democracy deficit' across the board.[34] Democratic rights might be seen as a nuisance and often refashioned or dismissed altogether when they get in the way of industrial expansion and extraction. The deficit plummets with respect to the supremacy accorded to nuclear industries and agencies apparent across the world. Whether it be for power or bombs, Indian nuclear developments are tied in with staunch secrecy and a special immunity enabled by a mélange of colonial and postcolonial laws: the Official Secrets Act (1923), the Atomic Energy of India Act (1962 with Amendments in 1986 and 2015), the National Security Act (1980) along with other clauses to exempt nuclear industries from public scrutiny. Even though on one level, fuel used for and derived from civilian reactors remains separate from military infrastructures, for all other purposes, the civilian and military sectors are acutely entangled. Both are treated as fundamental to India's national security that informs the basis for 'legitimate state violence' for anything or anyone that unsettles their domain.[35]

The onslaught of neo-liberal policies that saw its initiation in the 1980s and formalization in 1991 has come with a boost to global trade along with rising consumption, and corporate and industrial expansion in the country.[36] Nuclear power has become a massive player in this energy mix, particularly in the catalytic aftermath of the Indo–US civilian nuclear agreement (2005) and the waiver of the Nuclear Suppliers Group (NSG) in 2008. This waiver enables India to trade with the now 48 members of the NSG even though it is not a signatory to the Nuclear Non-Proliferation Treaty (1970). The accord was upheld with India's pledge to maintain its voluntary moratorium on nuclear tests, to not share nuclear material and technologies with others, and to open inspection of civilian reactors to the regulatory body, the International Atomic Energy Agency.[37] The agreements ended three decades of nuclear isolation after India's 'peaceful nuclear explosion' in Pokhran in 1974.

Consequently, the limited high-grade uranium supplies in the country have now been augmented by global trade. The Indian government made a definitive decision to upgrade its energy provisions, creating a boom in plans to construct new nuclear reactors. Glorious ambitions to get 'the Indian energy roadshow on the move' have come with goals to get to 470 gigawatts to meet India's growing economy and future needs by 2050.[38] This represents a hundred-fold increase from its current provision to the country's energy consumption.[39] Although the Indian government seeks to develop a diversity of energy pathways, the exorbitant costs of nuclear expansion are to the loss of investments in the research and development of alternatives. The nuclear option was heralded as the primary way to increase bulk power: catering to the shortage of electricity supplies as well as purportedly dealing with the excesses of climate change,[40] while nurturing, as the subcontinental metaphor goes, a nuclear, political, and economic tiger.

Such top-down development plans have a long history in grand nation-building models with centralized five-year planning and large-scale industrialization. Science and technology have played a premium role after the influence of India's first prime minister, Jawaharlal Nehru. They are now adapted for a neo-liberal age that has come with the dismantling of welfare provisions and deregulation of the economy. However, the Indian state has not entirely relinquished to market-driven forces. Rather, it continues to have a strong hand in orchestrating private and public sectors including the police (para) military, intelligence, and surveillance agencies in their market-led endeavours. It has led to what the anthropologist David Mosse calls 'a double-figure "centaur" state—neo-liberal at the top and penal at the bottom'.[41]

The social theorist Nikolas Rose adds: 'Neoliberalism does not abandon the "will to govern"'.[42] Instead, distance is created between formal political institutions and social actors who are seen 'in new ways as subjects of responsibility, autonomy, and choice, and seek to act upon them though shaping and utilizing their freedom'.[43] In other words, freedom is framed less through constitutional rights for expression and dissent, and more through a notion of individual choice and autonomy. This 'individualization' of freedom might be restated as the difference between the freedom to consume and the freedom to dissent. It is equivalent to saying that environmental degradation is best dealt with through individual choices to do with light bulbs, diets, and recycling rather than calls to change the industrial and political economy. As the environmental and political scientist Michael F. Maniates mentions:

> When responsibility for environmental problems is individualized, there is little room to ponder institutions, the nature and exercise of political power, or ways of collectively changing the distribution of power and influence in society.[44]

It is a similar process that attends the individualizing of freedom in the neo-liberal age. As we shall see in Chapter 3, constitutional rights have been overruled by consumptive rights: the right to consume, the right for electricity, the right to progress, come what may. These privileges are most apposite for middle-class urban subjectivities that stand apart from those people in the shadows of monolithic developments who do not or cannot subscribe to such aspirational models.[45]

At the launch of the 2016 Amnesty International report at the Press Club of India in New Delhi, it was concluded for the first time that India and China's rapid development had become another progenitor of human rights abuses. As the report reads for the two countries: 'Economic development did not prioritize realization of economic, social and cultural rights'.[46] In the race for supremacy, the juggernaut of forces came with the striking abrogation of citizens' rights that, in India, saw the increasing use of legislature including reviving acts from the colonial era to push through its agenda.

'Big Science, Big Development, Big Projects and Big Goals'—as the political theorist Achin Vanaik describes the Indian state's approach to development plans—comes with the considerable influence of market forces and reforms.[47] The attraction of 'ultra-mega power generation projects' as it applies to nuclear infrastructures and plans remain state-centric, but is now enmeshed in widening national and transnational corporate ties aided by amendments in 2015 to the Atomic Energy Act that allows Foreign Direct Investment (FDI) and private companies in the nuclear field, both energy and defence-related.[48] In this changing configuration of national–global relations, local residents had minimal say. They had become, in an adaptation of the two political philosophers, Giorgio Agamben and Achille Mbembe, 'bare life' slated to become subject to neo-liberal nuclear necropolitics.[49] Even the planning of nuclear plants engraves a national necropolitics, for the constructions are invariably located in the vicinity of those populations and regions deemed disposable:[50] in this case, the hinterlands of the deep south populated largely by fishing and farming communities and well away from metropolitan centres. Their lives mattered little when bulldozed by larger forces in the 'immortal timeline' of globalizing capital.[51]

Since the NSG waiver, trans/multinationals including the American General Electric and Westinghouse, and the French-based Areva, replaced later by Électricité de France (EDF), have stepped forward to collaborate with the Nuclear Power Corporation of India Limited (NPCIL) in order to build new nuclear reactors.[52] Under the aegis of the DAE, the NPCIL along with the Bhartiya Nabhikiya Vidyut Nigam Limited (BHAVINI) are responsible for the construction and operation of nuclear plants in India. In the case considered here at Kudankulam, NPCIL have liaised with Atomstroyexport, under the parent Russian company, Rosatom, which had first entered into agreements with the NPCIL in the 1980s. However, in 1991, the proposal was shelved for a decade due to the break-up of the former USSR and the assassination of India's then prime minister, Rajiv Gandhi. Agreements were renegotiated and finalized at the turn of the last century that brought the story of the Kudankulam Nuclear Power Plant into the new millennium.

Altering the National Nuclear Narrative

Nationalism is no less important now than when it fired the freedom struggle for India's national sovereignty against the British Empire.[53] However, in the current era, nationalism is deeply imbricated in nuclear security. 'Nuclear nationalism', as it has been termed, has become virulent in mainstream Indian political culture where the nuclear is heralded as a pre-eminent path for the postcolonial nation's defence and development.[54] With this has come the mobilization of a huge material, ideological, and surveillance apparatus in the effort for social control, not unlike the 'national security state' identified by Masco for post-World War II USA.[55] While he highlights the legacies of Cold War nuclear terror and the post-9/11 War on Terror as catalysts for the proliferation of the security state in the USA, in the case of India, counter-terrorism is to add to the public memory of colonial subjugation and threats to its progress and sovereignty. These forces have all been mobilized against those who dissent against development.

The benefits of big developments for the larger populace are aggrandized but mainly siphoned through trickle-down drips, much of which are marred by pollution, contamination, and the increased prospects of accidents, displacement, disease, and death. Nationalism therefore has become less about citizen's right and more about the state's rights. It has hijacked the language of democracy such that it has become a ruse to pursue a path in the name of the people even when the decisions act against the people as ulterior interests lie elsewhere.[56] A striking illustration of this cruel conundrum is when we consider state representatives who accuse certain citizens of being anti-national when they object against monolithic developments, while they themselves forge deals with trans/multinational agencies that invariably entail the exploitation of India's environment, resources, and residents.

People around this plant in Tamil Nadu read against mainstream rhetoric. With recollections of the 2011 Fukushima Daiichi nuclear reactor disaster in Japan firmly imprinted on their minds, more and more people realized how they too could be subjected to a nuclear calamity in an area that itself is not free from the risks of an earthquake or a tsunami as had besieged the Japanese plant. Grassroots resistance swelled across the southern peninsula and beyond. Consequently, they came together under the vanguard of the People's Movement Against Nuclear Energy (PMANE) with its peak of activities around 2011–2014 along with the support of hundreds of thousands of people.

Initially formed a decade earlier by the scholar and peace activist S.P. Udayakumar, PMANE set out to question, critique, and resist the opaque procedures of the nuclear authorities that were bulwarked by an increasingly intransigent and militarized state. According to a leaflet from the early 2000s, PMANE identifies 'nuclearism' as the main threat to an egalitarian and healthy life:

> PMANE does not see the nuclear program just as a matter of science, economics and development. It holds that nuclearism also has political, cultural, environmental, demographic sides and other externalities to it as a dangerous ideology and program.

Udaykumar was influenced by the founder of peace and conflict studies, Johan Vincent Galtung, who was a Distinguished Professor of Peace Studies at the University of Hawaiʻi where the former studied for his PhD in political science. Accordingly, Udaykumar draws out connections between India's nuclear industries and their danger to the fabric and future of the nation. By highlighting the structural violence around nuclear industries, PMANE sought to 'problematize the illegality, anti-democratic and anti-people nature of the Department of Atomic Energy (DAE)'. Rather than acting as public servants, he contended that it was as if the DAE make the public serve to their interests.

Environmental, political, and moral arguments converged on the nuclear plant. PMANE tried to intercept the dominant discourse of nuclear nationalism, not by resisting it altogether but first, by de-linking the *nuclear* from the *national,* and second, to question the role of the nuclear in the name of the nation. With this orientation, they were able to wrest the nation away from the state and revere the national as a discursive script that values the power, *kratos,* of the people, *demos*, along with the rights of the citizen, while encapsulating respect for their historical, cultural, and environmental heritage. Through reclaiming the national in this way, they resisted the nuclear, which they saw as part of a dominant 'nuclear-colonial' state imposing its agenda on them without their due consultation: as Udayakumar describes it 'nucolonization (nuclear + colonization)'.[57] They highlighted how 'spectres of colonialism' were more than just phantoms but continued to have material and physical repercussions.[58] By effective delinking, they aimed to circumvent accusations of being 'anti-national' and 'anti-development', and to try and fend off accusations of foreign infiltration and espionage. In this way, hegemonic meanings shared with the rest of the populace were destabilized.[59]

Drawing upon the works of M.K. Gandhi, initiatives to enable this reorientation included the revival and adaptation of India's non-violent freedom struggle that valorized the struggle as one for the people against colonization, and thus attempt to take on the higher moral ground in the name of the nation, a tactic that has been used in several other instances.[60] This entailed reasserting the integrity of non-violence and the environment to the nation. Campaigners emphasized their role as *nation protectors* more than they did as *anti-nation protesters*.[61] They advocated peaceful dissent over violent reproach. They sought to prioritize their custodianship of the planet, country, and people through a variety of practices and outlets in order to disseminate alternative views on the costs and dangers of the nuclear industries including the ills of radioactivity. They also made concerted efforts to connect with wider everyday issues such as the universal life value of fresh water, and its gradual depletion as thousands of litres are needed for a large power plant—this was a concern that spread to inland farming as well as fishing communities in their eco-nationalist campaigns. In the process, they mobilized striated conceptions of Indian and/or Tamil culture and anti-colonial history.

Instead of protesting against development per se, campaigners altered the terms of development to make it more people-orientated and environment-friendly—a modernized version of Gandhian self-reliance and relatively inexpensive, devolved development. In the midst of their campaigns, they made alternative proposals for the nation's electricity needs based on decentralized energy systems with provisions for bio-gas, mini-hydel (hydroelectric) plants, and wind and solar energy systems.[62] In tandem, they encouraged the tactical use of science, surveys, and statistics in a bid to create a parallel and locally informed knowledge base of 'citizen experts' with which to inform critical action and shake off accusations of them as irrational luddites or troglodytes. People from the movement entered into debate with representatives from central and state governments, and set up parallel expert committees with independent scientists and other experts to re-examine and question the NPCIL's claims about the site selection, geology, oceanography, emergency preparedness, and safety procedures to do with the nuclear power plant. This struggle for legitimacy encompassed all spaces—from the beaches to the court room.

In 2011, PMANE issued a 13-point agenda along with a series of mass hunger strikes.[63] It included appeals to follow due procedure with regards to plant construction regulations, public consultations, the dissemination of mandatory reports such as the Environmental Impact Assessment, Site Evaluation and Safety Analysis reports, and Emergency Preparedness Plans for the million or so people who live within a 30 km radius of the plant.[64] Their demands were extended to include nuclear waste management plans and liability agreements with the Russian supplier of reactors that were kept away from the public. The call for more information were also with the expectation that no information would be provided. The performative appleals then served the purpose of highlighting the non-procedural and opaque practices of the nuclear state, the disjuncture of democracy.[65] The main goal remained the cessation of nuclear activities at the plant altogether with a view to dramatize their undemocratic developments. As we will see from Chapter 8 onwards, by highlighting the 'lawlessness of the state' as against aspersions of their 'state of lawlessness', campaigners aspired to gain the higher ground as the true representatives of the people.

The clarion call was for social justice and a deepening of democracy that could extend to marginalized fishing communities, peasants, Dalits (Scheduled Castes), Adivasis (tribals), and the working classes in India's vast formal and informal economies.[66] While the movement was not against the political system per se, it raised several critical questions on their modus operandi. The struggle was for democratizing decision-making processes with more transparency while demanding accountability among the political elites. The demand was for a fair hearing and compensation for project-affected people. The movement unleashed multiple time-worn and new-fangled techniques to raise awareness and enlist people to their cause. In the process, campaigners sought to create, protect, and expand spaces for alternative people-centric development in what could be called a 'true democracy'.

The anthropologist Arjun Appadurai adds to his understanding of calls for deep democracy that their 'depth' are also based on 'the lateral reach of such movements with efforts to build international networks or coalitions of some durability with their counterparts across national boundaries'.[67] However, as we shall see in Chapters 9 and 10, the globalizing potential of an anti-nuclear movement is severely curtailed by allegations of anti-nationalism and sedition from the dark bastion of national security. Such a(nta)gonistic practices have led to fierce debates on the nuclear industries where issues to do with democracy, development, nation, state, economy, environment, health, science, and expert knowledge-claims as well as citizens' rights are thrown into sharp relief in a contested space of criticality.

Criticality

Drawn from the idea that a critical mass of radioactive material is required so that nuclear fission chain reactions become irreversible and self-sustaining as happens when a reactor is commissioned, criticality is invoked in this book in a multi-layered and multi-situated sense to connote a gamut of encounters. It indicates fraught and temporally contingent relations: from one perspective, that all is functioning smoothly and therefore an equilibrium that needs to be vigilantly maintained for more power; and from another that highlights various kinds of perils that need to be curbed in order to renegotiate the terms and relations of power.

Significantly, as with power so with the term, 'criticality': it is not just for the rarefied realms of reactor science and technology, but also how such processes are framed by, and impact on society—or, in other words, how scientific claims on behalf of national authorities are formed, engaged with, and interrogated by people in response to the nuclear industries. The struggle becomes one over the cultural meanings of nuclear power as well as the industries' biomedical, social, economic, environmental, and political consequences.[68] Criticality therefore has a politicizing resonance in that it highlights how historical, scientific, technological, and bureaucratic processes are never neutral, even if they pertain to be with the use of factual statements, statistics, and other forms of naturalized representations in the name of neutrality.

The political theorist Wendy Brown widens the scope to interrogate narratives of neutrality on critical conditions as 'a particular kind of call: an urgent call for knowledge, deliberation, judgment, and action to stave off catastrophe'.[69] While this sense of criticality implies the need for restorative action, criticality can also be precarious in that it seeks to upset the status quo in the effort to engage social action with new and alternative proposals. Theory and practice might be deployed to challenge dominant forces and reassess cherished 'truths' about the contemporary nation's nuclear development and security. Rather than a space of disinterest, criticality then becomes a space of vested interests where sanctioned truths and social realities become an issue of life and death, and subject to question and even transformation.

Resistance to a nuclear plant might be located in the intentions, motivations and meanings that people give to their actions. This could be by way of outright public challenges, rallies, marches, and other acts of opposition. However, dissent might also be much more diffuse or ambivalent as part of a low-level or broader critical orientation, whether it is every day or every other occasional day, that the wider encapsulating term of criticality enables us to accommodate.[70]

Much as spaces of criticality are instigated by domestic factors, global dynamics cannot be overlooked in such contact zones.[71] No matter how nuclear authorities present themselves under the umbrella of national security with ambitions to excel in 'indigenous' technologies, they are always inflected by transnational exchange in regulations, agreements, trade, materials, skills, and/or expertise.[72] A popular term for discrepant global interconnections is 'friction' that according to the anthropologist Anna Tsing describes a messy and uneven zone of 'awkward connection', leading to 'the unexpected and unstable aspects of global interaction'.[73] Through 'the 'makeshift links across distance and difference', she elaborates on how '[f]riction refuses the lie that global power operates as a well-oiled machine'.[74] In her study on uranium mining in African countries, the social historian Gabrielle Hecht adopts the term to consider the 'friction between transnational politics and (post)colonial power, between abstract prescriptions and embodied, instrumentalized practices'.[75] By comparison, the anthropologist Thomas Hylland Eriksen prefers to discuss fricative terrain in terms of the 'overheated world', which entails 'accelerated and intensified contact [that] leads to tensions, contradictions, conflict and changed opportunities in ways that affect identity, the environment and the economy'.[76]

Criticality accommodates their insights, but does not just rest on global–local interconnections and disconnections alone—whether they be described as fricative or overheated. In the people's movement considered here, the global might be delimited when the national is augmented in terms of sovereignty and territoriality as the anthropologists Aradhana Sharma and Akhil Gupta demonstrate in their analyses of the state in an era of globalization.[77] The enmeshment of universals and particulars in terms of Tsing's 'sticky engagements' are refracted here through a lens on the reified nuclear state with its Delhi- and Chennai-centric state circuits cross cut by a range of transnational currents that converge in the peninsular locality.[78]

Moreover, criticality is not just about the heat of movements, connections, interconnections, or disconnections, but it is also about *temporality*, the significance of timing and change that characterizes encounters: how a situation, the same situation even, may alter from moment to moment or from one perspective to another, and how the introduction of new entrants leads to ongoing transformations.

Spaces of criticality were sparked off almost immediately in reaction to the prospect of reactor criticality in the Kudankulam vicinity. A phantom icon that portends anxiety and catastrophe—in this case, a nuclear power plant—became a becoming event in the lives of project-affected communities. It gradually and sporadically became a contested object of emergence. It took on a phenomenal presence years before it was actually constructed. Criticality therefore has a historicizing aspect in that the declaration of plans for a nuclear power plant might become a turning or even breaking point in the region. Following the anthropologist Veena Das, it is another paradigm of a 'critical event' in that it generated new modalities of political action in the subcontinent.[79] This is with respect to her focus on seminal moments that have reconfigured conventional thought and social processes as with the partition of the subcontinent in 1947 and Bhopal's Union Carbide industrial disaster in 1984.

In the Kudankulam case, however, the critical event comes with starts and stops over a long-drawn-out period when plans were raised, negotiated, imposed, retracted, resurrected, halted, and resumed rather than in the singularity of a calamitous event. Even though the introduction of the nuclear plant to the south Indian landscape was episodic—an 'initiating event' to use a term by the sociologist Neil Smelser—the fateful moment surrounding people living in and around Kudankulam was not just around one singular instant, but a protracted period of chain reactions with several runs, alterations, hitches, and glitches along the way.[80] The idea of a nuclear plant triggered complex and unpredictable changes—immediate, spiralling, and/or long-drawn-out. It signalled an intensification that led to a concatenation of thoughts, becomings, and actions.[81] What exactly does the nuclear power plant involve? How many reactors would there be? How much radiation or contamination of the air, land, and sea would there be? How will the plant affect me? What can I do? What about my children? The 'event-effects', as the philosopher Gilles Deleuze puts it, prompted varied responses that altered with time and space like a palimpsest of wave graphs:[82] the ebbs and flows of suspicion, intrigue, indifference, pride, opportunity, belief, disbelief, risks, doubts, depression, anger, and collective dissent. [83]

There are therefore three main aspects to the way criticality is invoked in the book with respect to the nuclear industries in peninsular India. First, it is on account of the need to denaturalize the supposedly objective exercises of science, technology, and bureaucracy. Second, it is with respect to locating wider critical orientations that look into, but also go behind and beyond obvious displays of power and resistance—the everyday to the extraordinary, the ambivalent to the more explosive across sliding and fricative local–national–global scales. Third, it is in view of moving away from the singularity of events to how temporality sets the conditions of narrative possibilities in any kind of expectation, encounter, effect, or recollection.

In the process, criticality need be informed by self-reflexivity and careful ethnographic methods that question the positionality of the observer or researcher as well as of those around them who may have multiple and changing points of address and analyses on incidents and issues that affect their lives.[84] As will be clear from Chapter 8 on the post-2011 years, I talked, listened, and observed more than I participated in what I saw around me for too much of the latter could compromise representations of the growing movement as much as it could my safety. I therefore chose to adopt an understanding distance.

Whether they be scientists or social scientists, through the repercussions and partiality of vision, experiences and choices made, one cannot but be affected by the critical mix.[85] Representations are all narratives on ineffable truths—stories, or, as the title of the book goes, a story of stories.[86] The stories are by way of the narratives of fishermen and women, farmers, environmentalists, activists, scholars, teachers, authors, journalists, priests, children, as much as they are of lawyers, scientists, state officials, and myself as the researcher and writer drawing upon an interdisciplinary field as the subject compels. Characterized by varying truth-claims, idioms, and structuring devices, they are but representations of their realities. Indeed, the anthropologist and historian Carole McGranahan describes 'anthropology as theoretical storytelling' when one is 'engaged in explaining, understanding, and interpreting cultural worlds as well as in developing theoretical paradigms large and small for making sense of cultural worlds'.[87] While I present stories, I also weave them into a story. These stories are foundational even to those canons that choose to stand outside of stories, bowing as they do at the altar of objectivity.[88] Science and technology, facts and statistics: they too do not breathe outside of a 'universe of stories'—a proposition that will be further examined in Chapters 4 and 7.[89]

There is always a before and an after, a past and a future implicated in peoples' tales—whether they be conceived in terms that are euphoric or nostalgic, utopic or dystopic, or a restless combination of them all. Taking heed of the fact that every action has a reaction, and that there is nothing that remains constant but change, the stories presented in this book are also subject to new narratives, perspectives, and even radical revision. Criticality then encompasses transience and mortality, thus deepening the relation with the now, inscribed as it is by memories of the past, the contingencies of the present, and imaginaries of the future.

Altogether, the book provides ever-changing historical and ethnographic perspectives on spaces of criticality around (plans for) the Kudankulam Nuclear Power Plant since the 1980s. It recounts how residents contended with the prospect of one of Asia's largest nuclear power plant being built on their doorstep with the current proposal of 6,000 megawatts of power production in cahoots with state and central governments along with their vested collaborations with trans/multinational corporations. It delivers a *longue durée* and embedded account of the simmering sparks, and cycles of peaks and troughs of an anti-nuclear movement based in peninsular India. As it does so, it allows us to consider the larger politics of development, democracy and nationalism that has marked not just parts of India identified for large-scale developments, but also other regions of the world where state functionaries have much to gain from corporate collaborations at the cost of local residents who lose their livelihoods, and are forcibly displaced, persecuted, and even killed in order to execute governmental designs. It is yet another call for 'energy justice' that, according to the social theorist Benjamin K. Sovacool has to come with considerations to do with the availability of the energy, its affordability, good governance, prudence, intergenerational, and intra-generational equity, and responsibility.[90]

As such, this is one of the few books that have at its heart the many facets of a grassroots movement for energy justice in the global south centred on a nuclear power plant.[91] Nuclearism combined with nationalism and neo-liberalism is dictating the moral, media, and political economies of India. Strength is conceived in terms of numbers indexed in gross national and domestic products or the number of nuclear reactors and warheads. What fails to get as much notice is how this is at great costs to society and environment—both being seen as either utility or impediment in the bid for greater progress and power when they are its most important bedrock.

Notes

1. S. Anitha. 'Historic Day in Koodankulam, As People Plan to March towards The Nuclear Plant', *DiaNuke*, 8 September 2012. Available on http://www.dianuke.org/historic-day-in-koodankulam-as-people-plan-to-march-towards-the-nuclear-plant/, accessed 10 January 2013. I invoke Anitha's collation of comments for the post-2012 period throughout this book—a period when we jointly conducted fieldwork in Idinthakarai.

2. M.V. Ramana reports gaseous wastes released by stacks and the release of radionuclides such as tritium, argon 41, and fission products of noble gases into the atmosphere. He also notes low-level liquid wastes such as

tritium, caesium 137 and strontium 90. Ramana, M.V. *The Power of Promise: Examining Nuclear Energy in India.* New Delhi: Penguin Books, 2013, p. 225.

3. See Gregory, Robin, James Flynn, and Paul Slovic. 'Technological Stigma'. *American Scientist* 83, no. 3 (1995): 220–3.

4. This is a development of studies on the psychosocial effects of the Chernobyl nuclear accident that stops short of discussing political dimensions. See Bromet J. Evelyn 'Mental Health Consequences of the Chernobyl Disaster', *Journal of Radiological Protection*, 32(1): 71–5.

5. Ramana, M.V. 'Flunking Atomic Audits: CAG Reports and Nuclear Power'.*Economic and Political Weekly* 47, no. 39 (2012): 10–13; 'Supreme Court Moved against HC Go Ahead to Kudankulam'. *The Hindu*, 11 September 2012. Available on https://www.thehindu.com/news/supreme-court-moved-against-hc-go-ahead-to-kudankulam/article3885533.ece, accessed 10 February 2019.

6. Senthalir, S. 'Violence against the Non-violent Struggle of Koodankulam'. *Economic and Political Weekly* 47, no. 39 (2012): 13–15.

7. Hodge, Amanda. 'Anti-nuke Tamil Village under Siege'. *The Australian*, 22 March 2012. Available on https://www.theaustralian.com.au/news/world/anti-nuke-tamil-village-under-siege/news-story/0c7a2c22eca418019fbb150642a096e2, accessed 20 February 2019.

8. The term state when used in the abstract sense is shorthand for a complex of organizational structures, personnel, discourses, and practices that can be both dispersed and come together as a concrete force in people's lives. In Michel Foucault's words, it is the outcome of a *quadrillage* of performative processes: 'The state is at once that which exists, but which does not yet exist enough'. Foucault, Michel. *The Birth of Biopolitics: Lectures at the Collège de France, 1978–1979 (Lectures at the College de France)*. London: Picador, 2010, p. 4.

9. Available on http://www.nirs.org/international/asia/kudankulam2.html, accessed 10 January 2013. Reproduced in Rosa, Antonio C.S., ed. 2012. 'Media Blackout of the Ongoing Nonviolent Protest at Koodankulam Nuclear Power Plant', *TRANSCEND Media Service: Solutions-Oriented Peace Journalism*, 2 April. Available on https://www.transcend.org/tms/2012/04/india-the-ongoing-nonviolent-protest-at-koodankulam-nuclear-power-plant/, accessed 22 January 2019.

10. On the sovereignty of the 'people' as an organizing force of the modern state and political imaginaries, see Laclau, Ernesto. *On Populist Reason*, London: Verso, 2005.

11. See Jungk, Robert. *The Nuclear State*, translated by Eric Mosbacher. London: John Calder, 1979.

12. See Bidwai, Praful. 'People vs Nuclear Power in Jaitapur, Maharashtra,' *Economic and Political Weekly* XLVI, no. 8 (2011): 10–14; Kaur, Raminder. 'Nuclear Power vs People Power'. *Bulletin of the Atomic Scientists*, 9 July 2012. Available on https://thebulletin.org/2012/07/nuclear-power-vs-people-power/, accessed 22 January 2019.

13. Escobar, Arturo. 'Anthropology and the Development Encounter: The Making and Marketing of Development Anthropology'. *American Ethnologist* 18, no. 4 (1991): 16–40.

14. See Ferguson, James. *The Anti-Politics Machine: Development, De-politicisation and Bureaucratic Power in Lesotho.* Minneapolis: University of Minnesota Press, 1994; Mosse, David. 'The Anthropology of International Development'. *Annual Review of Anthropology* 42 (1993): 227–46, p. 230.

15. See Breckenridge, Carol A. and Peter van der Veer (eds). *Orientalism and the Postcolonial Predicament: Perspectives on South Asia.* Pennsylvania: University of Pennsylvania Press, 1993; Gupta, Akhil. *Red Tape: Bureaucracy, Structural Violence and Poverty in India.* Durham: Duke University Press, 2012; Yarrow, Thomas and Soumhya Venkatesan. 'Anthropology and Development: Critical Framings' in their edited volume, *Differentiating Development: Beyond an Anthropology of Critique*, pp. 1–20, Oxford: Berghahn, 2012.

16. See Gould, Jeremy and Henrik Secher Marcussen. 'Narratives of Democracy and Power', in *Ethnographies of Aid: Exploring Development Texts and Encounters*, pp. 15–44, Vol. 24, edited by Henrik Secher Marcussen. Roskilde: Roskilde University Press, 2004.

17. Gupta, Akhil. *Postcolonial Developments: Agriculture in the Making of Modern India.* Durham: Duke University Press, 1988. See also Chapter 6 in this volume.

18. Abraham, Itty. *The Making of the Indian Atomic Bomb: Science, Secrecy and the Postcolonial State.* London: Zed Books, 1998.

19. Wood, Richard. 'The World's 7 Largest Democracies—Where do America and India Fit in?' *HTC.* Available on https://www.hitc.com/en-gb/2017/10/20/the-worlds-7-largest-democracies-where-do-america-and-india-fit/, accessed 22 January 2019.

20. On the case of nuclear development in a democratizing Indonesia, see Sulfikar, Amir. 'Challenging Nuclear: Antinuclear Movements in Postauthoritarian Indonesia'. *East Asian Science, Technology and Society: An International Journal* 3, nos. 2–3 (2009): 342–66. Available on https://pdfs.semanticscholar.org/b633/0706fd657bd50fe994194b303baca1ed9b2e.pdf, accessed 22 January 2019.

21. This phenomenon is not unlike the 'deadly symbiosis' between unfettered markets and weakened social welfare that Loïc Wacquant describes for race-class social stratification in the neo-liberal era. Wacquant, Loïc. *Deadly Symbiosis: Race and the Rise of the Penal State*. Boston, MA: Polity, 2011. See also Güven, Ferit. *Decolonizing Democracy: Intersections of Philosophy and Postcolonial Theory*. London: Lexington Books, 2015.

22. Paley, Julia. 'Toward an Anthropology of Democracy'. *Annual Review of Anthropology* 31, no. 1 (2002): 469–96, p. 479.

23. Boutros-Ghali, Boutros et al. *The Interaction between Democracy and Development*. Paris, France: United Nations Education, Scientific and Cultural Organization, 2002.

24. Paley, 'Toward an Anthropology of Democracy', p. 484.

25. See Heller, Patrick. 'Degrees of Democracy: Some Comparative Lessons from India'. *World Politics* 52, no. 4 (2000): 484–519, pp. 487–8.

26. Paley, 'Toward an Anthropology of Democracy', p. 469.

27. Stromberg, Joseph. 'The Real Birth of American Democracy', 20 September 2011. Available on https://www.smithsonianmag.com/smithsonian-institution/the-real-birth-of-american-democracy-83232825/, accessed 22 January 2019. On a critique, see Gledhill, John. *Power and Its Disguises: Anthropological Perspectives on Politics*. Sterling, VA: Pluto, 2000, pp. 7–8.

28. Masco, Joseph. *The Theatre of Operations: National Security Affect from the Cold War to the War on Terror*. Durham: Duke University Press, 2014.

29. Güven, *Decolonizing Democracy*, p. 14.

30. See Verdery, Katherine. *What Was Socialism, and What Comes Next?* Princeton, NJ: Princeton University Press, 1996, p. 105.

31. Hansen, Thomas Blom. *The Saffron Wave: Democracy and Hindu Nationalism in Modern India*, New Delhi: Oxford University Press, 1999.

32. There is a large literature on democracy and development both on their own and with respect to each other—most say that the two are compatible, others disagree, and still others present mixed reports. On a useful overview, see Sirowy, Larry and Alex Inkeles. 'The Effects of Democracy on Economic Growth and Inequality: A Review'. *Comparative International Development* 25, no. 1 (1990): 126–57. The literature is invoked only so much as it aids an analysis of our focus on the nuclear plant in south India.

33. A revised EIA Notification was presented in 2006 to limit public participation at such public hearings in response to Department of Atomic Energy (DAE) complaints. See Ramana, *The Power of Promise*, pp. 88, 293–5.

34. Shore, Cris. *Building Europe: The Cultural Politics of European Integration*. New York: Routledge, 2000.

35. Abraham, Itty. 'The Violence of Postcolonial Spaces: Kudankulam'. In *Violence Studies*, edited by Kalpana Kannabiran. New Delhi: Oxford University Press, 2016. See also Weber, Max. 'Politics as a Vocation'. In *Rationalism and Modern Society*, translated and edited by Tony Waters and Dagmar Waters. New York: Palgrave Books.

36. See Ganti, Tejaswini. 'Neoliberalism'. *Annual Review of Anthropology* 43 (2014): 89–104.

37. On details of the negotiations, see Ramana, *The Power of Promise*, pp. 279–92.

38. Bagchi, Indrani. 'N-deal will be through in 36 Hours: Burns'. *The Times of India*, 8 December 2006, p. 15.

39. Ramana, *The Power of Promise*, p. xvii.

40. Kakodkar, Anil. 'Energy in India for the Coming Decades'. *Proceedings of an International Ministerial Conference on Nuclear Power for the 21st Century* 38, no. 25 (2005). Available on https://inis.iaea.org/search/search.aspx?orig_q=RN:38056572, accessed 10 October 2019; Raminder Kaur 'A "Nuclear Renaissance", Climate Change and the State of Exception', *TAJA: The Australian Journal of Anthropology* 22, no. 2 (2011): 273–7; Ramana, *The Power of Promise*, p. 264.

41. Mosse, 'The Anthropology of International Development', p. 237. See Ong, Aihwa. *Neoliberalism as Exception: Mutations in Citizenship and Sovereignty*. Durham, NC: Duke University Press, 2006; Padel, Felix and Samarendra Das. *Out of This Earth: East India Adivasis and the Aluminium Cartel*. New Delhi: Orient Blackswan, 2010.

42. Rose, Nikolas. 'Governing "Advanced" Liberal Democracies'. In *Foucault and Political Reason: Liberalism, Neo-Liberalism and Rationalities of Government,* edited by Andrew Barry, Thomas Osborne, and Nikolas Rose. Chicago: University of Chicago Press, 1996, p. 54.

43. Rose, 'Governing "Advanced" Liberal Democracies'.

44. Maniates, Michael F. 'Individualization: Plant a Tree, Buy a Bike, Save the World?' *Global Environmental Politics* 1, no. 3 (2001): 31–52, p. 31.

45. The Indian middle class is a complex, plural, and sometimes contradictory social formation. Most concur that since the 1990s in particular the old salariat version is being displaced by new consumerist and entrepreneurial middle classes. See Varma, Pavan. *The Great Indian Middle Class*. New Delhi: Viking, 1998; Mawdsley, Emma. 'India's Middle Classes and the Environment'. *Development and Change* 35, no. 1 (2004): 79–103; Mazzarella, William. 'Indian Middle Class'. 2005. Available on http://www.soas.ac.uk/ssai/keywords/; Brosius, Christiane. *India's Middle Class. New Forms of Urban Leisure, Consumption and Prosperity*. New Delhi: Routledge, 2010. On the middle class as it applies to the Indian

nuclear industries, see Kaur, Raminder. 'A Nuclear Cyberia: Interfacing Science, Culture and an Ethnography of an Indian Township Social Media'. *Media, Culture and Society* 39, no. 3 (2016): 325–40.

46. *Amnesty International Report 2006: The State of the World's Human Rights.* Oxford: Alden Press, 2006, p. 38.

47. Vanaik, Achin 'Ideologies of the State: Socio-Historical Underpinnings of the Nuclearization of South Asia' (paper presented at workshop, *Nuclear Understandings: Science, Society and the Bomb in South Asia,* Dhaka, Bangladesh, 17 February).

48. The Atomic Energy (Amendment) Act, 31 December 2015, Ministry of Law and Justice. Available on https://dae.nic.in/writereaddata/ ae_amend_2015_0116.pdf, accessed 9 October 2019. On market and military compulsions, see Raman, 'La Trahison des Clercs', p. 243.

49. On bare life, see Agamben, Giorgio. *Homo Sacer: Sovereign Power and Bare Life.* Stanford: Stanford University Press, 1995. On necropolitics, see Mbembe, Achille. 'Necropolitics'. *Public Culture* 15, no. 1 (2003): 11–40. With respect to their applicability to nuclear politics, see Kaur, Raminder. 'Nuclear Martyrs and Necropolitics'. *Countercurrents,* 16 September 2012. Available on http://www.countercurrents.org/kaur160912.htm. On a psychoanalytical interpretation of nuclear necropolitics, see Schwab, Gabriele. 'Haunting from the Future: Psychic Life in the Wake of Nuclear Necropolitics'. *The Undecidable Unconscious: A Journal of Deconstruction and Psychoanalysis* I (2014): 85–101. On 'the necro-politics of radia- tion', see Kohso, Sabu. 'The Age of Meta/Physical Struggle'. *Cultural Anthropology, HOT SPOTS, 3.11 Politics in Disaster Japan* (2011). Available on http://www.culanth.org/?q=node/422, accessed: 9 October 2019.

50. See Mbembe, 'Necropolitics'.

51. Jain, Lochlann S. *Malignant: How Cancer Makes Us.* Berkeley: University of California Press, 2013, p. 49. See also Patnaik, Prabhat. 'Neo-liberalism and Democracy'. *Economic and Political Weekly* XlIX, no. 15 (2014): 39–44.

52. Both transnational and multinational companies operate in multiple countries but multinational indicates a centralized management system in one of these countries. The latter include domestic companies such as Tata, Reliance Infrastructure Limited, and Larsen & Toubro who have been held to 'poach' NPCIL employees for their companies. Sudhakar, P. 'Private Sector Giants "Poach" NPCIL Personnel'. *The Hindu,* 20 November 2006, p. 3.

53. See Chatterjee, Partha. *The Nation and its Fragments: Colonial and Postcolonial Histories.* Princeton: Princeton University Press, 1993.

54. Mian, Zia 'Nuclear Nationalism'. *Nuclear Age Peace Foundation.* May 5. https://www.wagingpeace.org/nuclear-nationalism/, accessed 9 October

2019. See also Bidwai, Praful and Achin Vanaik. *South Asia on a Short Fuse: Nuclear Politics and the Future of Global Disarmament*. New Delhi: Oxford University Press, 1999.

55. Masco, *The Theatre of Operations*.

56. See Whitecross, Richard W. 'Intimacy, Loyalty and State Formation: The Spectre of the "Anti-National"', pp. 68–88, in *Traitors: Suspicion, Intimacy, and the Ethics of State Building*, edited by Tobias Kelly and Sharika Thiranagama. Pennsylvania: Pennsylvania State University Press, 2010, p. 68.

57. Vishwanath, C.K. 'National Convention on "The Politics of Nuclear Energy and Resistance"'. 4–6 June 2009. Available on http://www.mail-archive.com/greenyouth@googlegroups.com/msg08208.html, accessed 8 October 2009. See also Valerie L. Kuletz who refers to it as 'nuclear colonialism', Ward Churchill and Joseph Masco on 'radioactive colonization' of the lands of Native American communities, and Navajo activists describing it as 'radioactive colonialism' as reported by Gabrielle Hecht. Kuletz, Valerie. *The Tainted Desert: Environmental Ruin in the American West*. Hove: Psychology Press, 1998, p. xviii. Churchill, Ward. *Acts of Rebellion: The Ward Churchill Reader*, New York: Routledge, 2005, p. 103. Masco, Joseph. *The Nuclear Borderlands: The Manhattan Project in Post-Cold War New Mexico*. Princeton: Princeton University Press, 2006, p. 101. Hecht, Gabrielle. *Being Nuclear: Africans and the Global Nuclear Trade*. Massachusetts: MIT Press, 2012, p. 177.

58. Güven, *Decolonizing Democracy*, p. 59.

59. See Jameson, Frederic. *The Political Unconscious: Narrative as a Socially Symbolic Act*. Ithaca: Cornell University Press, 1981.

60. Hardiman, David. *Gandhi: In His Time and Ours*. New Delhi: Permanent Black, 2003; Nilsen, Alf Gunvald. *Dispossession and Resistance in India: The River and the Rage*. New Delhi: Routledge, 2012.

61. See '"We Are Protectors, Not Protesters": Why I'm Fighting the North Dakota Pipeline'. *The Guardian*, 18 August 2016. Available on https://www.theguardian.com/us-news/2016/aug/18/north-dakota-pipeline-activists-bakken-oil-fields, accessed 9 October 2019.

62. See Srikant, Patibandla. *Koodankulam Anti-Nuclear Movement: A Struggle for Alternative Development?* Working Paper 232, Bangalore: Institute for Social and Economic Change, 2009.

63. 'Thirteen Reasons Why We Do Not Want the Koodankulam Nuclear Power Project'. 25 August 2011. Available on http://www.dianuke.org/thirteen-reasons-against-the-koodankulam-nuclear-power-project/, accessed 4 July 2018.

64. A 'plume emergency planning zone' applies to those living within 10 miles from the plant, and an 'ingestion planning zone' to those within

50 miles from the plant. Udayakumar, S.P. *Emergency Unpreparedness at Koodankulam Nuclear Power Project*. Nagercoil: People's Movement Against Nuclear Energy, National Alliance of Anti-Nuclear Movements, 2010, p. 22.

65. On 'disjunctive democracy', see Holston, James and Teresa P.R. Caldeira. 'Democracy, Law, and Violence: Disjunctions of Brazilian Citizenship', pp. 263–96, in *Fault Lines of Democracy in Post-Transition Latin America*, edited by Felipe Agüero and Jeffrey Stark. Miami: North-South Cent, 1988.

66. Appadurai, Arjun. 'Deep Democracy: Urban Governmentality and the Horizon of Politics'. *Environment and Urbanization* 13, no. 2 (2001): 23–43. Similarly, Rosemary Coombe proposes 'dialogic democracy' as an alternative to rational argument in 'deliberative democracy' that ends up excluding socially marginalized groups. Coombe, Rosemary. 'Introduction: Identifying and Engendering the Forms of Emergent Civil Societies: New Directions in Political Anthropology'. *PoLAR: Political and Legal Anthropological Review* 20, no. 1 (1997): 1–12.

67. Appadurai, 'Deep Democracy', p. 42.

68. See Escobar, Arturo. 'Culture, Practice and Politics: Anthropology and the Study of Social Movements'. *Critique of Anthropology* 12, no. 4 (1992): 395–432.

69. Brown, Wendy. *Edgework: Critical Essays on Knowledge and Practice*. Princeton: Princeton University Press, 2005, p. 6.

70. Accordingly, theories of resistance need to go beyond a reliance on singular notions about state dominance and sovereignty, on the one hand, and ideas to do with unitary subjectivities, hegemonic blocs, and metaphors to do with raising politicized consciousness, on the other. Since the 1970s in particular, what may be loosely defined as resistance studies, encompassing largely anthropological, sociological, and historical works have taken root (which in the Indian case has been termed subaltern studies after the influence of Antonio Gramsci). Inspired by the works of Michel Foucault and Michel de Certeau, they moved away from binaries of the powerful and powerless to consider everyday forms of resistance where small acts of defiance suggest dissatisfaction with the system. An emphasis on veiled tactics or 'hidden transcripts' to cite James C. Scott has led to studies of 'cultures of resistance' entailing gossip, possession, song, poetry, festivals, and even love letters that challenge notions of subservience. Albeit contentious as to what extent such mundane activities alter the status quo, in their particular ways, the studies have enabled a means by which the terrain of power and political resistance can be reconceptualized. The problem then arises, first, of over-romanticizing

resistance, and second, identifying when everyday practices constitute resistance that breaks out of the cyclical equation that multiple forms of power necessitates multiple forms of resistance. See Foucault, Michel 'The Subject and Power', *Critical Inquiry* 8(4): 777–95; Certeau, Michel de. *The Practice of Everyday Life*. California: University of California Press, 2002; Ong, Aihwa. *Spirits of Resistance and Capitalist Discipline: Factory Women in Orders*. Chicago: University of Chicago Press, 1987; Scott, James C. *Domination and the Arts of Resistance: Hidden Transcript*. New Haven: Yale University Press, 1990; and Abu-Lughod, Lila. 'The Romance of Resistance: Tracing Transformations of Power through Bedouin Women'. *American Ethnologist* 17, no. 1 (1990): 41–55.

71. On linguistic and cultural encounters in contact zones, see Pratt, Mary Louis. 'Arts of the Contact Zone'. *Modern Language Association* (1991): 33–40.

72. Abraham, 'The Violence of Postcolonial Spaces', pp. 334–5.

73. Tsing, Anna. *Friction: An Ethnography of Global Connection*. Princeton: Princeton University Press, 2005, pp. 11, 3.

74. Tsing, *Friction*, pp. 2, 6.

75. Hecht, *Being Nuclear*, p. 46.

76. Eriksen, Thomas H. *An Overheated World*, 2013, p. 1. Available on https://www.sv.uio.no/sai/english/research/projects/overheating/publications/overheated-world.pdf, accessed 22 January 2019.

77. Sharma, Aradhana and Akhil Gupta. 'Introduction: Rethinking Theories of the State in an Age of Globalization', pp. 1–42, in *Anthropology of the State: A Reader*, edited by Aradhana Sharma and Akhil Gupta. Oxford: Blackwell, 2006, p. 6.

78. Tsing, *Friction*, p. 6.

79. Das, Veena. *Critical Events: An Anthropological Perspective on Contemporary India*. New Delhi: Oxford University Press, 1997.

80. Smelser, Neil. *Theory of Collective Behaviour*. London: Routledge and Kegan Paul, 1963, p. 196. Anthony Giddens discusses similar experiences in terms of 'fateful moments'. Giddens, Anthony. *Modernity and Self-Identity: Self and Society in the Late Modern Age*. Stanford: Stanford University Press, 1991, p. 113.

81. See Badiou, Alain. 'The Event in Deleuze'. *Parrhesia* 2 (2007): 37–44. Available on http://www.parrhesiajournal.org/parrhesia02/parrhesia02_badiou02.pdf, accessed 10 January 2019.

82. Deleuze, Gilles. *The Logic of Sense*, translated by Mark Lester and Charles Stivale, edited by Constantin Boundas. London: Athlone Press, 1990, p. 12. Deleuze focuses on '*événement*' as a disjuncture implicated in time and meaning such that what made sense before the event becomes distant

or opaque to be supplanted by new meanings after the *événement*. Alain Badiou discusses the event in terms of the 'Truth of power' that becomes suddenly visible. Borrowing from the works of both Gilles Deleuze and Félix Guattari, Bruce Kapferer elaborates on the temporality of the multi-faceted event as 'the critical site of emergence, manifesting the singularity of a particular multiplicity within tensional space and opening toward new horizons of potential' (p. 15). Laurent Berlant prefers to distinguish the drama of an event from *situation* as a 'genre of unforeclosed experience' that could then become an event'. Badiou, Alain. *Being and Event*. Chicago: A&C Black, 2007; Badiou, Alain. *The Rebirth of History: Times of Riots and Uprisings*. London: Verso, 2012; Kapferer, Bruce. 'Against the Case as Illustration: The Event in Anthropology'. *Social Analysis* 54 (3): 1–27, p. 15; Berlant, Lauren. *Cruel Optimism*. Durham: Duke University Press, 2011, p. 5.

83. Criticality recalls other theoretical methods about socio-political challenges and changes. One of these is critical methods reflection that tends to focus on transformative moments in the individual's biography and how they are experienced and remembered as significant for their lives. Thomson, Rachel, Robert Bell, Janet Holland, Sheila Henderson, Sheena McGrellis, and Sue Sharpe. 'Critical Moments: Choice, Chance and Opportunity in Young People's Narratives of Transition'. *Sociology* 36, no. 2 (2002): 335–54, p. 335. The focus in this book is less on the individual's growth, but a *distributed biography* of a nuclear power plant and its imbrication among diverse communities.

84. On postmodern anthropology, see Clifford, James and George E. Marcus. *Writing Culture: The Poetics and Politics of Knowledge*. Berkeley: University of California Press, 1986; Marcus, George E. and Michael Fischer. *Anthropology as Cultural Critique: An Experimental Moment in the Human Sciences*. Berkeley: University of California Press, 1986.

85. On hegemonic claims to objectivity with respect to scientists, see Latour, Bruno and Steve Woolgar. *Laboratory Life: The Construction of Scientific Facts*. Princeton: Princeton University Press, 1988.

86. On an analysis of the use of narratives in politics, see Polletta, Francesca. 'Contending Stories: Narrative in Social Movements'. *Qualitative Sociology* 21, no. 4 (1998): 419–46.

87. McGranahan, Carole (2015) 'Anthropology as Theoretical Storytelling', 19 October, Fall, Savage Mind Writers' Workshop Series, *Savage Mind: Notes and Queries in Anthropology* https://savageminds.org/2015/10/19/anthropology-as-theoretical-storytelling/, accessed 18 January 2019.

88. See White, Hayden. 'The Value of Narrativity in the Representation of Reality'. *Critical Inquiry* 7, no. 1 (1980): 5–27; and Rosaldo, Renato. *Culture and Truth: The Remaking of Social Analysis*. Boston: Beacon Press.

89. Accordingly, I do not subscribe to Polletta's view that narrative stands outside of referential modes for the latter too are marked by structuring conventions that are historically and culturally specific. Polletta, 'Contending Stories'.

90. Sovacool, Benjamin K. *Energy and Ethics: Justice and the Global Energy Challenge*. Basingstoke: Palgrave Macmillan, 2013, p. 15.

91. Most studies on anti-nuclear movements are based in the global North. They include Kitschelt, Herbert. 'Political Opportunity Structures and Political Protest: Anti-Nuclear Movements in Four Democracies'. *British Journal of Political Science* 16, no. 1 (1986): 57–85; Rudig, Wolfgang. *Anti-nuclear Movements: A Survey of Opposition to Nuclear Energy*. Harlow, Essex: Longman Group UK Limited, 1990; Lofland, John. *Polite Protesters: The American Peace Movement of the 1980s*. Syracuse, NY: Syracuse University Press, 1993; Flam, Helena (ed.). *States and Antinuclear Movements*. Edinburgh: Edinburgh University Press, 1994; and Welsh, Ian. 'Anti-nuclear Movements: Failed Projects or Heralds of a Direct Action Milieu?' *Sociological Research Online* 6, no. 3 (2001). Available on http://www.socresonline.org.uk/6/3/welsh.html. On the campaign to stop the construction of a nuclear power plant in Kerala, see Chapter 3 and Varghese, M.P. *A Critique of the Nuclear Programme*. New Delhi: Phoenix Publishing House Pvt Ltd., 2000. On more recent stories of women's experiences in Idinthakarai, see Vaid, Minnie. *The Ant in the Ear of an Elephant*. New Delhi: Rajpal and Sons, 2016. On anti-nuclear weapons campaigns across South Asia, see Bidwai and Vanaik, *South Asia on a Short Fuse*.

2

A Nuclear Paradise

Visible by its glimmer, [the mountain] is thickly covered
With wild rice, sugarcane with severed joints,
Millets ready for plucking, ragi that grows
On fertile soil, garlic, turmeric,
Lovely kavalai vines, plantains,
Arecas, bunches of coconuts hanging low,
Mangos, and jackfruits.
(Cilappatikaram)[1]

A crow will conquer the owl in broad daylight
(Thirukkural)[2]

Many myths envelop the south Indian peninsula. According to one of them, India's Malabar coast, on its western side, is a part of the rich and verdant land reclaimed from the sea by Parashurama, a warrior sage incarnation of the Hindu deity, Vishnu. On the southernmost tip of the district is a temple dedicated to the virgin deity, Kumari, after which the district of Kanyakumari is named. Otherwise known as Devi Bhagavathi or Kanya Devi, she is an avatar of the Hindu deity, Parvati, who was supposed to have married Shiva. However, he did not turn up for the ceremony, casting her in the permanent role of a virgin goddess. According to another myth, one of the prominent hillocks, Marunthuvazh Malai, was an accidental by-product of the monkey-headed god, Hanuman. On one of his aerial journeys, he dropped a piece of earth from a Himalayan mountain that he was carrying. It harboured a life-saving herb to tend to the wounded in Ram's war against Ravan in Sri Lanka. The hillock where the mound landed is now reputed to be home to about sixty species of medicinal herbs.

These shrouds of mythical mists envelop a land of lavish vegetation with lowland growth of coconut, areca nut, neem, *amla*, rice, and plantain (*musa x paradisiaca*) of which there are remarkable fifteen varieties, ranging from yellow and slim to pink and plump. In the midlands, there are tapioca, groundnut, and vegetables; and in the highlands, tea, coffee, rubber, cocoa, pineapple, cloves, and pepper. Scattered around the fertile forests are orchards of mango, citrus fruits, cashew nut, jackfruit, and striking flame-coloured flowers. Fourteen types of forests traipse the level lands all the way up the rugged ghat mountains, an extension of the Deccan plateau (Figure 2.1). Some of these forests have been converted into rubber and teak plantations, and others are threatened by ongoing road-widening and development projects.

Figure 2.1   Kanyakumari District, 2006.
Source: Raminder Kaur.

Opposite the benevolent gaze of Devi Bhagavathi are two rocky islets known as 'twin rocks'. Keen observers on a boat might spot whales, dolphins, or even the endangered dugong when the season permits. One rock is situated about 200 m from the shore, with a monument dedicated to Swami Vivekananda who famously came here to meditate and visualize 'One India' before his trip to the World's Parliament of Religions in Chicago in 1893. The shrine was erected in 1970 and consists of a building containing the Swami's statue and a small temple facing it. The second rock further out from the harbour is host to a colossal 95-foot high statue of the Tamil sant, Thiruvalluvar, a saint reputed to have been born in 31 BC.[3] He stands tall on a plinth with verses from his work, *Thirukkural*, carved onto the interior stonework of the pedestal. This monument was erected more recently with its inauguration in 2000 (Figure 2.2). From these rocks, in the far distance to the east, about 30 km away, ethereal cream and ochre domes are barely visible. They could be modernist Buddhist stupas but the more perceptive ones know that these are the definitive new structures of the Kudankulam Nuclear Power Plant.

Figure 2.2 Vivekananda Rock Memorial and Thiruvalluvar Monument, Twin Rocks off Kanyakumari Town on Cape Comorin, 2006.
Source: Raminder Kaur.

This natural treasure trove has now become a national atomic trove. To add to the enterprise of mining sands for the nuclear fuel thorium, the southern peninsula has become arraigned to India's ambitions to career ahead with its nuclear industries. Inevitably, the decision to build a nuclear power plant in the area generated tremendous changes that triggered a sea change in people's awareness of themselves regionally, nationally, and internationally.[4] Residents of the southern districts were compelled to join a larger drama whether they wanted to or not. They were mainly located across three administrative districts— Kanyakumari, Tirunelveli, and Thoothukudi. Going beyond a 'not in my backyard' (NIMBY) protest, their struggle for recognition went on to question the entire political apparatus of nuclear energy and its implications for the Indian nation, democracy, and development.

Along with the adjoining state, Kerala, Kanyakumari District was a part of the Travancore–Cochin state along the Malabar coast during the colonial era. This was before language-based campaigns resulted in the district merging with the rest of Tamil Nadu in 1956. The area now adjoins Tirunelveli District, the location where Kudankulam is sited, whose eponymous capital lies further away from the nuclear plant than Nagercoil, the capital of Kanyakumari District. It is these two districts that form the hotspots for this book. Nagercoil is a local centre of transport, communication, and activism. The nearby Manavalakuruchi is where Indian Rare Earths Limited is located, a major site for the mining and processing of thorium from monazite beaches. Kudankulam is where the nuclear plant is sited in Tirunelveli District; and, later in the book, the adjoining coastal village, Idinthakarai, becomes our focus as the hub of a large-scale anti-nuclear movement not too long after the Fukushima Daiichi nuclear disaster in 2011.

Modern nuclear India sees itself on a three-stage program, a plan that was conceived in the 1950s by the founding chairman of India's Atomic Energy Commission, Homi Jehangir Bhabha, and chemical engineer and bureaucrat, Nuthakki Bhanu Prasad. The program is influenced by considerations to do with India's limited amounts of high-grade uranium as against a large percentage of the world's thorium present in monazite beaches in southern India. A thorium-based nuclear fuel cycle is, therefore, held as 'the only way to insure a stable, sustainable, and autonomous program'.[5]

The stages include, first, to use natural reserves of uranium for power production, much of which is the isotope uranium 238 with a small amount of the fissile uranium 235 to produce plutonium 239. The second stage is to process the spent fuel (mixed oxide or MOX made from plutonium 239) along with natural uranium in fast breeder reactors that are now located in Kalpakkam along the eastern coast of India. The third is to utilize the abundant supplies of monazite on the beaches of Kerala and southern Tamil Nadu to produce thorium 232 that, once introduced to the plutonium 239, can be transmuted to uranium 233, a fissile material that can produce energy in a thermal breeder reactor.[6] After the initial fuel charge, this thermal breeder reactor, in principle, can be refuelled with naturally occurring thorium.

The available evidence suggests that the Indira Gandhi Centre for Atomic Research (IGCAR) in Kalpakkam is yet to master fast breeders for it to make any progress onto the third stage, a route that other nuclear countries have said is a futile one for the high-risk and costly yet poor performance record of the fast breeder.[7] However, India's Atomic Energy Commission pride themselves as being more advanced in nuclear technologies, even more so than the US.[8] After several delays, a recent announcement decrees that commissioning of the prototype fast breeder reactor *may* be in 2019 but, to date, there has been no further news.[9] Great expectations such as these are repeatedly spouted by the nuclear authorities but not shared by all, not least by the many coastal village residents of Kerala and Tamil Nadu, for it would spell nothing but disturbance and further health hazards in places where thorium extraction would take on a war footing.[10]

Taking such contentions on board, in this chapter, we ask what are the ecological, economic, and socio-political implications of nuclear industries in peninsular India. Our scope ranges from the history of the region, the potentials and constraints of the environment, and the larger impact of a nuclear power plant on surrounding communities.

There are some overlaps to note with a space of criticality and Gabrielle Hecht's notion of 'nuclearity' as 'a contested technopolitical category': 'It shifts in time and place. Its parameters depend on history and geography, science and technology, bodies and politics, radiation and race, states and capitalism'.[11] The categorization is contingent on human interventions in industries such as uranium mining, milling, constructing, and commissioning and how they make places, objects, and hazards 'nuclear'. As she elaborates: 'Nuclearity is not so much an essential property *of* things as it is a property *distributed over* things'.[12]

The space of criticality takes Hecht's insightful analysis on board while spreading its remit to the emergent and attendant contexts of a place marked with the growth of atomic industries. As with nuclearity, this space is itself locked into broader historical, regional, national, and transnational dynamics to do with human interventions, technologies, materials, electricity production, trade, security, the politics of their representation, and socio-political tensions and hierarchies that attend these developments. In peninsular India, the Kudankulam Nuclear Power Plant wore the hard hat of nuclearity.

However, in comparison to nuclearity, criticality does not just focus on the coming into being of the 'nuclear'. It has a larger purview that also accommodates the potential, the enveloping, the overlapping, and the nuclear-essential referring to certain fundamentals of a nuclear plant at Kudankulam.

By *potential*, I allude to postcolonial state ambitions to channel the copious supplies of thorium that could produce the fissile isotope uranium 233, and thereby hope to demonstrate techno-scientific capacity and capability.[13] Not too dissimilar from the early days of uranium, the status of thorium remains aspirational.[14] Without as yet the rank of nuclearity that could produce energy, thorium is at best a second-rate contender in material terms although there is no slackening when it comes to its first-rate role in fantasies of nuclear and economic benefits for sand-miners, their families, and the nation at large.

Enveloping alludes to social and ecological elements that predispose this area to heightened sensitivities around anything atomic, as with natural residues of alpha radiation from thorium in monazite sands and their consequences in terms of the inordinately high levels of background radiation and low levels of health along the coast. Living in high radioactive zones that are not due to human intervention adds another penumbra to man-made nuclearity that defines only those things that have attained visibility and value in national and international politics, markets, and regulatory frameworks.[15]

Overlapping ties in related industries, formal and informal, that transcend nuclearity as with rare mineral mining, transport networks, other constructions, and commercial opportunities. This perspective owes to the knock-on effects of plans for a nuclear plant in the region that might act as a magnet for a plethora of other businesses along with a variety of state, corporate, entrepreneurial and casual labour actors.

The *nuclear-essentials* owe to certain foundational imperatives that vary from region to region. These are strictly outside the ambit of national security, but without them nuclearity cannot proceed. In peninsular India, it was water and land that was most at issue in highly charged contexts of depletion and degradation, and with the dawning realization amongst local residents of contamination and displacement due to the nuclear plant.

Following the sociologists Scott Lash and John Urry's account on disorganized capitalism, we have here a case of 'disorganised nuclearism'—powerful yet uneven processes that span the aspiration and the reality, the official and the unofficial, the conspicuous and the clandestine, and the peripheral and essentials that have their own vectored logics and circuits of value.[16] The space of criticality is then less a concept or classification, unsettled and ambivalent as it may be, but the grounds for its emergence or, in words, the background story to nuclearity in the region. In our overview of Kanyakumari and Tirunelveli Districts, we will explore the dissipating dimensions of this space, covering environmental and demographic features, nuclear and sand-mining operations, the personnel that work for them, issues to do with water and land, and implications for the inhabitants of the inland village, Kudankulam, after which the nuclear plant is named.

The Place and the People

Enjoying both tropical and subtropical climates, Kanyakumari District is appreciated for its weather, hot but invariably tempered by a breeze and occasional storms. The plains are saturated with many natural water tanks along with a series of rivers, tributaries, canals, and ponds that help irrigate agricultural fields. This is unlike the neighbouring district of Tirunelveli where Kudankulam is located in Radhapuram taluk—harsh, hot, and barren and therefore, on its own, unable to

cater to the mammoth water needs for the construction and maintenance of a nuclear power plant.[17]

There are several dams in Kanyakumari District, the most significant of which is the Pechiparai Dam constructed in 1906, a dam
that was once earmarked to provide vital fresh water supplies to the
nuclear plant. Although the district is blessed with water, its future
needs are not guaranteed. While water supplies have been over-uti-
lized by construction projects, a crumbling infrastructure and a lack
of maintenance has undermined their capacity to provide succour.[18]
Sporadic rainfall in the region has compounded the problem such
that the region has undergone severe drought where water may not be
available for weeks at a time. This is to add to the fact that coastal wells
have become saline, degraded further by the tsunami of December
2004 that ravaged the shores of south India along with other coastal
populations around the undersea Pacific Ocean eruption. Ground
water tables too are depleting, not least because litres of water were
used for the construction of the nuclear plant. Much of it was through
illegally dug bore wells to provide for the contractors: the Ground
Water (Regulation and Control of Development and Management)
Act (2005) stipulates that you cannot sell water dug out from a well
but its lack of implementation meant that plenty of landowners in
neighbouring villages and towns reaped dividends regardless (see
Chapter 6). This is all the more reason why fresh water is such a valu-
able entity in the region, and a deeply disputed one as thousands of
litres are required, not just for the construction, but also the continual
running of the nuclear power plant as moderator and as coolant with
sufficient reserves at hand to deal with any emergency. Due to the
controversy over siphoning water from the dam, original plans to
use dam waters were dropped and the water is currently provided by
costly desalination units at the plant.[19]

To give a sense of demographics, according to the 2011 census, the
peninsular population in Kanyakumari District is constituted by about
49 per cent Hindus, 49 per cent Christians, 4 per cent Muslims, with
the remainder consisting mainly of Buddhists, Jains, and Sikhs.[20]
This compares with about 79 per cent Hindus, 11 per cent Christians,
and 10 per cent Muslims in the neighbouring Tirunelveli District.[21]
Education is highly valued across the board. As with the adjoining
state of Kerala, literacy rates are high in Kanyakumari District relative

to the rest of Tamil Nadu (91.96 per cent as compared to an average of 80 per cent across the southern state with Tirunelveli District at 82.5 per cent).[22] Despite the statistics, several residents boasted that there was a 100 per cent literacy rate in Kanyakumari District. A perception of high literacy rate owes to a history of educational institutions established by Christian missionaries in the Malabar region which did not exclude girls and low-caste children. The missions were established after the princely state of Travancore was annexed to the British Empire in the early nineteenth century.[23]

Traditionally, most employment has been in fishing, farming, and weaving. In 1980, the district was officially declared as 'industrially backward' and new industries were eligible for a 10 per cent subsidy towards their fixed capital investment subject to the maximum of Rs. 10 lakh (100,000).[24] This has encouraged a degree of industrial growth—a soft drinks plant, cashew processing, cement works, and the manufacture of safety matches, leather goods, plastic articles, wooden articles, and tiles, to name but a few. This is a scenario that has dramatically changed with the introduction of the nuclear power plant in the region that spawned a range of industries and construction projects. Commercial developments were then boosted with the nation-wide initiative of Sagarmala from 2016. Literally meaning 'necklace of the sea', Sagarmala comes with the tagline, 'port-led prosperity' with a swathe of multimodal infrastructure development projects that span from coasts to inland, from the northern to the southernmost posts of the country.[25]

Due to the relatively high levels of literacy, Kanyakumari District has hitherto suffered from a case of *underemployment* in which the highly educated take on more menial employment. Moreover, as with the lower middle classes in the rest of the country, there is a certain level of prosperity, prestige and permanence attached to 'government jobs'.[26] Despite possible hazards to his health, for instance, a manual worker at the government-owned corporation, Indian Rare Earths Limited, was earning 10,000 rupees a month in 2006 before he retired. Compare this to about 4,000 rupees for a taxi driver (before expenses), 1,000 rupees for a teacher (in a private school), and 3,000 rupees (in a state school). Earlier there had been some cachet around the idea that government departments and public–private companies were involved in the nuclear plant, and therefore skilled work opportunities could be increased for the educated middle classes.

A prime site for such middle class aspirations was the Kanyakumari District capital, Nagercoil. Lying about 10 km from a monazite-rich coastline and about 30 km away from Kudankulam, the town has a growing population of around 225,000 people.[27] It is a central hub from which to reach surrounding villages and towns, yet it is a fairly humdrum town with its most famous monument being the Nagaraja Temple dedicated to the serpent god. In its gardens, it contains the very unique Naga flower (*Couroupita guianensis*), a symbolic representation of Nagaraja, the 'king of snakes', after which the town is named.

One of the town's other conspicuous features is the number of medical facilities. It has been dubbed 'the city of hospitals': according to the scientist P.D. Mercy, they 'are flourishing like an industry in Nagercoil'.[28] This growth owes in part to missionaries such as Charles Meath who were also trained as doctors.[29] Nowadays, there are about 130 medical establishments in Nagercoil, all private, except one government-run hospital. Just outside the town, there is a medical college to train the next generation of experts.

The highly disproportionate number of hospitals in a small town such as Nagercoil acts as both cure and panacea for what may, with certain reservations, be described as 'a culture of hypochondria'. Cancer is one of these abiding fears and the subject of intense medical specialities. It is a slippery enigma that compels an exceptional drive to understand, explain, and manage mutation. Even though causation is controversial and cannot be accurately determined, it is common knowledge that cancer incidents owe a great deal to high background radiation in the region. According to the United Nations Scientific Committee on the Effect of Atomic Radiation (UNSCEAR) and a team of international scientists, the western coastline of Kanyakumari District along with the contiguous shoreline of Kerala 'contains the world's highest level of natural radioactivity in a densely populated area'.[30] The flipside to this was a high rate of radiation-induced mutations and diseases.

However, data on cancer was not centralized or available—and even if it was, say for instance, in the Regional Cancer Centre (RCC) in Kerala's capital Trivandrum (aka Thiruvananthapuram), it could not always be relied upon. Many such public institutions were dependent on funding from government sources and minimized the role of the environment in cancer causation preferring to stress genetics and lifestyles as the more significant factors. Moreover, as we shall see in Chapter 7, it was not in their interests to publicize data that might inform an anti-nuclear drive, which highlighted the already high levels of cancer in the region.

Rare Minerals

Commercially exploitable levels of mineral deposits include a 7 km stretch between Kadiapattinam and Colachel, and a 2 km one in Midalam on the south western Malabar coast. The coast here contains high concentrations of heavy minerals such as ilmenite, sillimanite, rutile, garnet, zircon, and monazite. The relatively poor radioactive cousin to uranium, thorium, is extracted from the abundant supplies of monazite for the nuclear industries.

Before the onslaught of the atomic age, the monazite sand was most appreciated for its ilmenite, an ore for titanium that was previously used for incandescent lighting in gas mantles. Monazite was first identified by a German scientist, Charles Schomberg, who saw minerals stuck to coir imported from Manavalakuruchi in a German godown in 1906. This was due to the coconut husks that were processed and converted into coir on the sands. The scientist traced the coir to Manavalakuruchi, and a factory was established in 1911, which was taken over by the British at the onset of World War I.[31] The sand mining industry was picked up again after 1945 once the phenomenal power of atomic minerals was realized soon after the bombs in Hiroshima and Nagasaki were dropped on those fateful days in August 1945.[32]

Imperial Chemical Industries, a company that was mining sands in Travancore, had another role with the British government in the attempt to develop nuclear weapons during the period of World War II. Codenamed Tube Alloys, it was the first British nuclear weapons programme in collaboration with the Canadian government and was later subsumed into the US-led Manhattan Project until the end of the war.[33] Hopes began to be expressed for thorium as well to step up to the radiant red carpet of nuclearity. An advertisement by the

Imperial Chemical Industries (India) in 1947, for instance, high-lighted the large deposits of thorium on the Travancore coastline that 'has attracted world-wide attention':

> For more than twenty years monazite ilmenite has been exported, mostly to America, and only recently has the strategic importance of thorium, with the discovery of the atomic bomb, been realized as a source of atomic energy. Together with uranium, thorium can pro-duce a chain-reaction to manufacture a material which is an atomic explosive.[34]

Post-war claims to having exported thorium to USA in the adver-tisement appear like a misplaced bid for strategic greatness, as it was with uranium from the then Belgium colony of Congo (now the Democratic Republic of the Congo) that had the dubious hon-our of supplying the Manhattan Project scientists for the making of atom bombs.[35] Nevertheless, even a vague connection to the atomic enterprise in USA triggered princely ambitions of strength and even separation from the rest of independent India.[36]

In the twilight hours of colonial rule, C.P. Ramaswamy Aiyar, the Dewan of the state and the president of Sri Chitra Council, was enthusiastic about the exploitation of thorium, first for the Travancore princely state, and in June 1947, hand-in-hand with the Indian gov-ernment to pursue conjoint mineral sands and atomic research.[37] Ambitions were channelled into a dual approach to development: that India needs 'industrialisation, not as contra-distinguished from or opposed to cottage industries, but side by side with it'.[38] It was felt that nation-orientated amends needed to be made for the fact that huge quantities of mineral sands had been shipped abroad to produce other items used for industry such as ilmenite, sillimanite, and zircon.

The monazite in the south Indian region became the lottery ticket to fuel imaginaries of an independent, developed as well as atomic state. In an address to the Sri Chitra Council, Aiyar stated:

> Members must have read about the atomic bomb which was used at present as an instrument of destruction. But this source of energy might be diverted to the production of power in peacetime in addi-tion to uranium. Experts have held that thorium is only next in

importance to uranium and this thorium is found in our mineral lands. It thus becomes all the more necessary for us to conserve our mineral resources on which so much of our future welfare depends and steps in that direction have already been taken by the Government.[39]

In 1950, after an extractive venture with the French company, La Socéité des Terres Rares, Indian Rare Earths Limited (IREL) was established to mine and process the heavy mineral sands as a government of India undertaking under the administrative control of the Department for Atomic Energy. Manavalakuruchi is one of their sites of operations.[40]

Literally meaning 'taking the sand away', Manavalakuruchi is a village that comprises several hamlets on its coast where the fishing community resides. The beach is punctuated here and there by houses among coconut groves, roads, and the IREL factory. Minerals on the beach are the result of the disintegration of the Western Ghats brought to the river by rains and then flushed into the sea. Black rivulets of sand form as the tide returns to its oceanic forebears leaving the heavier sediment behind. Around the IREL plant are mountains of sand including 'waste sand'—that is, sand with monazite taken out of it. It is shored up by truck and by hand at the behest of sub-contracted fishermen who move radioactive sands in baskets on their heads, a practice that has not much changed since 1957 when John Bugher of the US Atomic Energy Commission made similar observations.[41]

With the input of Australian expertise, the plant is a wet concentration, one that enriches the heavy mineral content from 50 to 95 per cent. Daily, about 1,200 tonnes of beach sands are dredged and 600 tonnes of heavy mineral concentrates produced.[42] The sand mining is done by manual methods and dredging with machinery while processing of the sun-dried sand is done by electromagnetic and electrostatic operations. First, silica and other waste is removed from the sand, Then the concentrate is processed separately for minerals while the waste sand is dumped back on the beach. After processing the residue, radioactive monazite tailings are covered with sand and stored in designated sites according to rules by the Department of Atomic Energy. These need to be periodically checked by the Atomic Energy Regulatory Board to ensure that radiation levels are within permissible limits.[43] Monazite, along with thorium oxide, is processed at the

Bhabha Atomic Research Centre in Trombay in Greater Mumbai for the manufacture of rare earth oxides and nitrates.

Currently, IREL employs about a thousand people, an estimated 90 per cent of whom are from the locality according to local residents. The rest are scientists and engineers drafted in from other cities. IREL is a prohibited zone to the general public. Initially there was resistance to sand mining from fishing communities immediately surrounding Manavalakuruchi. Concerns were raised that mining would hamper fishing, and that they may not be able to keep their catamarans on the beach. At that time, they were not aware of the pollution and contamination that sand mining for atomic minerals could cause to the environment.[44] By mining the coastal sands, radioactivity is released from the monazite. This is to add to the uranium released into the air by the many granite quarries in the district, some of which supplied the construction of the Kudankulam plant.[45]

Controversy has also sparked around those working for the IREL colluding with private ventures such as V.V. Minerals, a powerful family-run company based in the port town of Thoothukudi on the southeastern coastline of Tamil Nadu.[46] According to V.V. Minerals publicity, the company is India's first and largest mining, manufacturer and exporter of garnet and ilmenite with the logo, 'Grit Strength—Inside Out'. In these beaches of southern India, it is mainly a sulphateable ilmenite with low constituents of uranium, thorium, and chromium (III) oxide that is extracted for titanium for use in anything from the paint industry to the manufacture of nuclear submarines. This ilmenite is distributed globally at the behest of the Australian company, the Australasian Minerals and Trading Pty. Limited.

Beginning their operations in 1989, V.V. Minerals is run by three brothers, S Vaikundarajan, S Jegatheesan, and S Chandresan. They boast on their website about their 'control of a 15 kilometres beach area with continuous placer mineral deposits plus another 1,200 acres of heavy mineral-rich land'. Certainly, their appetite for buying up land is unremitting.[47] They continue: 'At the global level, we are poised to rise to the position numero uno' [sic].[48]

V.V. Minerals' official script veers quite dramatically from ground realities that include *goondha* (gangster) tactics in the peninsular region.[49] Otherwise known as the 'sand mafia', the company is renowned for their strongman tactics to keep village residents to heel in areas that they operate.[50] When situations become recalcitrant, they send out their henchmen to 'get the work done', as one local resident put it.

Those living near sands that are mined or in their separation facility in Midalam have become sensitized to health issues as a result of radiation from the minerals. Questions have been raised as to what V.V. Minerals does with atomic residue after mineral separation and whether instructions by the Atomic Energy Regulatory Board have been adequately followed for its safe storage. According to news reportage:

> Church records show an average of 10–15 cases of cancer in every village. 'It's absolutely not true that this area already had high levels of cancer. It has increased in the past few years since the increase in illegal sand mining,' says a church official. Cases of Down's Syndrome and impotence have also been reported.[51]

On many occasions, fights have erupted between V.V. Minerals supporters and fishing communities if the company tried to make a bid to expand their mining enterprise near their homes. One local observer stated:

> What V.V. Minerals do is strictly illegal because no private company can indulge in such practices [on government-owned land]. But they say they have a High Court order which allows them to proceed. This stops the District Collector from doing anything. They have such power that they have bought out MPs and MLAs [Members of Legislative Assembly] even in central government to continue their trade ... Strictly it's not allowed, but they are so powerful they are above the law. They are basically uneducated musclemen.

On top of their strongman tactics, an activist recounted the time that he was subtly warned not to direct their anti-nuclear campaigns against sand mining:

> Once I was in Radhapuram with a friend. I was at a hotel [a café] and there must have been some V.V. Minerals manager there. I went to pay the bill and owner said 'It's OK. It's already been paid' ... I insisted

on paying. And he said: 'Sir, why do you want to cause problems for me? It has been paid and that's it!' Basically V.V. Minerals were subtly threatening me. They were hinting that you may be known for your anti-Kudankulam work, but don't start anything against my factory.

Paying for someone else put the recipient in a position of gratitude and therefore inferiority. It also implied that a relationship of debt had been created that the beneficiary should not upset.[52] In a case of what may be termed *belligerent benevolence*, the expectation was that the recipient of the 'gift' should not interfere in V.V. Minerals operations. Through such benev(i)olent ploys ranging from the aggressive to the manipulatively benign, V.V. Minerals have become a power to be reckoned with at the edges of powerful networks that include the nuclear industries.[53]

The Kudankulam Nuclear Power Plant and Township

By the turn of the millennium, Kudankulam, a village of about 13,000 residents, was indelibly marked on India's map of nuclearity.[54] Over 929 hectares of land were taken for the project and another 150 hectares for the township, Anuvijay, located about 10 km further down the coastline.[55] Security, intelligence, and related operations firmly tagged this hitherto hinterland to central government. The gates to the nuclear plant and township are akin to national border security. Central Industry Security Forces, with their signature red berets and their deep khaki uniforms, patrol the imposing fence and gate. A large sign with a neat picture of the nuclear plant, a 'Government of India Enterprise', lies at the edge of the main Kudankulam junction surrounded by a small clinic, a medical store, a police station, various other tea stalls and shops, and the Sri Ram Orthopaedic Hospital. By 2011, the space was bulwarked with cordons of security forces and police when local resistance against the plant picked up. There is a board at the gate to the plant reminding visitors that this is a prohibited zone, and any offence would be punishable under the Atomic Energy and the Official Secret Acts. Just outside the gate, there is a booth that, during the heyday of construction from 2002, was attended to by a long line of men waiting to get daily labour with a parallel entrance for trucks and lorries carrying water, sand, cement, and rocks, each driver showing his specially made identification card to the officers in this tight net of security.

The virtually immaculate surroundings of the nuclear plant and township recall the environs of a premium resort albeit with staid touches that characterize most Indian governmental ventures. Sponsored by Larsen & Toubro, an Indian multinational conglomerate and contractor for the plant, signs in English, Tamil, and Hindi about safety flank the large road on both sides as one approaches the main reactor buildings. This is one of the few places in this southern tip of Tamil Nadu where the Devanagari script is out in the open, a sure sign that Hindi-speaking north Indians had come to stay (Figure 2.3). 'Safety first or be the first', one sign warned the nuclear employees. Others proclaim the importance of safety, a message that they have inculcated to the point of obsession: 'Kudankulam Nuclear Power

Figure 2.3   Signpost for Kudankulam Nuclear Power Plant, Kanyakumari District, 2006.
Source: Raminder Kaur.

Plant motto: safety in toto'; 'Safety is a state of mind. Accident is an absence of mind'; 'Always alert. Nobody hurt'; 'One man's careless-ness is a 100 family's torture'; 'Check wires. Avoid fires'; and 'Work safely...Go home with a smile'.

The reactors and buildings girdling the coastline are a good 2 km drive from the main gate. The two 1,000 megawatts VVER reactors stand tall among the other compound buildings. VVER refers to pres-surized water reactor designs and is from the Russian *Vodo–Vodyanoi Energetichesky Reaktor;* translated as Water–Water Power Reactor. Nearby, a modernist–styled, smoky glassed three-storey building, the Nuclear Training Centre, houses most of the employees' offices. Around this complex is a clinic, a meteorological laboratory (in asso-ciation with the Bhabha Atomic Research Centre), an anti-fire unit, and a 50-foot high observation tower.[56]

Opposite to the Nuclear Training Centre are a few steps that lead up to a large abstract humanoid statue swooping its arms up at 45 degrees from a swirly white cascade of white stone. The erection is called 'Power of Wings' and stands about 12 feet high with hands held aloft around an egg-like head. The statue was raised in 2002 after the laying of the foundation stone. The sculpture faces east, the direction of the sunrise, where the egg-like reactor domes can be seen through the vista between the two outstretched hands. Humanity seemed to cradle the sun, the source of all energy, in between its hands. The plaque underneath the statue endorsed a metaphysical message writ-ten in English:

> We do not come with anything in this world. We do not go with any-thing either. What we leave back are some thoughts and ideas that we generated with the power and energy we got from this world.

Beguilingly humble, the message veils an unbending belief in the sanctity of nuclear power. It is an ode to the fundamental nature of nuclear matter, while the human condition is a mere ephemeral shadow. Nuclear energy is re-scripted as the eternal font of inspiration and creativity. As public servants the message is to be stewards to this fundamental power of life itself.

The public servants work for the Nuclear Power Corporation of India Limited (NPCIL) whose headquarters are in Anushakti Nagar (the Vikram Sarabhai Bhavan), in Trombay, a north eastern suburb of Greater Mumbai.[57] They live in Anuvijay, a residential complex about 10 km from the plant built to house the permanent workers and visitors for the Kudankulam Nuclear Power Plant. Literally meaning 'victory of the atom', Anuvijay is a contemporary update on Le Corbusier's model Indian city, Chandigarh—a nondescript, uniform yet salubrious and well-catered for place marked by the linearity of modernist ambition.[58]

In local eyes, the gated township appeared to be a desirable abode. The main site of Anuvijay, with ongoing expansion plans, houses about 700 regularly spaced apartments for families, and flanks the beach on the southern side of the Tamil Nadu coast. Efforts were made to turn the township green with the planting of many saplings and plants. Nutritionally rich reddish coloured soil was bought in and pipelines were installed for water sprinklers. But they barely kept the greenery alive—fledgling coconut trees looking sorry for themselves stood cowering in the harsh sunlight of this starkly dry and desolate scrubland.

In the centre of the residential complex, a large roundabout is crowned with a gigantic tetrapod. Standing like an epic god of atomic science, it is surrounded to the south by a brightly painted building that houses the Atomic Energy Central School for the township's children. Around the arterial roadways flank several buildings with about 10 flats in each and car-parking space below, mainly for a fleet of white Ambassadors that have been gradually replaced by Marutis and other cars made possible by the deregulation of the economy over recent decades. On occasion, women are visible on the balconies putting out washing on the lines. It appears to be just like any other suburban area of an Indian heartland.

There is a guesthouse and a restaurant facing the sea, but this is only for VIP visitors. Opposite to it is a brightly coloured Hindu temple. It stands out as a brash reminder of the resilience of religion even at the heart of a techno-scientific enterprise.[59] Around the corner, in Russian doll-like style is a smaller gated community within the larger gated community. This compound is for Russian employees who have stayed to oversee the construction and maintenance of the plants. These buildings along with a few VIP houses, a swimming pool (behind high walls) and a bar said to be a 'Russian-only zone' are among the areas that are out of bounds for other residents and their visitors.

There are a couple of commercial zones on the main site consisting of a sparse array of shops in a circular building with a courtyard as a canteen. Not far from one of them is a flour-grinding area called Anu Atta (literally atomic flour) servicing the North Indian palette of roti or chapati. Air-conditioned buses are parked at several places that carry staff to and from the plant. Some vehicles also drive township residents to Kanyakumari town and Nagercoil for recreational purposes. Residents occasionally venture out to have their beloved over-brewed Indian chai outside the sanitized confines of the township to a stall that also doubles up as a bus shelter. In the last few years, the *chaiwallah* stall has expanded alongside a line of small retail shops catering to those coming in and out of the township. Apart from this string of shops on the doorstep, Anuvijay denizens are for most purposes screened off from the rest of the populace, resting secure in their air-conditioned atomic bubbles cordoned off by central security forces whether it be at work or at home. Indeed, some residents have hardly seen life out of such bubbles, with life journeys that start from their township home and school, pass through a specialist government college and end up with one of the nuclear organizations and townships again.

Neo-Brahmanism

A new(clear) middle class had firmly implanted itself in peninsular south India. Many of them are even visibly different from the majority of local Tamilian populations, conspicuously north Indian in their relatively tall and fair-skinned looks and dress. Personnel from Chennai and the state of Kerala had also joined the raft of north Indians. They all conveyed the slickness of a savvy scientist, engineer, or bureaucrat: wearing crisp pairs of trousers and stripy shirts with expensive pens tucked away in their chest pockets, they signalled the introduction of distinctly different hierarchies in the region.

Discipline and docility are deeply injected in the veins of a nuclear authority employee. Such features are structured by individual perks and structural constraints.[60] Well-paid jobs (often for life), housing, and other dividends keep employees and their entire family comfortable and compliant.[61] Unlike other departments that are severely underfunded, the nuclear agencies are the mollycoddled darlings of a Nehruvian legacy. No one has quite stepped into the hallowed shoes of Homi J. Bhabha, a close friend of Prime Minister Jawaharlal Nehru and the pioneer of the atomic industry in India. However, the top ranks of the Department for Atomic Energy, the Atomic Energy Commission, and the Nuclear Power Corporation of India Limited continue to enjoy a cosy relationship with the decision-makers in power.

Nuclear scientists and bureaucrats in India see themselves at the frontier of techno-scientific knowledge, torchbearers for a modern nation, and trailblazers for an imagined hypermodern India.[62] Nuclear technology is presented as the cutting edge of pioneering work, even though it is a technology that is well over half a century old. It is the synaptic shortcut for India to prove its superpower status that belies the country's high rate of poverty and malnutrition amongst its billion plus population.

It comes as little surprise that the late rocket scientist (aerodynamics engineer) and former head of the Defence Research and Development Organisation, A.P.J. Abdul Kalam, assumed India's presidency only two years after the 1998 nuclear tests at Pokhran by the Bharatiya Janata Party-led alliance. Otherwise known as 'the missile man' or a 'tech-savvy prez', Kalam remained incredibly popular to the point of becoming 'incredible' as go eulogies for the country itself.[63] Described as selfless and fair, he was a champion for the disabled, poor, and youth, and an eminently learned yet humble man who had little ambition other than to teach after his presidency. For some, Kalam was the only island of hope in political waters of corruption and neglect. In an oft-cited incident, many people wrote a letter to him after a case on the murder of the model and society host Jessica Lal ended with the accused, Manu Sharma (the son of the then Union Minister, Venod Sharma) walking away scot-free. After popular protest took off against the miscarriage of justice, the case was re-opened in 2006, not least due to Kalam's intervention, and the accused in the murder case was duly sentenced.[64]

In another local incident that was relayed to me in the same year, a cook from Nagercoil borrowed Rs. 8,000 from her bank. She repaid Rs. 3,000 of the sum, but then became widowed. She was also a weaver and, to add misery to tragedy, a flood ruined all her material. A representative from the bank came to see the damage and got the proof they needed to write off the loan. However, eight years later she got a letter from the bank manager saying that if she did not pay the loan off—which had built up to Rs. 8,000 inclusive of interest —she would be penalized and even arrested. She panicked and showed the letter to her domestic employer who suggested that she write to Kalam. After writing to him, the president personally phoned up the bank manager, requesting him to look into the matter and if the circumstances were correct, that the bank should write off the loan, which they promptly did. The manager was very embarrassed, and admonished the woman, complaining: 'Why didn't you call me first?' The woman thought this was the height of duplicity, her view being that officials do nothing for the likes of her: 'They have all these organisations and departments, but they are simply namesake. India is full of good-for-nothings'. Kalam, however, was deemed as one shining exception. He was 'a Tamil man done good'. The fact that he grew up in Rameswaram on the eastern coast of Tamil Nadu, from where the legendary god-king Ram is reputed to have travelled to Sri Lanka to defeat the unscrupulous Ravan, did not go unremarked.

This irreproachable 'missile man' became one of the biggest propaganda elements in the state's armoury for nuclear power in the region. His selfless and non-corrupt personality lent a certain aura to nuclear and defence organizations. On more than one occasion, he has been called out to state that he was 'fully satisfied' with the plant's progress and safety features.[65] Nevertheless, as critique grew against the Kudankulam Nuclear Power Plant, so did it against even incredible Kalam. Apologists suggested that essentially he was a good man but that he needed to be 'educated' into seeing how nuclear policies kill people, especially the poor, the weak, and the children. Others were less sparing. A cultural worker from a village in Kanyakumari District commented:

Kalam is a 'good terrorist'. Good terrorist in the sense that the only thing he has invented is for the damage of India—missile, Agni and what have you. Not for genuine and peaceful development of country. Secondly, he meets many young students, but the view he's giving them is a dream of a future India, 2020 India, which does not talk about education for all, upliftment of the poor, development for all. His development [referring to his support for the nuclear power plant] is anti-people.

Recasting Kalam's charm offensive as an offensive charm, he thought that this missile man was overrated. He concluded: 'He loves children, but his pro-nuclear policies actually kill and deform children. Even though he goes on about helping the disabled, doesn't he realise many of these children are disabled because of radiation?'

Another critic in Nagercoil pointed out the misfortune of being subjected to problems that essentially emanate from north India:

This [The Kudankulam Nuclear Power Plant] is all happening here because of power crisis in Delhi which is dictating the terms in areas like Tirunelveli. India is seen to have an energy crisis and nuclear is the answer without enough investment and research in alternative energy sources. It is due to the strength of the Brahmanic lobby.

The strength of the Brahmanic lobby wedded well with what many saw as the strength of the nuclear authorities. Kalam was a token Muslim in the echelons of central power, an honorary Hindu in an otherwise Brahmanic fold in the nuclear–military establishment.[66] As go arguments about whiteness, that this is a discourse about the institutional power of white hegemony, rather than white people being inherently powerful, here we are talking about the institutional power of Brahmanism in the nuclear establishment—neo-Brahmanism.[67] Similarly, the physicist M.V. Ramana describes the nuclear enclave as part of an 'elite priesthood', noting that the structure of the DAE is hierarchical and not conducive to dissent.[68]

It is true that nuclear departments along with intelligence and paramilitary units are exclusive and consensual. Neo-Brahmanism characterizes many of the powerful institutions that are seen to constitute the fulcrum of national security and where reservations are not in place, as is also the case with high-security think thanks, intelligence bureaus, and the like. Unlike other governmental departments and institutions, they are not open to quotas for Backward and Scheduled Castes (SCs). According to the political scientist Sankaran Krishna, it is as if 'upper castes are uniquely fit to govern India and any dilution of their presence could only mean an impoverishment of quality'.[69]

Indeed, as we will go on to see in Manavalakuruchi amongst many other places noted in Chapters 4 and 7, there seems to be an utter disregard for the poor and lower castes who work as daily labourers in the nuclear industry, exposing them to various health hazards. Safety and monitoring measures are lacking, many of the casual labourers using porous bags for the transport of sand instead of plastic ones that ought to be used to contain the alpha radiation in the monazite. In addition, the sheer weight of them as they carry them on their heads led to several of these workers developing spondylitis, spinal, and neck problems. M. Ahmed Khan, an advocate from Friends of Nature Society, aptly points out: 'These workers will be used like curry leaf and put in the dust-bin after extracting the essence'.[70]

In a variation of a term proposed by the critical race theorist David Theo Goldberg, 'neo-liberal racism'—one that promises equality and social inclusion while at the same time contributing to the 'passive extermination' of racial minorities—here we have a case of neo-liberal elitism, a paper democracy over passive extermination that has its roots in earlier caste hierarchies and fears of the numerical expansion of the subaltern.[71] Krishna expands how this elitism comes with a (metaphorical) murderous impulse when upper caste and middle class Indians bemoan Indira Gandhi's aborted attempts at sterilizing the poor and quite happily look the other way if the 'masses' disappeared:

This self-imposed distance between the middle class and the 'masses' sometimes partakes of a genocidal impulse, as is indexed in many milieus—everyday expressions of desire for a country with a smaller population; the occasional wild-eyed scheme for secession from the rest of India by momentarily prosperous enclaves such as the IT sector in Bangalore or parts of Mumbai or Gujarat or Punjab; the oft expressed idea that it may not have been a bad thing if Sanjay Gandhi had had a relatively freer hand for a few more years back in the mid-1970s [with his compulsory sterilization programmes]; urban planning schemes that fantasise bypassing slums through freeways, sub-ways, hovercrafts and helicopters—but is more often indicated by a simple wish for the masses to simply, magically, disappear.[72]

Krishna identifies a liberal discourse that loves the masses in the abstract but not in the actuality, as is evidenced in his analysis of the atomic scientist Raja Ramanna's autobiography, *Years of Pilgrimage*.[73] In the neo-liberal era, it would seem that even the abstract sense of love for the masses has virtually disappeared.

More specifically, fishing communities have been targeted as a beneficiary of the National Planning Commission since independence. Authorities repeatedly state that the communities are 'largely of a primitive character, carried on by ignorant, unorganised, and ill-equipped fishermen' to cite the anthropologist Ajantha Subramanian.[74] Mainstream views deem fishermen as lowly, unruly, drunkards, and as with the sea, unpredictable and dangerous.[75] They are dismissively characterized by 'indolence, lack of thrift, resistance to change and violence, and itself, a product of social isolation'.[76] We hear of '*jai jawan, jai kisan, jai vigyan*' (hail to the soldier, farmer, and scientist), but nothing for the fisherman. They are literally and conceptually on the cusps of the nation in a state of permanent precarity. On grounds of minority religious affiliations as many in the peninsular region are Christians, they are again outside the metanarrative of India.

In the context of Kudankulam, where the sea is a source of livelihood and sustenance for many, neo-Brahmanic philosophy and practice is both threat and sorely out-of-place. Traditionally, high-caste nuclear workers have little respect for the sea: it is *kala pani*, black waters that pollute the human body if you cross them. However, it would seem that the reverse is closer to the truth for it is the human who is polluting the environs. The Environmental Impact Assessment report produced by the National Environmental Engineering Research Institute (NEERI) in 2003 for the proposal of more reactors at Kudankulam, point to the release of water coolant in Zone 2 of the sea, away from the shoreline where the water and any radionuclides can be dispersed and diluted. Not unlike other parts of the world, the sea becomes like a vast rubbish tank where all kinds of items are dispensed with impunity. Based in Canada, the anthropologist Zoe Todd observes that the ocean is treated like 'a dead zone for us to pollute, damage, and poison. It emboldens us to be the very worst kind of human kin to the vast waters of the world'.[77] She advocates a contrary view:

> ... the ocean is not a massive 'unknown' territory *out there*, nor is it an earthly watery outer space for us to mine and dive ..., but is in fact an intimate familiar.[78]

Coastal communities have a deep sense of this intimate familiar that saturates their 'oceanic consciousness'. Others are less respectful, however. 'What do they care? These people are vegetarian only!' retorted one fisherman. He alluded again to the privileged neo-Brahmanic status of the NPCIL that had 'landed' on their shores. Another woman from a coastal village testified to their regional ignorance by telling a story of how a man from the NPCIL advised not eating fish because the bones could get stuck in the throat! Her cautionary tale was shared as a sign of neo-Brahminic incomprehension and stupidity with much mirth.

An Anxious Accolade

Not only were divisions between the migrant middle class and fishing communities made more evident with the construction of a nuclear plant along the coast, but it also entrenched local divisions between Nadar communities, who saw their small landholdings rise in value, and the coastal fishing community who worried about the plant's likely impacts upon their lives and livelihood. Categorized as Backward Castes, Nadars are a constellation of various sub-castes. Traditionally known as cultivators of palmyra trees and jaggery, Nadars have since made great economic strides through education and entrepreneurship to form middle-ranking groups. Despite measures for social reform, caste alliances and hierarchies remain robust. They often provided strong reasons for not getting involved in the anti-nuclear movement, several inland residents, initially at least, dismissing the nuclear plant as merely 'fishermen's problem'.[79]

For Kudankulam residents who lived about 2 km away from the coast, the nuclear plant was initially a mixed blessing.[80] Formerly a relatively insignificant and impoverished place, the inland village consists of landlords, labourers, cultivators, and petty business traders mainly from Hindu Nadar backgrounds and SCs who traditionally worked as beedi twisters (thin rolled up cigarettes), shepherds, palmyra palm tree climbers, and labourers in salt-pans.

After a sustained publicity campaign by the nuclear authorities promising employment and development, inland village residents began to warm to the idea of the nuclear power plant. Many of the petty landowners had sold their land at paltry prices to the NPCIL, not fully realizing the gravity of their decisions. Compensation was as little as Rs. 2,000 per acre and Rs. 100 per tamarind or cashew tree, but even this was not guaranteed.[81] The NPCIL promised that at least one person from each family would get employment in the compound, recalled several people with a note of cynicism: 'Big promises from big people', as one woman in Kudankulam put it.

While the plant indicated an anxious imminence, it also made some local people feel important by association: 'everybody knows where Kudankulam is' as one resident put it. Through regional development and contracts with the NPCIL, many of them prospered. In the mid-2000s, people in the area described it as going through a 'boom time': 'Before buses only stopped there for a second. Now they come all the time, and they have built a bus shelter', remarked one man underlining how central this area had become to the region. Since revived measures to build the plant commenced around 2002, the national highway has also been widened and strengthened and a string of other commercial operations have mushroomed around Kudankulam.

On one level, the nuclear plant was a marvel of development in a relatively downtrodden area. Before plans for the nuclear power plant materialized, Kudankulam consisted of about 2,200 houses of which 1,700 were palm leaf-roofed shacks. Nowadays, as an indication of their level of prosperity, signs of old thatched houses remain, but most of the residences have now been constructed out of bricks and mortar, pukka houses. One woman elaborated how in 1989: 'No one had a refrigerator, or electronic articles like television, or VCR. We also have less power cuts now. Before it was all the time'. In the early days, the ability to enter consumer culture fuelled fanciful ideas, with one person believing that Kudankulam residents would eventually move to the Anuvijay township. He knew how well people were looked after in the township, was aware of their relatively substantial salaries, and was quietly envious of their privileges. Some people had swallowed the government line—seeing it as a positive sign for development and jobs. 'Perhaps my son can get a job there', admitted one hopeful father.

On another level, as the construction proceeded, disillusionment spread in alongside a caustic view in the fact that there were not as many jobs allocated for native residents as the nuclear authorities had promised. About 5,000 jobs were reputed to have been created in the nuclear plant complex—but not all of them were for local denizens for they lacked the professional training and experience. One locally based computer operator at the plant commented:

> The government publicity is such that they promise 95 per cent of jobs to local people from Tamil Nadu. And 5 per cent for outsiders. Actually, it must be the other way round...There's only a couple of local people who have permanent jobs with residence [at the Anuvijay township].

With little exception, those who gained employment through the nuclear plant development tended to be casual labourers and entrepreneurial individuals. In one Kudankulam household, a lorry driver had gained much business from the development and would regularly go inside the plant to deliver his load for construction. In another, a young woman was a sewing machinist who along with her relative had rented out a small shop in the Anuvijay Township. Even though business in the township was slow, the local entrepreneurs nurtured ambitions of increased clientele especially after the settlement expanded. All of them were tenuous beneficiaries of the nuclear state that had implanted itself in the region. They were cautiously grateful for their gains and the accolade that conferred to their village growing into a town, but also anxious that changes were precarious and may not work out in the long run where official claims did not match their reality.

Once requests to extend or make new houses had been refused by local authorities, worries were raised about being displaced from the region to make room for the expansion of the Kudankulam plant. In 2006, talk was rife of letters to residents saying that they will have to move out of their homes for a period of a month, and be temporarily relocated elsewhere with full provisions including schooling for their children. However, residents became suspicious of NPCIL's motives, thinking that if they left their homes, perhaps they would never be able to return. After all, the proverbial saying held traction: 'possession is nine tenths of the law'.

As the reporter Gopal N. Raj generalized on nuclear plant operations in India:

> When, as in Narora [the site of a nuclear power plant in the state of Uttar Pradesh in north India], the locals find that their sleepy hamlet

with a little more than a ramshackle shop has metamorphosed into a small township where everything from textiles to television sets is available, they are likely to be well-disposed to nuclear power plants. When, as in Kaiga [a nuclear power plant in the state of Karnataka], there are fears that the Government will not keep its promises, the advantage is bound to shift the other way.[82]

In Kudankulam, the shift over the years was going the same way as Kaiga where people's resistance to a nuclear plant had earlier been mobilized. The lack of transparency amongst nuclear quarters only added to their growing disaffection.

Rumours abound in cultures of secrecy where the authorities disclose little. This is not to say that secrets or conspiracies do not circulate in apparently transparent contexts, only that where opacity is the norm, transparency becomes the ideal of good governance. This is even in the knowledge that the 'paradoxical tendency of transparency measures ... yield, in practice, new opacities', to cite the anthropologist William Mazzarella.[83] Moreover, laws and national security discourse can never render nuclear authorities transparent. They stand at an opaque apex of what can be called an 'irregular governmentality': a twilight zone where authoritarian *government* follows on from colonial regimes combined with modern *governmentality*—the 'art of government' that has as its aims the (re)production of a content and stable society under the rubric of a democracy.[84]

Kudankulam residents learnt about this analysis at the sharp edges of their experiences. They were in a suspension, not only content about their material gains but also cautious about measures that were imposed on them. As one auto-rickshaw driver put it in 2006: 'I don't like it, but it's good for my business'. This boom in business meant dividends for family, which was his primary concern (Figure 2.4). Others seemed to be apathetic, unsure of anything that was outside of their domain but assertive about what they considered as their priorities. One cynic proffered, 'We're all going to die anyway so what does it matter?' Another labourer stated, 'I would rather die with a job than without a job'. His comment reflected the official stance of Communist Party members of parliament who prized employment over radiation. As reported by Itty Abraham, their view was that 'even if the nuclear reactors were dangerous, it was better to work and die in

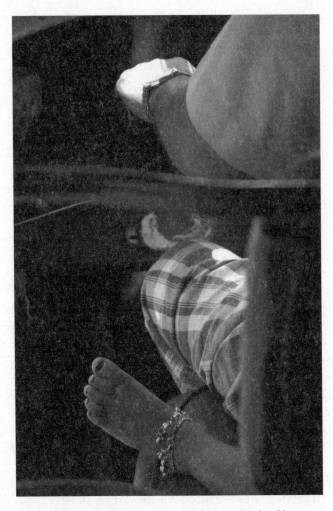

Figure 2.4   Autorickshaw Driver with His Child Seated Behind him,
Tirunelveli District, 2006.
Source: Raminder Kaur.

10 days (from radiation poisoning) than to starve from never having
worked at all'.[85] Prior to 2011 when anti-nuclear civil disobedience
reached its heights, economic necessities and family responsibilities
tinged with a fatalist pragmatism summed up the mood of the people
in Kudankulam.

Deep Vision

If we were to reverse the direction of the gaze and stand near the Kudankulam Nuclear Power Station on the coast, looking out to the west, the faint outlines of Vivekananda Rock and the monolith of Thiruvalluvar is visible in the dewy oceanic distance. One wonders what such sages would have thought of this nuclear plant across the sea of consciousness had they been alive. One verse by Thiruvallavar engraved on the pedestal states: 'All our actions should be pure and should be free from blame by others'.[86] In view of the amount of disruptions and distortions, secrecy and violence that is implicit in the planning, construction and operation of the nuclear plant, there is a fundamental disconnect with such a message. For the nuclear state, the 'power and energy that they got from this world' is far from blame-free. The nuclear industries had become a feverish source of ambivalence and anxiety, particularly as power in all its shades—electrical, political, social, and muscular—was foisted on others.

This paradisiacal peninsular has so many treasures as well as terrors lurking beneath and around its landscape that sand mining companies, state-backed and private, are keen to exploit to turn to tradeable riches. The establishment of the Kudankulam Nuclear Power Plant and Anuvijay township have added to these concerns. They have become dominant players in the region that herald both boom time and broken time depending upon who one talked to and when. Whatever their views, local residents saw their presence as a major departure from what in comparison was a relatively ordinary place.

The space of criticality was informed by the potentials of the region in terms of material benefits for various social actors including aspirations and anxieties about monazite sands. It was enveloped by heightened health sensitivities around ionizing radiation in the environment. It was marked by formal industries that overlapped with informal incursions in terms of casual employment, rogue traders and gangster harassment, directly and indirectly. It was grounded by an adherence to valuing the essentials—water, land, sustenance, and livelihood differently perceived by inland and coastal communities. This is the back story to nuclearity in the region crowned as it is now by the domed towers of power at the Kudankulam Nuclear Power Plant.

Temporality plays a significant part throughout. Over the last four decades, local reactions to the plant have vacillated between the positive and negative, with the balance tipping heavily to the latter. As we shall see in the following chapter, locally based dissent followed a variety of paths but most of them tended to centre on the power of peoples' numbers, peaceful protest, the protection of the environment, the freedom of expression, and the right to know about the impact and assessment of projects that would affect their lives and livelihoods.

To develop an oceanic consciousness is no tall order. It comes from certain privileges and years of embodied practice, learning, and reflection. It became the basis for sustenance and seeking truths. It became a non-sectarian repository for the light and force that informed people's non-violent satyagraha to come. It was a scenario not entirely different from the colonial era—relying upon the numerical strength and moral force of the indigenous population and the inspiration of Mohandas Karamchand Gandhi who continues to yield influence in the region as an enduring ambassador of non-violent resistance.

Notes

1. Parthasarathy, R. translated. *The Cilappatikaram of Ilanko Atikal: An Epic of South India*. New Delhi: Penguin Books, 2004, p. 112.
2. Cited in Kalairasan, M. 'Thiruvalluvar's Observations on Nature: A Study on the Classical Tamil Text Thirukkural'. *Jai Maa Saraswati Gyandayini* 3, no. IV (2018): 417–19, p. 418.
3. Kalairasan, 'Thiruvalluvar's Observations', p. 417.
4. On 'myths of paradise lost' as a response to modernity, see Deckard, Sharae. *Paradise Discourse, Imperialism and Globalisation: Exploiting Eden*. London: Routledge, 2009. In peninsular India, it was not so much an anxiety about modernity that was at issue but *nuclear* modernity.
5. Busher, R.G. 'India's Baseline Plan for Nuclear Energy Self-Sufficiency'. *Argonne National Laboratory*, ANL/NE-009/03, January, p. 1.
6. Busher, 'India's Baseline Plan'.
7. Ramana, M.V. *The Power of Promise: Examining Nuclear Energy in India*. New Delhi: Penguin Books, 2013, pp. xxiv–xxv, 63–70; Subramanian, T.S. 'Indo-US Nuke Deal Will Help Launch Thorium Reactors'. *The Times of India*, 25 February 2006. Available on https://timesofindia.indiatimes.

com/india/Indo-US-nuke-deal-will-help-launch-thorium-reactors/articleshow/1428328.cms?referral=PM, accessed 25 September 2018.

8. 'Identifying a Civilian Nuclear Facility in India's Decision'. *The Hindu*, 12 August 2005, Thiruvananthapuram edition.

9. 'Kalpakkam Fast Breeder Reactor May Achieve Criticality in 2019'. *The Times of India*, 20 September 2018. Available on https://timesofindia.indiatimes.com/india/kalpakkam-fast-breeder-reactor-may-achieve-criticality-in-2019/articleshow/65888098.cms, accessed 22 January 2019.

10. Other reactor designs are also being pursued that could utilize the thorium directly as S.K. Jain reports in his address. Jain, S.K. 'Nuclear Power—An Alternative'. Available on https://docplayer.net/21365593-Nuclear-power-an-alternative.html, accessed 20 November 2018.

    John Stephenson and Peter Tynan estimate that if the three stages are to be followed to completion, the full exploitation of thorium could take place around 2050. (2007) 'Will the US-India Civil Nuclear Cooperation Initiative Light India?', in *Gauging U.S.-India Strategic Cooperation*, edited by Henry D. Sokolski, Army War College (U.S.) Strategic Studies Institute.

11. Hecht, Gabrielle. *Being Nuclear: Africans and the Global Nuclear Trade*. Massachusetts: MIT Press, 2012, p. 14.

12. Hecht, *Being Nuclear* (author's emphasis).

13. As Itty Abraham writes: 'Thorium remains a value *in potentia*'. Abraham, Itty. 'Geopolitics and Biopolitics in India's High Natural Background Radiation Zone'. *Science, Technology and Society* 17, no. 1 (2012): 105–22, p. 110.

14. Hecht, *Being Nuclear*, p. 55.

15. Hecht notes a similar tension, referring to agencies who monitor radiation only do so for those in workplaces with man-made sources that then become designated as nuclear, rather than those sites with natural sources of radioactivity. *Being Nuclear*, p. 14.

16. Lash, Scott and John Urry. *The End of Organized Capitalism*. Oxford: Polity Press, 1988.

17. Tirunelveli District covers 145,000 hectares of which less than a fifth is cultivated land in comparison to Kanyakumari's 167,000 hectares of which over half of it is under cultivation. Gopalakrishnan, M. (ed.). *Gazetteers of India, Tamil Nadu State, Kanyakumari District*. Madras: Government of Tamil Nadu, 1995, p. 202.

18. Mohan, R.S. Lal. 'Report of the Seminar on the Water Problems of Nagercoil with Recommendation'. *Water Problems of Nagercoil*, edited by R.S. Lal Mohan. Nagercoil: Conservation of Nature Trust, 1998, pp. 63–7, p. 65.

19. Padmanabhan, V.T., R. Ramesh, and V. Pugazhendi. 'Water Balance Sheet of Kudankulam Nuclear Power Plant (KKNPP)'. 2012. Available on http://www.dianuke.org/water-balance-sheet-of-koodankulam-nuclear-power-plants-kknpp/, accessed 10 January 2019.

20. Available on https://www.census2011.co.in/census/district/51-kanniya-kumari.html, accessed 10 January 2019.

21. Available on https://www.census2011.co.in/data/religion/district/50-tirunelveli-.html, accessed 10 January 2019.

22. Available on http://www.census2011.co.in/census/district/51-kanniya-kumari.html, http://www.census2011.co.in/census/state/tamil+nadu.html, https://www.census2011.co.in/census/district/50-tirunelveli.html, accessed 10 January 2019. It is not clear according to what criteria, literacy numbers are measured.

23. Anglican and Protestant denominations from the colonial era came together as part of the Church of South India (CSI) after independence. Catholicism in the region goes back to the times of Francis Xavier, a Portuguese Jesuit, who visited India in the sixteenth century. There is also an earlier tradition that goes back to the Syrian Christians of the first century onwards. With the latter, Saint Thomas the apostle is reputed to have converted the indigenous people of India before his martyrdom in Tamil Nadu in CE 72.

24. Gopalakrishnan, *Gazetteers of India*, p. 383. At the time of writing, there are about Rs. 60 to a US dollar, and Rs. 80 to the sterling pound.

25. Available on http://sagarmala.gov.in/media/press-release/pms-inaugu-ral-address-maritime-india-summit-2016, accessed 10 January 2019.

26. See Jeffrey, Craig *Timepass: Youth, Class and the Politics of Waiting in India*. Stanford: Stanford University Press, 2010.

27. Available on https://www.census2011.co.in/census/city/490-nagercoil.html, accessed 10 January 2019.

28. Mercy, P.D. 'Socio-Economic Impact of Tsunami'. In *Tsunami and Its Impact*, edited by R.S. Lal Mohan. Nagercoil: Conservation of Nature Trust, p. 129.

29. Early Protestant Christian missionaries founded the London Mission Society hospital in Neyyoor in 1830 and in Nagercoil in 1838. Gopalakrishnan, *Gazetteers of India*, p. 1067.

30. Forster, Lucy, Peter Forster, Sabine Lutz-Bonengal, Horst Willkomm, and Bernd Brinkmann. 'Natural Radioactivity and Human Mitochondrial DNA Mutations'. *Proceedings of the National Academy of Sciences* 99, no. 21 (2002): 1–5, p. 1; Ajithra, A.K., B. Venkatraman, M.T. Jose, S. Chandrasekar, and G. Shanthi. 'Assessment of Natural Radioactivity and Associated Radiation Indices in Soil Samples from the High Background

Radiation Area, Kanyakumari district, Tamil Nadu, India'. *Radiation Protection and Environment* 40, no. 1 (2017): 27–33, p. 33.

31. Gopalakrishnan, *Gazetteers of India*, pp. 347–8.

32. See Abraham, Itty. 'Rare Earths: Travancore in the Annals of the Cold War, 1945–47'. Unpublished paper, University of Texas.

33. Available on http://www.atomicarchive.com/History/british/, accessed 10 January 2019.

34. Imperial Chemical Industries advertisement, *Bombay Chronicle*, 10 December 1947, p. 2.

35. See Hecht, *Being Nuclear*, p. 3.

36. Abraham, 'Geopolitics and Biopolitics', pp. 108–10.

37. On a fuller discussion of C.P. Aiyar's peregrinations on atomic resources, see Abraham, 'Rare Earths'.

38. Sundararajan, Saroja. *Sir C.P. Ramaswami Aiyar—A Biography*. New Delhi: Allied Publishers, 2002, p. 425.

39. 'Thorium Found in Travancore', *Bombay Chronicle*, 13 August 1945, p. 1.

40. Others include Chavara in the Kollam district of Kerala, and Chathrapur in the Ganjam district of Odisha to the north east of India.

41. Cited in Abraham, 'Geopolitics and Biopolitics', pp. 110–11.

42. Gopalakrishnan, *Gazetteers of India*, p. 12.

43. Ravishankar, Sandhya. 'Exclusive: Atomic Minerals Found in Tamil Nadu Beach Sand Samples Meant for Export, Says Report', *Huffington Post*, 10 May 2018. Available on https://www.huffingtonpost.in/2018/05/10/ exclusive-atomic-minerals-found-in-tamil-nadu-beach-sand-samples-meant-for-export-says-report_a_23431545/?guccounter=1, accessed 10 July 2018.

44. See 'The Number Game: Occupational Health Hazards at Indian Rare Earths Plant'. *Economic and Political Weekly* 21, nos 10–11 (1986): 443–52; Padmanabhan, V.T. 'All within Limits: Radioactive Waste Disposal at IRE'. *Economic and Political Weekly* 22, no. 10(1987): 419–25.

45. Murali Das, S., V. Sasi Kumar, and S. Sampath. 'Atmospheric Electrical Conductivity Measurements as an Indicator of Natural Radioactivity'. *The Journal of Indian Geophysical Union* 5, no. 2 (2001): 93–101, p. 97.

46. Chellappan, Kumar. 'Rich N-fuel at TN Mafia's Mercy'. *The Pioneer*, 1 January 2016. Available on https://www.dailypioneer.com/2016/india/ rich-n-fuel-at-tn-mafias-mercy.html, accessed 10 January 2019.

47. Available on http://www.vvminerals.com.vvminerals_frm.html, accessed 10 January 2018.

48. Available on http://www.vvmineral.com/content/about, accessed 10 January 2018.

49. See Arasu, Sibi. 'Battling India's Sand Barons', 11 May 2016. Available on https://news.mongabay.com/2016/05/battling-indias-sand-barons/, accessed 10 January 2018.

50. Chellappan, 'Rich N-fuel at TN Mafia's Mercy'.

51. *Tehelka*. 'There's Nuclear Gold in this Sand. And it's being sent out with Impunity'. 16 October 2010. http://archive.tehelka.com/story_main47. asp?filename=Ne161010Therenuclear.asp The original article is no longer available on the *Tehelka* archive but can be accessed here http://radiationstudy.blogspot.com/2015/05/theres-nuclear-gold-in-this-sand-and.html, accessed 10 January 2019.

52. See Peebles, Gustav. 'The Anthropology of Credit and Debt'. *Annual Review of Anthropology* 39, no. 1 (2010): 225–40.

53. Such is its power that the company has even managed to lure IREL employees and twist the arm of authorities to persist with their ventures unimpeded. The Atomic Energy Act (1962) strictly prohibits individuals or private enterprises from undertaking mining activity related to atomic material, but with latter-day amendments, the government has relaxed its policy to allow select private enterprise. According to a recent court-appointed Sahoo Report, this privilege has been flouted such that there is cause to suspect that V.V. Minerals are also trading in radioactive monazite. Available on https://ejatlas.org/print/beach-minerals-sand-mining-in-tamil-nadu-india; Ravishankar, 'Exclusive'.

54. Available on http://tirunelveli.nic.in/population.html, accessed 10 January 2018.

55. D.N. Moorty cited in Srikant, Patibandla. *Koodankulam Anti-Nuclear Movement: A Struggle for Alternative Development?* Working Paper 232, Bangalore: Institute for Social and Economic Change, 2009, p. 4.

56. On an animation of the main site, see 'Kudankulam Nuclear Power Plant—India's Nuclear Pride' produced by Anthelion Technology in 2017 for the Hindustan Construction Company. Accessed 20 February 2018.

57. Kaur, Raminder. *Atomic Mumbai: Living with the Radiance of a Thousand Suns*. New Delhi: Routledge, 2013.

58. Abraham, 'The Violence of Postcolonial Spaces', p. 332.

59. See Kaur, Raminder. 'A Nuclear Cyberia: Interfacing Science, Culture and an E-thnography of an Indian Township Social Media'. *Media, Culture and Society* 39, no. 3 (2016): 325–40.

60. Ramana, M.V. 'La Trahison des Clercs: Scientists and India's Nuclear Bomb'. In *Prisoners of the Nuclear Dream*, edited by M.V. Ramana and C. Rammanohar Reddy. New Delhi Orient Longman, 2003, p. 215. However, there are a few exceptions to this picture of compliance among nuclear

scientists and bureaucrats, most of whom are retired and therefore not directly answerable to the state.

61. In a growing economy, however, some of the employees have been attracted by better packages in the private sector particularly with software firms. In 2006, for instance, the 'nuclear brain drain' saw nine engineers and three financial experts leave NPCIL's base in Kudankulam with many more leaving its Trombay headquarters. Sivan, Jayaraj. 'Nuclear Brain Drain'. *Mumbai Mirror*, 13 December 2006, p. 18.

62. See Vanaik, Achin. 'Unravelling the Self-image of the Indian Bomb Lobby'. *Economic and Political Weekly* 39, nos 46–47: 5006–12.

63. Dasgupta, Debarshi. 'Masala Dosa at Midnight', *Outlook India*, 2 July 2007. Available on https://www.outlookindia.com/magazine/story/masala-dosa-at-midnight/234983; Chatterjee, Kaushiki. A.P.J. Abdul Kalam: Reminiscing the 'Missile man' of India', *Deccan Herald*, 27 July https://www.deccanherald.com/national/apj-abdul-kalam-reminiscing-the-missile-man-of-india-749977.html, accessed 6 November 2019; Bhushan, Gopal (ed.). *Memories: Incredible Kalam*. New Delhi: Ministry of Defence, Defence Research & Development Organisation (DRDO) Defence Scientific Information & Documentation Centre (DESIDOC), 2016. Available on https://www.drdo.gov.in/drdo/pub/Memories_Incredible_Kalam.pdf, accessed 21 January 2019.

64. O'Flaherty, Brendon and Rajiv Sethi. 'Public Outrage and Criminal Justice: Lessons from the Jessica Lal Case', 2019. Available on http://www.columbia.edu/~rs328/Jessica.pdf, accessed 21 January 2019.

65. *India Strategic*. 'India's Kudunkulam Nuclear Plant Safe: Kalam', November 2011. Available on http://www.indiastrategic.in/topstories1241_Kundakulam_nuclear_plant_safe.htm, accessed 21 January 2019.

66. Laxman, Srinivas and Siddhartha D. Kashyap. 'Kalam set to Shatter Sound Barrier Today'. *The Times of India*, Mumbai, 8 June 2006, p. 15. By his own admission, Kalam saw himself as well-tuned to Hinduism, having many Hindu friends who he endearingly writes about in his biography. See Tiwari, Arun and A.P.J. Abdul Kalam. *Wings of Fire: An Autobiography*. Himayatnagar, Hyderabad: Universities Press, 1999.

67. See Dyer, Richard. *White: Essays on Race and Culture*. New York: Routledge, 1997.

68. Ramana, 'La Trahison des Clercs', pp. 207, 213.

69. Krishna, Sankaran. 'The Bomb, Biography and the Indian Middle Class'. *Economic and Political Weekly* 41, no. 23 (2006): 2327–31, p. 2328.

70. 'Koodangulam: A Nuclear Graveyard'. *Indian Express*, 20 June 1989.

71. Goldberg, David Theo. *The Threat of Race: Reflections on Racial Neoliberalism*. Malden, MA: Wiley Blackwell, 2009.

72. Krishna, 'The Bomb, Biography and the Indian Middle Class', p. 2327.

73. Krishna, 'The Bomb, Biography and the Indian Middle Class', p. 2329.

74. Subramanian, Ajantha. 'Mukkuvar Modernity: Development as a Cultural Identity'. In *Regional Modernities: The Cultural Politics of Development*, edited by Arun Agrawal and K. Sivaramakrishnan. Stanford: Stanford University Press, 2003, p. 264.

75. According to a study conducted by the National Institute of Nutrition and the All Indian Institute of Medical Sciences, 92.2 per cent of people in south of India are non-vegetarians, the highest in the land. The main exceptions to this habit are orthodox Brahmins. See 'Over 64 p.c. People are Non-vegetarians: Report', *The Hindu*, 17 October 2006, p. 7, Thiruvananthapuram edition.

76. Subramanian, *Mukkuvar Modernity*, p. 3.

77. Todd, Zoe. 'Protecting Life below Water: Tending to Relationality and Expanding Oceanic Consciousness Beyond Coastal Zones'. *De-Provinicializing Development Series* http://www.americananthropologist.org/2017/10/17/protecting-life-below-water-by-zoe-todd-de-provincializing-development-series/, accessed 16 December 2017.

78. Todd, 'Protecting Life Below Water'.

79. Along the Kanyakumari coast that stretches across into Kerala, the Mukkuvar caste predominates. They are viewed as a sea-faring sub-category of More Backwards Castes, and entitled to welfare support and government reservations. Paravar are conceived as even lower down in the social hierarchy—Scheduled Caste Christians and Hindus who are scattered to the south across peninsular Kanyakumari and the Tirunelveli District shorelines.

80. On a comparison with ruses deployed to build nuclear power plants in the French countryside from the 1950s, see Hecht, Gabrielle. *The Radiance of France: Nuclear Power and National Identity*. Massachusetts, MIT Press, 2009, pp. 241–70.

81. See Srikant, *Koodankulam Anti-Nuclear Movement*, pp. 6–7; Bhawna. 'Nuclear Energy, Development and Indian Democracy: The Study of Anti-Nuclear Movement in Koodankulam'. *International Research Journal of Management Sociology and Humanity* 7, no. 6 (2016): 219–29, p. 224.

82. *Frontline*. 'Issues Nuclear—The Safety Concerns'. 4–17 March 1989, pp. 89–95, p. 90.

83. Mazzarella, William. 'Internet X-Ray: E-Governance, Transparency, and the Politics of Immediation in India'. *Public Culture* 18, no. 3 (2006): 473–505, p. 476. See West, Harry G. and Todd Sanders. 'Power Revealed and Concealed in the New World Order'. In *Transparency and Conspiracy:*

Ethnographies of Suspicion in the New World Order, edited by Harry G. West and Todd Sanders. Durham: Duke University Press, pp. 1–37.

84. Foucault, Michel. 'Governmentality'. In *The Foucault Effect: Studies in Governmentality*, edited by Graham Burchell, Colin Gordon, and Peter Miller, translated by Rosi Braidotti. Chicago: Chicago University Press, 1991, pp. 87–104. However, this is not to overlook how government and governmentality overlap—for instance, in the mobilization of authoritarian populism. Alexander Dunlap argues 'hard' coercive force vs 'soft' social/political technologies of control are never exclusive to each other. Dunlap, Alexander. 'Counterinsurgency for Wind Energy: The Bíi Hioxo Wind Park in Juchitán, Mexico'. *The Journal of Peasant Studies* 45, no. 3 (2018): 630–52.

85. Abraham, 'The Violence of Postcolonial Spaces: Kudankulam', p. 322.

86. See Ramasubrahmanyan, R. *Thiruvalluvar*. Kanyakumari: Hari Kumari Arts, not dated.

3

Cultures of Dissent

Do you really need nuclear energy to boil water?

(Matthew, Tirunelveli District)[1]

Globally, anti-nuclear campaigns came to a head as part of peace movements from the 1960s against Cold War arms proliferation.[2] They received an extra shot in the aftermath of nuclear plant accidents and with the spread of environmental awareness particularly from the 1970s. While those who decried anti-nuclear weapons were already present in pockets of urban India, social movements against nuclear plants emerged elsewhere on the ground in the southern states of Kerala, Karnataka and Tamil Nadu from the 1980s. This grassroots upheaval came about in large part due to awareness of the problems that beset nuclear power plants, especially in view of major disasters at the Three Miles Island plant in the USA in 1979, and at the Chernobyl plant in the former USSR in 1986. However, the agitations were also in league with a growing culture of entitlement for project-affected people responding to the emergence of, as the political scientist Sanjay Ruparelia terms it, a 'new rights agenda'.[3]

More often than not, people's movements—the more common term in use in the subcontinent—stand apart from the predominantly middle-class movements against nuclear power plants in the global north. Environmentalism, as it emerged in India, was intricately embedded in social justice and concerns about habitat and livelihood.[4] A lack of adequate public consultation, compensation, and the prospect of living with health hazards or displacement from their homes rang loud as distrust led to discontent that spread and sharpened into dissent. If they were not at the helm themselves, people's movements could not thrive without engaging the support of the subaltern sectors involving the likes of farming, coastal, and tribal communities.[5]

The movement against the plant in Kudankulam began with fishing communities, activists, and environmentalists who spanned a range of professions within and outside of non-governmental organisations (NGOs) including the likes of doctors, scholars, teachers, lawyers, journalists, and priests. The ignition point was around 1988 when plans for an Indo-Russian nuclear power plant were announced amid secretive meetings and token public relations exercises. Lessons were learnt, and ideas and practices shared from earlier anti-nuclear campaigns in the neighbouring state of Kerala. It was a gathering of minds and bodies that took on pressing urgency in light of the Chernobyl disaster only two years earlier.

The second phase of the movement was about a decade later from around 1998 when Kudankulam plans were revived and construction of the plant began in neo-liberal India's post-nuclear weapons declared status. The government nurtured grand globalizing ambitions with increased efforts to forge international agreements between the nuclear lobby and corporations. As the political stakes were raised, a revitalized anti-nuclear movement came with an expanding base. This encompassed farming communities worried about the siphoning of water supplies to the plant, along with others who might have initially supported the plant for employment and other dividends, but turned against them when nuclear realities hit the fan. They too began to join ranks against threats to ecology, safety, and health as well as the prospect of displacement.

Correspondingly, the anti-nuclear movement in south India took on a more national platform by engaging with people's movements in other parts of the subcontinent in the effort to ensure social justice was integral to processes of globalization.[6] This included those affected by nuclear operations, as has occurred around uranium mining and tailings in Jadugoda in the northeastern state of Jharkhand.[7] People's movements stepped up a concerted evidence-based series of actions while availing of the proliferating communication channels to make emboldened calls for citizen rights, equity, transparency, and accountability. Following the terms proposed by William Mazzarella, the 'politics of immediation' and a 'righteous realism' were pitched against the caverns of political opacity, corruption and neglect.[8] Several of these movements went beyond a struggle against state irregularities alone to pose a threat to the very premises of neo-liberal governance.[9]

The third phase of the anti-Kudankulam struggle came to light in 2011 in the aftermath of the Fukushima Daiichi nuclear disaster in Japan when two reactors had been constructed, one of which was about to go critical. This 'Fukushima effect', combined with domestic factors, saw a dramatically intensive period of campaigning that drew hundreds of thousands to the cause.[10] This was not least due to the enduring appeal of non-violent civil disobedience and its intensification through digital technologies to spread news about the struggle while relaying alternatives to nuclear developments from its 'epicentre' in the coastal village of Idinthakarai, which is the focus of the last four chapters in the book.[11] Before we concentrate on a history of the struggle in south India, we turn to an overview on social movements and the significance of non-violence to the subcontinent.

Movements in Perspective

There has been a discernible growth over the twentieth century from movements that were based largely on economic or resource mobilization as accompanied (post-)industrialization, to ones that are characterized by more of an identity or issue perspective, where social positions and cultural 'framings' take on more prominence.[12] Emerging in the 1960s, these 'new social movements' included the struggle for peace and nuclear disarmament, and civil, human, women's and LGBTQ+ rights.[13] A related development from the 1970s was the rise of 'risk movements' that set out to highlight health and environmental hazards.[14] Many of these were driven by 'crisis narratives' based on a sense of emergency, disaster, or impending catastrophe.[15]

While there are several views on how social movements bring people together to address a grievance and make collective claims, most agree that they are quite apart from the relative inflexibility of political institutions, parties, and advocacy groups.[16] Although this is not to say that they do not interact with them, for the establishment might determine, be challenged, or alter official policy in response. Largely based on informal networks, social movements are essentially fluid and open to anyone who sympathizes with the cause. This 'libidinal economy of movements', in the words of political and social scientists Donatella Della Porta and Mario Diani, is marked by increasing cross-cultural/group communication, and may lead to an umbrella of alliances with other organizations.[17] The loose coalitions are formed on the basis of identifying commonalities with one another that has the potential to cross social boundaries and hierarchies to increase the movement's life.[18] Accordingly, there is a premium placed on faith, loyalty, and the dedication of time and energy to sustain the movement.

In terms of their actions, Alex Shankland and his team of social theorists propose 'unruly politics'. This they argue encompasses a manner of 'political actions which escape, exceed or transgress "civil" forms of civic and democratic engagement'.[19] They add that such actions often 'take forms that are juridically illegible, extra-legal, disruptive of the social order, strident or rude'.[20] They might range from what might be described as 'politely provocative', tongue-in-cheek and theatrical to the more subversive and militant.[21] Unruliness implies not quite reckless driving, but calculated transgression, in the course of which new imaginaries, vocabularies and tactics are envisaged and mobilized so as to overturn hegemonic regimes.[22]

Collective rituals play a component part in (re)creating an embodied commitment to the cause.[23] Dominant social relations may be suspended, displaced, or even neutralized with 'areas of anti-power' as the sociologist and philosopher John Holloway puts it—an anti-power that is hinged upon a *power-to* change through doing, rather than merely usurp and replace the structures of *power-over*. The aim then becomes to change the world but without taking power. Holloway elaborates: 'Anti-power, then, is not counter-power, but something much more radical: it is the dissolution of power-over, the emancipation of power-to'.[24]

Related perspectives are borne out by what has been described as 'prefigurative politics' that came to centre stage in the alter- or anti-globalization movements from the 1990s.[25] This vanguard seeks to create, protect, construct, and expand spaces for an alternative society configured by sustainable economies and ecologies based on ideals of non-hierarchical, consensual decision-making. Such movements seek to offer 'a glimpse of what real democracy might be like' through processes of a direct democracy, to cite the anthropologist and anarchist David Graeber.[26] The latter-day phenomenon is different from socialist struggles against global capitalism in that it prefers process over outcome, and largely rejects the assuming of roles in established structures of law, politics and representative democracy.[27]

Social movements that call for environmental, livelihood, and health justice might be abbreviated as *right-to-lives* movements. The term is in the plural to respect the diversity of supporters and their life worlds, and not to be confused with pro-abortion or protection from danger 'right to life' groups. Right-to-lives movements could be influenced by most if not all of the features identified above. More often than not, they are driven by a philosophy and practice of peaceful dissent.

Non-violence is not diametrically opposed to more overt kinds of violence, and has to be seen both relationally and contextually.[28] It could, as the political scientist Gene Sharp outlines, cover a number of civilian-based challenges that could be highly disruptive.[29] The preference for non-violence as against brute force compels imaginative interventions that highlight and invert the logics of oppressive regimes while drawing more support from those across the board. Releasing what the historian Madeleine Hurd describes as the 'emotional energy of hope', creativity is unleashed to convey their moral and political protest in numerous ways that fall short of physical violence.[30] For maximum leverage, the aim is to exert steady pressure on public opinion and the authorities where independent media plays a crucial role. Creating an autonomous expert base and the documentation of human rights abuses, environmental assessments, and technical irregularities are also instrumental. This evidence might form the basis for educational material that could aid further mobilization, all features that mark the growth of resistance against large-scale developments as more and more people learn about their consequences on their lives.

Right-to-lives movements depart somewhat from the classic understanding of rights and life as applying only to the human species. Here, the environment is also a social actor in the sense that it is fundamental to worldviews and fiercely contested. At one extreme, it is a 'natural resource', a field for profit, and held of inferior value unless it can be utilized for these purposes.[31] On the other, it is intimately associated with identity, habitation, a repository of daily sustenance, even spiritual significance, and of supreme value in and of itself. Life from these perspectives has to be construed not just about the biopolitical management of populations, but one where the population is interconnected with life forms on and including the earth—an eco-governmentality that takes on board the intergenerational depth and diversity of life forms.[32] As Zoe Todd contends: 'It is crucial, in times of environmental/socio-ecological upheaval, to underscore and impress the interconnectedness of lands, waters, space, people, and time'.[33] This dictum was easier to appreciate by fishing and farming communities in the subcontinent than it was by nuclear technocrats and their law-enforcing entourage.

Subcontinental Struggles

Throughout the decades, we have seen the rise of issue-based people's movements to add to party-affiliated, resource-based, and redistributive movements in India. They have borrowed from international trends as much as they have from the history of struggle against colonialism. They have been interlaced and configured throughout by a blend of unruly and 'identity politics', the latter based on a mixture of class-caste, gender, ethnicity, regional, and linguistic vectors.[34] Many of them were influenced by 'non-party political formations' that were established in the wake of the repressive Emergency years of Indira Gandhi's rule (1975–7) to mobilize issues, tactics, and groups that were sidelined by the mainstream Left.[35] Some of them went on to establish themselves as NGOs. However, even though they might join forces, people's movements stand apart from NGOs for being a larger multi-canopied vehicle that cannot stand accused of being co-opted, nor scrutinized by governmental bodies on organizational procedures.[36]

Non-violent resistance has been integral to most people's movements, largely due to its resounding legacy as part of civil disobedience against British colonial rule from the 1920s. It owes its popularity to an ethical commitment and a flexible vocabulary with which to communicate with and raise awareness across diverse communities using an accessible, transferable, creative, and often witty repertoire of remonstration.[37]

In the face of mounting public–private ventures, right-to-lives movements are fundamentally affected by livelihood matters to do with shelter, sustenance, and frequent lapses in compensation. The Chipko movement led by hill women against state-sanctioned deforestation is one of the earliest escalations from the 1970s.[38] Campaigns against industrial hazards in the wake of Bhopal's Union Carbine pesticide industry disaster in 1984 is another prominent one that has also foregrounded health, environmental, and compensation concerns particularly as they apply to slum-dwelling communities who lived next to the toxic plant.[39] The Narmada Bachao Andolan (NBA, Save the Narmada Campaign) from the mid-1980s is one other movement renowned for the struggle of tribal and other marginalized people against the construction of a series of dams on the Narmada River in western and central India. Along with its most prominent spokesperson, Medha Patkar, NBA campaigned for equitable and sustainable development with proper rehabilitation and compensation for displaced people along the river that snakes through the states of Gujarat, Maharashtra, and Madhya Pradesh.[40]

With the neo-liberal entrenchment of national and trans/multinational industries into tribal, agricultural, and coastal lands, threats of displacement and attendant health and environmental hazards are another vindication of 'accumulation by dispossession' to draw upon the work of the anthropologist and geographer David Harvey.[41] Take the cases of Dongria Kondh tribal groups against Vedanta's plans to mine their sacred lands and the Pohang Iron and Steel Company's (POSCO) expansions onto forest lands in Odisha; and the rise of resistance against Mahan Coal's excavations in Madhya Pradesh, and other nuclear projects as with proposals for Areva (later EDF France) to work with NPCIL to construct an even larger plant with up to six 1,600 megawatt reactors in the biodiverse region of Jaitapur in Maharashtra.[42] As in Kudankulam, agitations against the plans were primarily based on a critical stance against top-down models of development imposed on people and places.

With the emphasis on bringing different groups together to protest against an Indian nuclear plant, there are elements that are shared with anti-nuclear movements in other parts of the world.

A few features, however, make the Kudankulam case qualitatively distinct.

In terms of who was involved: prominent were the number of sub-altern communities from fishing backgrounds, some of whom had not completed their schooling, and therefore markedly different from the class–caste constituency of the nuclear state as outlined in the previous chapter. By 2011, subaltern groups had become a core part of the movement. Even while the educated middle classes were involved, they invariably gave the bow to project-affected representatives along the coast whose lives were to be most forcefully affected. As with other people's movements in India, the anti-nuclear campaign was strongly moored in social justice on the basis of which multipronged alliances were formed. A related point of note is that, while the main spokespeople were men, women formed the most determined major-ity who steered the anti-Kudankulam struggle on to other levels.[43]

In terms of motivating visions, the nation remained supreme as a sanctuary from accusations of foreign funding and outside interven-tion.[44] While its hegemonic terms were reworked and inverted, most acts were done in the name of the *nation as people*, and not in terms of the official coupling of *nation to state*. The latter was laced with the profanity of politics that could not be trusted nor bode well for the future of the country.

Using the vocabulary of the nation was also instrumental in main-streaming marginalized narratives. Many in the movement wanted to prove that they were knowledgeable and that their proposals were beneficial to the country. They did not want to be cast as primitive, irrational, Luddite, or anti-development as was the metropolitan elite's dismissal of coastal and rural communities. Rather they wanted to rework the terms of the future development of the country so as it is justly pursued and complementary rather than detrimental to their lives and environments. Theirs was a future-orientated vision that transcended governmental five-year plans and elected terms.

Prefigurative politics has had less of a pull in India.[45] While alter-native visions of the nation ran rife throughout such right-to-lives movements, they were not then a vindication of autonomy in that they did not altogether reject social and political convention. Compared to those in the global North, their horizon of political possibilities was more guarded. Campaigners were extra vigilant about keeping the higher moral ground and towing the khaki line of lawfulness even while they challenged the agents of law enforcement. For the majority, non-violence was an ethical imperative to challenge a state that had more firepower and the authority to use it.[46] When it came to any arrests, it did not mean a couple of days in custody and/or a relatively small fine as might be the case in the global North.[47] Due to a slower and weaker justice system, charges and incarceration could go on for lengthy periods of time in twisted interpretations of the legislature. As we shall see in Chapter 10, the drawn-out process could atrophy physiques, funds, and willpower in a seemingly endless and unsavoury maze where the central hub becomes just another waiting room that descended into another labyrinth. To keep out of jail while holding onto a placard for peaceful dissent was paramount. Lead activists therefore aspired to work *through* the imagined state rather than *against* it, knowing full well that any suggestion of anti-state violence as was associated with the guerrilla tactics of Naxalite or Maoist groups mobilizing for land reform in North-East and south-central India, was to invite policing and (para)military advances that could end in the arrest and death of people, if not the movement itself.[48]

However, this is not to say that the anti-nuclear movement did not nurture alternative suggestions for the political economy. Visions of an alternative to capitalist societies associated with the paradigmatic West have had a long history in the subcontinent. This has led to alternative proposals for community life and has been repeatedly revived in different ways in the modern era to uneven effect. In peninsular India, the ultimate motive was to move away from centralized models of nuclear power in India, but not to altogether usurp extant structures of law and politics. Rather, they sought to almost 'cleanse' the state, to pull it to task, so to speak, so that it did not prevaricate from regulatory protocols and values enshrined in the sublime canopy of the Indian Constitution.[49] They rallied, campaigned, negotiated, renegotiated, and ironically, made demands on protagonists of the state to keep to the law *themselves* and due process regarding accountability and public consultation over the nuclear power plant. Therefore, while they fought against the nuclear plant, they also fought for a more functioning, 'effective', and 'substantive' democracy, to draw upon Patrick Heller's analysis.[50] As the social scientist proposes, the 'formal democracy' to which India subscribes through universal suffrage in periodic elections since 1950, the constitutional framework, along with the ideals of state accountability and legal and rights-based codifications falls short in several ways. India has neither become an 'effective democracy' with values diffused throughout the entirety of society, nor has it become a 'substantive democracy' that can ensure that all social groups are fully involved in political and economic matters. People's movements have periodically arisen with a view to give substance to formalities across the subcontinental spectrum.

The Initial Phase

The anti-Kudankulam struggle began soon after an intergovernmental agreement was publicly announced under the Congress leadership of Rajiv Gandhi and sealed with the former USSR president Mikhail Gorbachev on 20 November 1988. High-level talks had been underway since the 1970s. After India's 'Peaceful Nuclear Experiment'—the term used to describe the tests in the deserts of Rajasthan in 1974—global sanctions and the abortion of supplies for the Tarapur Nuclear Power Plant led to the Indian government conferring more and more with the USSR. Collaborations for Kudankulam had commenced as early as 1979 under Morarji Desai's premiership when the Soviet statesman, Alexei Kosygin, offered to supply India a reactor.[51]

While central government entertained ambitions of cheaper reactors on favourable terms in the wake of a nuclear disaster in the former USSR, those in the peninsular drew upon the post-Chernobyl global momentum against nuclear power to lend both an understanding of and urgency to the effects of a nuclear power plant on health and environment. Such was the opposition to the project that, amidst security concerns, the proposed foundation-laying ceremony on 19 December 1988 was postponed. With an umbrella organization called the Samathuva Samudaya Iyakkam (Social Equality Movement), people from the three southern districts of Kanyakumari, Turunelveli, and Thoothukudi organized massive rallies and marches. Several organizations formed empathetic networks and had come together. They included the National Fishworkers Forum, the Tamil Nadu Fishworkers. Union, the Social Action Movement, the Palmyra Workers' Development Society, the Peace Association for Social Action, and Group for a Peaceful Indian Ocean amongst others.[52] Catholic priests in coastal dioceses in the three districts also got involved. At a meeting in 2006, the activist priest, Reverend Y. David, recalled these early days of anti-nuclear resistance with pride:

It was a spontaneous growth. It included women's groups, environmental groups, human rights people, and fisher people's groups. There was a strong awareness amongst the people of the dangers of a nuclear power plant. On one occasion, we had 10,000 people in a long procession.

'Spontaneous' implied that dissent came from the people themselves, and not merely orchestrated by NGOs or external agents as the latter were often held by officials to be the instigators, the 'troublemakers'. A show of people's strength was memorably marked by a coastal march on May Day, 'Protect Waters, Protect Life' that covered both eastern and western coastlines and converged in Kanyakumari in 1989. The march was to raise awareness about the pollution of coastal waters and the destruction of mangrove, estuary, and wetland ecosystems.[53] Those in and around Kudankulam channelled the water issue to worries about the effects the plant would have on sea life and fishing livelihoods.

The march was, however, broken up by a local bus driving into it. What was widely reported as a 'road accident' by state authorities resulted in the injury of six fishermen and trumped-up criminal charges. Many believed that this 'accident' was at the behest of authorities acting in collusion with henchmen. One man that I had met in Kanyakumari District in 2006 recalled how:

...the police came down at them with force and brutality resulting in several injuries, including for the women. They were slapped with very heavy charges. Some of them are only coming out of prison now, after about twenty years.

The high-handed response only convinced campaigners of the need to adopt the higher moral ground. By contrasting the *peaceful and lawful* in their acts of civil disobedience, they could further highlight the *violence and lawlessness* of nuclear state conduct.[54] By August of that year, people had come together under the banner of over 120 organizations in Kanyakumari to form the short-lived Anumin Nilaya

Ethirpu Iyakkam (Nuclear Power Project Opposition Movement). This was followed by further rallies and several cultural programs against the plant. Fishing and farming communities united with the prospect that not only would sea water be affected but water from the Pechiparai Dam could be channelled towards the plant in a context of depleting ground water for agricultural use.

Subaltern populations drew strength and vigour from experts in the field. As was reported in a 1989 newspaper article:

> Dr Indira Surendran, vice president of FONS [Friends of Nature Society], has categorically ruled out the possibility of allowing water to be taken from Pechiparai dam. Even this meagre water source was created a few decades ago when the population of Nagercoil was almost 50 per cent of what is now, and the district is perennially water-starved. Under these circumstances, 'We won't allow even a single drop of water to trickle down to Kudankulam'.[55]

Against such bold resistance, by the mid-2000s, the NPCIL was forced to declare that they would not draw water from the dam, and instead would depend on a 'high capacity desalination unit and also the discharge water into the sea which would not contain any radioactive material'.[56]

Despite official assurances, health concerns added to the turbulent mix. Along with sand mining of monazite sands on the southern coasts, having nuclear reactors in the region added to resident's radiation burdens. N.D. Jeyasekharan, who had been conducting health studies in Kanyakumari District since the 1980s, observed:

> This radiation already poses a grave threat to the people's health, and adding any more to it will be disastrous. It may go beyond the control of man or machinery to set right things.[57]

S. Nagaraja Pillai underlined the cumulative effect of year-round radiation while also referring to the recent cancer deaths of nine employees at Indian Rare Earths Limited at Manavalakuruchi: 'The radiation will certainly be more when the nuclear plant is commissioned and more people falling prey to cancer will become a necessary evil'.[58]

Such experts called upon those based in the neighbouring state of Kerala who too had a history of agitation against a nuclear power plant from the early 1980s. This recourse acted as a font of both inspiration and tactics against the Kudankulam Nuclear Power Plant.

It was in the regional daily, *Malayala Manorama*, on 29 March 1982 that an announcement was made that investigations were being conducted with a view to establish a nuclear plant near the Periyar River at Bhoothathankettu near Kothamangalam. One of the Kerala movements' key protagonists, M.P. Varghese, a former lecturer in economics and Principal at the Mar Athanasius College, lived only four miles from the proposed site. Forty years on, recollections of this struggle are limited, therefore attributing Varghese's book on the struggle a great deal of significance. Initially, he organized a meeting and prepared two reports about the inadvisability of establishing a nuclear plant. A 'society' was formed, the Organisation for Protection from Nuclear Radiation, and seminars and agitations were organized in and around the proposed site.

Referring to the Three Miles Island nuclear plant accident, Varghese recalled how 'these very monuments of social disaster continue to haunt me in many a nuclear nightmare'.[59] Together with others, he campaigned and appealed to the judiciary. Eventually, they won their case.

Reflecting on their success, the activist K. Ramachandran recalled that one of their main objectives was to cast any aspersions of them as 'primitives or monkeys' averse to the modern bounties of electricity understood as the kingpin for development:

> The campaign consciously desisted from exaggerations, cooked up statistics, and sentimental methods appealing to religious, caste or party loyalties. It wanted to be objective, logical and efficient in its presentation of facts as far as possible and democratic in implementation of decisions.... It was a sincere attempt to remove nuclear illiteracy which existed even among the educated.[60]

Led by the middle classes, the anti-nuclear movement in Kerala sought to convince through persuasive and logical argument for the 'nuclear illiterate' across the spectrum. With a surge of support, Varghese claimed to have even altered the view of K. Babu Joseph, a renowned scientist and erstwhile nuclear power advocate based at Cochin University of Science and Technology. Varghese elaborated on how they had earlier 'held opposite views regarding the safety and advisability of the nuclear plant' but remarkably, Joseph had now 'converted himself into an anti-nuclear proponent'.[61] As a tribute to his loyalty shift, Joseph wrote the foreword to Varghese's book where he declared:

> The uranium sources in the country totalled to about 30,000 tonnes. Today thorium reserves in the country are estimated to be around 300,000. With such a picture depicting an abundance of nuclear fuel reserves in the country, the case against nuclear power is as good as lost!.... Varghese's arguments and especially this book have influenced me a great deal in undergoing a change of heart. I now concur with his opinion that nuclear power is anti-people, dangerous and must be eschewed at all cost. He has succeeded in demolishing the myth of nuclear power as a panacea for our energy famine.[62]

Joseph continued, 'One thing is clear—it is a calculated move towards mass killing, not only the people but also life in all forms in the environment'. He concluded by advocating that:

> It is high time that these handful of scientists were prevented from their onslaught on the nation. The time has come for all sensible people of this country to wake up and react vigorously.[63]

In an astonishing departure from the conventional narrative of the post-colonial nation, nuclear scientists were held not as sensible builders of the nation but rather its senseless destroyers by a scientist himself.[64] Varghese's appeal was to 'common sense' which, in the end, was shared by the judiciary. He beamed, 'perhaps the first time in the world—a popular agitation against the establishment of a nuclear plant has succeeded'.[65]

Plans were also made for a nuclear power plant in Peringome to the north of Kerala that too were defeated by collective agitation. When government plans for the nuclear plant were shifted to Kudankulam in Tamil Nadu with ambitions to attempt to supply electricity to two states, like-minded activists and representatives of fishing communities from across the state border would regularly come to attend meetings. Before 2011, most of these were held in the central town of Nagercoil in Kanyakumari District only a two-hour drive away from Kerala's capital, Trivandrum (now renamed Thiruvananthapuram).

Activated by telephone and in the early days, one-line telegram communication, many wanted to share their experiences and embolden
the movement in the south. After all, radiation does not respect the
lines of administrative borders.

However, the attendants from Kerala would complain about the
difficulties of organizing a similarly effective anti-nuclear movement
in Tamil Nadu due to what they saw as the idiosyncratic culture of
populist politics in the state.[66] They parodied it in terms of 'item politics' where political difference lay in the electioneering of consumer
handouts such as television sets and gas burners rather than burning
issues that affected people's health and welfare. One Kerala-based
activist decried, 'In Tamil Nadu, everyone can be bribed. It is more
corrupt than Kerala'.

It is certainly true that Kerala's history of left-wing politics, high-
density demography, and the dominance of educated and professional classes played an influential hand in repelling a nuclear plant.
Hardly any political party in Tamil Nadu was interested in taking up
environmental issues, let alone a nuclear plant issue, seeing it as a
vote loser.[67] It was recognized that many, including the state's ruling
party had made personal gains from the Kudankulam Nuclear Power
Plant in their engagements with the central government. Ultimately,
whether it be through the two main parties, Dravida Munnetra
Kazhagam (DMK, Dravidian Progress Federation) or the Anna
Dravida Munnetra Kazhagam (AIADMK, All India Anna Dravidian
Progress Federation), the Tamil Nadu state government was brought
or, more to the point, *bought* into the bargain with the centre-state in
the politics of purchasing power.

The political Left was considered bankrupt and confused: ' ... it
cannot decide whether they're pro- or anti-nuclear', observed one
activist in Kanyakumari District. With few exceptions, left-wing politicians provided little support or inspiration for an anti-nuclear energy
project, gripped as they were by their repeated calls for jobs, electricity, and the country's industrial development, whatever that industry
happens to be. The Russian connection to Kudankulam continued
to recall an erstwhile socialist era to which the Left were favourably
disposed even though contemporary capitalist realities did not match
their weathered fantasies.[68] Comparing their latter-day rejection for
the French 'capitalist reactors' at Jaitapur to their ambivalence and

even welcome to the nuclear power plant in Kudankulam, the author and social activist A. Muthukrishnan mocks their stance: 'It's as if the Russian reactors are Communist in nature'.[69]

Timing played a critical part. By the late 1980s, the Communist bloc had collapsed, the USSR economy was in turmoil, and the main signatories disappeared—Mikhail Gorbachev lost power and Rajiv Gandhi was assassinated, both in 1991. Kudankulam project plans were left to rot on the NPCIL shelves. On reflection, it is not so much effort and efficacy that decided the success or failure of the anti-nuclear movement in Tamil Nadu, but how larger national and global crises play into local circumstances such that nuclear plans no longer have traction on the ground. Whether it was through a nuclear disaster, economic collapse, or an assassination, world events came to the rescue, and for a moment at least, campaigners enjoyed a joyous moment of victory.

The Neo-liberal Era

The 1990s marked a dramatic change in the Indian political economy with the rapid growth of trans/multinational ventures and consumer society. Development projects had also taken on breakneck speed as deals were struck up and down the country with the ambition to make India into a regional superpower. As an era that foregrounded the free market and consumerism, it also saw the hollowing out of the modern subject through a gradual process of democracy diminishment.[70] As the neo-liberal state dismantled its welfare provisions in public and social services and deregulated the economy, corporate ventures grew in scale and ambition. This was matched in intensity by a slide into more and more repression of marginalized people, particularly if they presented an obstacle to such undertakings. Consequently, while a minority prospered, liberalization policies began to wreak havoc on the lives of many.

With these changes that orientated themselves more and more to the urban middle classes came a thirst for power, electricity being the lamp that could lead the way to expansive plans. Even though a diverse energy portfolio was sanctioned by the Indian government, and that nuclear produced less of the country's power than any of the others, state compulsions geared by military–industrial interests meant that nuclear technology continued to be portrayed as a beacon for bulk electricity. Power projects became not only more pressing, but also increasingly oppressive. It was difficult to oppose the desire for more uninterrupted supplies of electricity—most people if not all wanted it, and if it did not affect them directly, were equally ambitious for more power projects. To have it any other way was quite literally a return to the dark ages.

Kudankulam fell subject to the exigencies of another moment. In 1997, the then Indian Prime Minister H.D. Deve Gowda and the Russian President Boris Yeltsin instigated further talks and the final agreement was signed by their successors, Atal Bihari Vajpayee and Vladimir Putin, four years later. India entered into renewed arrangements for two standard high-pressure VVER-1000 reactors that became connected to other defence deals including the supply of T-90 tanks, Sukhoi Su-30 planes, and the Admiral Gorshkov-class frigate.[71] After a revised agreement on loan interest, repayments in dollars, and rearrangements regarding the spent fuel and radioactive waste to be kept in India, construction of the nuclear plant began.[72] By 2006, an Indo-Russian joint task group was set up that announced plans for further reactors that had been approved by the central government.

The development in Kudankulam was paralleled with reinvigorated ambitions to construct civilian nuclear power plants up and down the subcontinent, particularly in the aftermath of the Nuclear Supplier Groups waiver in 2008. Other plans were proposed for Haripur (West Bengal), Mithi Virdi (Gujarat), Madban (Maharashtra), Pitti Sonapur (Odisha), and Kovvada (Andhra Pradesh), a couple of which have been subsequently dropped.

In the volley of arguments about battling climate change and global warming fuelled by the Kyoto Protocol of 1997, the environment itself was co-opted as an alibi for the nuclear industries. It was claimed that nuclear power could reduce up to almost a half of India's carbon emissions over the next few decades.[73] In other words, a twisted environmentalism emerged—one where nuclear developments could be seen as an ecologically sound and safe solution to fossil fuel-induced climate change. Conservationist Samuel Lal Mohan remarked in a seminar in Nagercoil that:

... an amendment is suggested in the Kyoto protocol (which prescribes reduction of Carbon emission by 5% of 1990 level) that building of Nuclear Power Plants, in poor and underdeveloped countries by the G-7 countries like USA, UK, Canada, Russia and Japan, as an effort to reduce Green House Gases like Carbon-di-oxide. The global environment fund, and World Monetary Fund and World Bank may support such funding for building Nuclear Reactors at the insistence of the USA. So Russia may utilise this 'Pro-Environment' or 'Environment friendly' funds of G-7 countries to build Nuclear Power Plants in India, and escape the Kyoto protocol by back door. We should not fall prey to the International game played on the poor countries by the rich countries [sic].[74]

Local inhabitants felt they were pitched in an uphill battlefield that had taken on vastly global dimensions to do with climate change compulsions. As the environmental anthropologists Alexander Dunlap and James Fairhead argue, terms such as 'green', 'sustainable', 'clean', and 'climate friendly' have become part of a worldwide vocabulary to legitimate extractive, exploitative, and/or toxic industries.[75] Casting a blind eye to their intensive construction and radioactive waste nuclear power plants were rebranded as green and clean.

Anti-nuclear campaigners upped the ante to cement their bond with other grassroots movements across the country that had grown as a result of India's proliferation of large development projects related to the mass production of energy and other goods and materials. This included what had become a force to be reckoned with, the National Alliance of People's Movements (NAPM), that coordinated campaigns across the country led by those threatened by large-scale and non-procedural developments. The NAPM organized a four-day workshop in Nagercoil in 1998 that sealed this need for national grassroots solidarity. After the setting up of the PMANE in 2001, another group emerged in south India, the National Alliance of Antinuclear Movements (NAAM) with 'The Kanyakumari Declaration' in June 2009.[76] This included embracing other social justice organisations that tended to the marginalization of the poor and minorities, displacement, sanitation, healthcare, and the scarcity of potable water. Fishing populations joined ranks with people of the land including farming and tribal communities across the country.[77] To cite the anthropologist and historian Partha Chatterjee, they vindicated a differentiated 'moral community' of people from different walks of life.[78] They came together to oppose what they saw as unjust and immoral forces orchestrated by a bandwagon of national and transnational elites. There was a cyclical and dynamic diversity to the anti-nuclear momentum, that gained more ground but also declined at intervallic moments due to various factors to be elaborated in the rest of the book.

Nuclear energy's close association with weaponization cannot go unremarked even though the Kudankulam Nuclear Power Plant is designated a civilian operation. Focusing on the case of Britain, Emily Cox, Phil Johnstone, and Andy Stirling propose that several 'perceived imperatives' are extant in 'military and civilian nuclear sectors in terms of high-level policy processes around supply chains, skills and expertise'.[79] The science and technology policy analysts hold that these connections are part of the state's 'deep incumbency complex' that is orientated towards gaining political purchase through nuclear weapons in global affairs.[80] With regards to south Asia, 1998 was the year when both India and Pakistan declared their nuclear bomb ambitions through their spate of underground tests. This was an outcome of the deep incumbency complex fused with nationalist machismo through a demonstration of nuclear and scientific strength. The complex is both binding and gripping, deeply incumbent, and compulsive.

Spurred on by the tests in 1998, an embryonic movement had developed against the nuclearization of south Asia that saw attempts to form peace alliances between activists in the global South. This anti-weapons movement was largely urban-based, led by activists and intellectuals in north India in arms with those based in Pakistan and other regions in South and South East Asia, as evident with the Coalition for Nuclear Disarmament and Peace—an umbrella organization for about 200 other grassroots groups, movements, and advocacy groups.[81]

South Indian peninsular campaigners also tapped into this broader anti-nuclear sentiment who shared resources and tactics with them. They likened the Kudankulam reactors to nuclear bombs, and speculated how the fuel could be used for more nuclear weapons and not its stated purpose for electricity. So as not to be cast out as anti-national Luddites, protesters foregrounded alternative proposals for electricity production as with the promotion of solar, wind, and mini-hydel power projects, each of which were summarily dismissed for one reason or another by the political elites. In the words of the ecologist Madhav Gadgil and historian Ramachandra Guha, the latter were bound to a 'pattern of development [that has] inevitably favoured the dinosaurs'.[82] Nuclear and, to a lesser extent, grand hydroelectric dams were regarded as the most impressive dinosaurs of electricity.

By the mid-2000s, two broad streams to the movement in peninsular India became evident. Those akin to the conservationist Lal Mohan preferred to gather data and information for seminars, conferences, court cases, and the like, hoping to affect policy and decisions but also fully aware of the power of people's pressure (Figure 3.1). Coming from a government scientist background as a marine biologist, Lal Mohan spent his time running several organizations and initiatives such as the Nuclear Power Awareness Committee, the Conservation of Nature Trust, the Nagercoil Citizens Welfare Council, and the Nagercoil

Figure 3.1   Dr Samuel Lal Mohan, 2006.
Source: Raminder Kaur.

chapter of the Indian National Trust for Art and Cultural Heritage (INTACH). His joined-up critique was 'side on'—that is, he was interested in collecting 'scientific' facts and figures and focusing on the environmental and health impacts of nuclear power plants—rather than a head-on attack of the nuclear industry. His preference was for a challenging civil cooperation rather than civil disobedience where even the terms, 'anti-nuclear' or 'environmentalist', were eschewed.

On a second level, the tactics of another influential resident of Nagercoil, Udayakumar, were much more direct with declarations to step outside of the 'civility of society' but in a peaceful manner (Figure 3.2). From his years of scholarly work in historical and peace studies, he had come to the view that civil disobedience was the need of the hour.[83] In an ardent speech on the Kudankulam Nuclear Power Plant at a meeting in Nagercoil in 2006, he announced, 'We are ready to die. It is unacceptable. It is anti-democracy, anti-future, anti-children, anti-human'. As he spoke inconvenient truths to power, his

Figure 3.2   Dr S.P. Udayakumar, 2012.
Source: Raminder Kaur.

speeches were informative and persuasive, rational yet impassioned displaying sparks of rage against his experience of the intransigence and duplicity of nuclear authorities.

Lal Mohan and Udayakumar were not averse to each other's styles of awareness-raising and often they worked together along with others in multipronged strategies. Challenging civil cooperation and civil disobedience were enfolded in each other. Joining up with others, they organized 'inreach' public seminars for officials, NGOs, students, and the like to spread more awareness about related problems such as environmental pollution, water scarcity, and the hazards to health by ionizing radiation. There were court proceedings in another attempt at challenging civil cooperation from within. There was also 'outreach' attempts to reach out to wider publics as with marches, lecture tours, sit in protests (*dharna*), and cultural activism where drama and film shows were used to disseminate their anti-nuclear message across the land.

Both protagonists produced literature in Tamil and English. With a talent for fictional writing, Udayakumar occasionally engaged in imaginative work in addition to his informational outputs. In 2004, he had written a satirical one-act play, *Anushakthi Amma* (*Atomic Mother*). Set in a dystopic future, the Kudankulam Nuclear Power Plant is managed according to ludicrous logics, and treated like divinity even while people suffer and die around it. Another local drama group, Murasu, explored themes such as the problems with Kudankulam, the problems of sand mining, and threats such as tourist development firms threatening to take over the beaches and people's homes. They visited various community centres and villages with their performances (Figure 3.3). Two of their plays were also taken to Mumbai as part of the Peace and Justice Festival in 2006. Using a rousing style of performance, song, and drumming, the group inventively created skits of fishermen against government and Indian Rare Earths Limited officials to take over their homes and beaches. Packed within its multilayered and entertaining plot were criticisms of duplicitous nuclear officials. Despite the fact that it was performed in Tamil, its wit, message, and energy enlivened Tamil, Hindi, and English-speakers alike. Following the philosopher Jacques Rancière, such interventions were an attempt at a 're-distribution of the sensible' with a reconfigured politicized aesthetic on nuclear industries.[84]

Figure 3.3   Performance by the Cultural Group, Murasu, in a Kanyakumari District Village, 2006.
Source: Raminder Kaur.

Even though traversing a national platform, solidarity with international campaigns had to be more cautious when compared to the political climate after the Chernobyl accident in the 1980s. This was an era when anything nuclear was arraigned in a strong national security discourse that itself was entrenched in neo-liberal economics and an increasingly muscular sovereignty.

As the Indian government sought to empower itself, insurgency had become more widespread. Briefly, terror attacks had arisen most markedly in the 1980s with respect to the regions of Punjab, Jammu and Kashmir, and the continuing ferment in the north-eastern states from the 1970s. It spiked again in the 2000s with Islamic extremism across the subcontinent and Maoist Naxalites spreading to the south-central states that formed a 'red corridor' with the north east. After a reactive period of economic sanctioning in the wake of the nuclear tests, Western governments renewed initiatives to engage the Indian government. Sanctions were relaxed only five months later with their lifting in 2001 under the George Bush Junior administration. Compelled by the 11 September aircraft attacks in USA, the post-2001 period saw the execution of the 'War on Terror' and the birth of an international US-led alliance that included India. Several knowledge, trade, and security initiatives began between India and USA that led to talks about a civilian nuclear agreement.[85] The climate of terror with its national and transnational components ricocheted into hardened and securitized political cultures within which nuclear deals were embedded.

While the Indian government engaged US officials, it continued nuclear collaborations with Russia in a recalibrated post-Cold War neo-liberal world. Drawing from a public memory of British colonial rule, anti-nuclear critics compared the post-1990s period to a form of neo-colonialism where people's lives were ruthlessly oppressed and their rights abrogated. A leaflet from the Conservation of Nature Trust in the early 2000s declared:

> The Russians may be happy over the [Kudankulam] pact as it will boost its dollar hungry economy with an inflow 360 crore of US dollars (Rs 173,000 crores) work as this is entrusted to 300 industrial units in Russia which will manufacture parts for the VVER-1000 Reactors. It is like manufacture of cloth in Manchester, England for India in the Pre-independent India. The much publicised job opportunity for the local labourers is only a mirage [sic].[86]

As with the extraction of raw materials from the subcontinent to serve others, and then selling it back at extortionate prices in the colonial era, so with the Kudankulam contract between India and Russian authorities that were seen to be pitched against Indian people.

Some residents in peninsular India went as far as to describe the construction of reactors as the 'rape of their land'. Aside from coastal considerations, thousands of tons of rocks had to be brought to the nuclear power plant to push back the sea and for other construction purposes. It was as if entire mountains had been cut up, crushed, and transported to keep the ocean at bay as another patch of about half a kilometre square was pumped clear off seawater in front of the reactors. The mind rationalized but, for many in the region, there was also the visceral reaction to the devastation of the environment, as if

Mother India had herself been 'disembowelled' for nuclear ambitions. The feminine principle had been overrun by the masculine sense of exploitation, brutality, and control.[87]

As Wendy Brown has proposed, such phenomena vindicate a defining feature of the power and privileges accorded to a masculinist state: 'it entails both a general claim to territory, and claims to, about and against, specific "others"'.[88] In this gendered governance paradigm, the environment is effectively feminised and defaced with modern man's ambitions to conquer nature and the communities that depend upon it.

Udayakumar too underlined the point that 'rape of nature and women [is] analogous'.[89] The land has no voice, no rights except on febrile paper. It is a part of Mother India, a feminine principle of commercial and extractive value only until its productive life ends.[90] After that, it is expendable other than as a glorified yet eviscerated icon for the nation. Udayakumar elaborates:

> The state is masculine, inhuman and obsessed with power and violence. We can't match its muscle. That's why we have insisted our people be non-violent but courageous and persistent in our struggle.[91]

The anti-nuclear movement's emphasis on non-violence suggests a mix of Mohandas Karamchand Gandhi's political strategy dovetailed with eco-feminist movements adapted for the contemporary era.

The post-1990s period therefore proved to be a very different and difficult era that saw the toughening of the Indian state when it came to national security and development. The non-violent and anti-nuclear stance associated with Gandhi had fallen out of favour in the mainstream to be replaced by the jouissance of consumption and the bomb compatible with imaginaries of a muscular and masculinist nation.[92] Anyone who sought to protect the environment and human rights was cast as a suspect insurrectionist against the country's interests. Any resistance group denouncing nuclear power in India was tarnished with a terrorist or Naxalite brush, seen also to be threatening national security. Critics and agitators became more wary, emphasizing that their struggle was about the rights to water, livelihood, and dissent for people of a democratic India, rather than a direct affront at the nuclear industries—an assault that could easily backfire with accusations of national treachery. Inevitably, the discourse of democratic rights was eroded in favour of individualist and aggressive enterprise and, as we shall see now, the overriding principle of policy.

The Tyranny of Technicality

Virtually all avenues for social justice in the country were explored to the extents possible—the rights to lives, to peacefully dissent, for information and transparency, accountability, and the need to follow mandatory procedures to do with environmental assessments, public hearings, safety and evacuation measures, and the reliability of VVER reactors later alleged to be of substandard quality.[93] One of the primary initiatives was to stake a claim on peoples' fundamental rights as defined by the Constitution of India.[94] Specifically, Article 19(1)a refers to 'the right to freedom of speech and expression' and 19(1)b to assemble peacefully and without arms. Article 21 decrees 'no person shall be deprived of his life or personal liberty except according to procedure established by law'.[95] This article in particular has been interpreted as an essential aspect of rights-to-lives movements including rights to water, a livelihood, health, education, and a speedy trial. With Article 32, individuals could seek remedy for any violations of their fundamental rights.[96]

One memorable case was in 2002 when people came together in Nagercoil to start a court case under Article 32 of the Constitution of India. It went to the Supreme Court of India but, in a country where the average duration of a court case is five to ten years, the case was fast-tracked and dismissed in a matter of weeks. The verdict:

> There is an inordinate delay in filing the writ petition. Secondly, the question of setting up the power station is a matter of policy. There is no reason as to why this court should sit in appeal over the Governmental decision relating to a policy matter more so, when crores of rupees having been invested.[97]

Pointing out the delay of the petition and reminding the complainants of the cost of government policy made the appeal redundant.

The petition was dismissed with 'costs of Rs 1,000 which shall be paid to the Supreme Court Legal Services Committee'.[98] All in all, the case cost the complainants not just Rs 1,000, but about Rs 40,000 plus flights to and from Delhi—by train, one journey alone would take three days, a time that many of those who worked could little afford. Lal Mohan recalled his feelings at the time of the Supreme Court verdict:

> When the SC (Supreme Court) verdict was announced, I felt saddened. My heart sank. I felt that there was no justice in this land. And that we had no other recourse. The SC is the highest in the land. We thought about turning to international agencies but the problem they will identify to act on our cause is national security.

Nevertheless, exasperated by regulatory shortcomings and the incestuous drawing of ranks between Indian authorities, The People's Movement against Nuclear Power extended its appeal outwards.[99] On 23 October 2002, they submitted a letter to the Executive Director of the United Nations Environment Programme (UNEP). The letter stated:

> The Indian Central Government is proceeding with the construction of the Kudankulam Nuclear power plant in spite of all the protests that has been made by numerous bodies and organisations. We have now decided to make a plea to your good self having no other recourse [sic].

After a list of the main problems, the letter ended with allegations of lax governmental regulations and corruption that could have impact on wider regions of the world, and concluded:

> Finally, we believe, that with your good office having responsibility for the environment of the whole world, that you will be able to intervene on this monstrous project and make our area a safe haven to live in nuclear pollution free environment. [sic]

A month later, a response arrived. The Regional Director and Representative for Asia and the Pacific recommended 'that this matter goes beyond the mandate of UNEP's activities in the region, and perhaps, it would be best if you were to consult the International Atomic Energy Agency (IAEA)'. Knowing full well the futility of approaching an international organization whose very raison d'être was, according to their website, to 'promote the safe, secure and peaceful uses of nuclear science and technology', campaigners felt that they were between a rock and a hard place that left them no room to manoeuvre.[100]

Campaigners turned to India's National Human Rights Commission. In this appeal, they again stressed the urgency of a public hearing and an environmental impact report for the initial proposal of two reactors as per India's Environment Impact Assessment (EIA) Notification (1994), along with a summary of nuclear hazards to nearby communities.[101] The response received on 14 November 2005 was a point-by-point rebuttal. At heel with the governmental position, it declared that the project was first cleared in 1989 and therefore lies out of the Act's purview.[102] It furnished select evidence from the NPCIL and the Atomic Energy Regulatory Board to conclude that 'the plants under construction will not pose any unacceptable risk to the plant personnel, public and the environment'. It emphasized the importance of dealing to the nation's 'energy famine' and the 'need to produce electrical power by all available generation technologies, that is, hydro, thermal, nuclear or renewable energy sources'. Comparing the situation in France, where 'more than 75% of the total power requirements of France are met by nuclear power', it asserted that 'Nuclear power plants are environmentally benign with respect to carbon dioxide emission and therefore cause no global warming'. In its detailing of somewhat questionable evidence regarding health hazards, the letter continued:

> Cancer Research Institute, Thiruvananthapuram has undertaken in-depth study of effect of low level radiation on the population in high natural background area of Kerala. The study revealed that the prevalent high natural radiation background does not contribute to any abnormal increase in cancer/early abortions and mental retardation.... The Nuclear Power Plant like Tarapur Atomic Power Station, Madras Atomic Power Station have been operating in coastal areas for about thirty and twenty years respectively and fishermen have not observed any adverse effect on their catch. In view of the factual position detailed above, it has been concluded that the complaint does not warrant any consideration, being devoid of merits [sic].

Finally, the response concluded: 'In view of the facts and circumstances detailed above, report of Undersecretary, Department of

Atomic Energy, Govt of India, is taken over the record and matter is closed' [sic].

An expressly open and shut case: the word of the highest court in the land carried authority. A 'factual position' was engineered by the nuclear state and all authorities were in agreement with it. Even the National Human Rights Commission conferred more rights to 'electrical power' citing reports and studies that themselves had been conducted by government-backed organizations.

Making appeals to state departments to broker and lever regulation by nuclear agencies was a fraught strategy due to the especial status of the latter enshrined in several exemption acts and clauses. It was also an indication of a judicial cul-de-sac: that Supreme Court judges do not have the technical expertise to judge the matter at hand and therefore need to call upon nuclear specialists, all of whom are employed or sponsored by the Indian government.[103] The irony is quite astounding: state-backed experts are brought into a case complaining against state-backed conduct, a catch-twenty-two that allowed little openings into the realm of officialdom for contrary views on nuclear developments. Even the Atomic Energy Regulatory Board, supposedly an independent overseeing body, is funded by the Department of Atomic Energy (DAE), and answerable to the Atomic Energy Commission, whose Chairman, to seal the loop, is the DAE Secretary.[104] Although top Indian judges were on the peripheries of circuits of corruption, they fell into other ones to do with the sanctity of the state apparatus and policy.

International agencies did or could not take any meaningful interest due to the nuclear installation being outside of its remit and heavily embroiled in national sovereignty and security. The obstacles were formidable. While complainants nurtured a modicum of optimism in navigating through a country they believed to be a democratic one, in a matter of months if not days, downhearted cynicism set in.

To similar ends, Alf Gunvald Nilsen notes how the Supreme Court dismissed a very credible case on the Sardar Sarovar Dam in the Narmada River valley in 2000 with technicalities such as a lapse in time and the significance of governmental policy. This was a very different kind of response to an earlier appeal to the Supreme Court when the fundamental rights of people who lived in the region were honoured by the judiciary.[105] The Supreme Court verdict justified its latter-day verdict on the Sardar Sarovar Dam by saying that whether hydro, thermal, or nuclear power projects in India:

> ... the electricity has to be generated and one or more of these option exercised. What options to exercise in our constitutional framework, is for the government to decide keeping various factors in mind.[106]

Nilsen observes that this is one typical trap of 'judicial activism'.[107] Once a resistance movement enters into the fray of the legislature, then it is already compromised. True as this may be, there were earlier cases that had been won even on the nuclear issue, and this avenue that combines moral arguments with legal technicalities continues to be pursued by campaigners in what Erica Bornstein and Aradhana Sharma describe as 'technomoral politics'.[108]

An institution such as the Supreme Court is held to orbit above corruption and culture. However, the Supreme Court's decisions are not just about the specificity of individual cases and social actors. Rather, in the adoption of recalibrated goals, the terrain has changed.[109] The Court relies upon the structural constraints of government-framed policy that poses as neutral but ends up acting like tyrannous technicalities effectively debilitating right-to-lives movements. Conceivably, while justice in India had not been patently politicized, it had become more and more technocratic, hemming in appeals on nuclear plants to matters of punctuality, policy, and organizational structuring. As the anthropologist James Ferguson has demonstrated in his classic work on the 'anti-politics machine' of development and bureaucracy, policy was screened away from politics despite the fact that guiding principles were steeped in highly politicized or contested contexts.[110] The rights of Indian citizens as framed in the Indian constitution and related laws became diminished in the superior light of India's policies on development, electricity and, with respect to the Kudankulam case, 'nuclear exceptionalism'.[111] This then became the yardstick of rights across the establishment.

Relatedly, in the neo-liberal era, corporate–state actors too have adopted the language of human rights, social responsibility, and even ecological care for their large-scale projects.[112] Against this balance with its constantly shifting loads, anything to do with the *human* or the *environmental* as understood by activists could maintain little weight on the scales of justice.

However resolutely policy might be held in the courts, retired nuclear engineer R.V.G. Menon, observed that nuclear authorities have a 'dilute and disperse' policy towards radioactive waste in a seminar on Kudankulam to mark Chernobyl Day in 2006. He continued:

Safety mechanisms cost more money. So where possible the nuclear authorities cut corners. In the US, plant authorities had to show that they had evacuation plans for people within a 30 km radius of the plant. They failed in providing a blueprint, so the plant was not given approval. We have to ask the same question in India.

However, nuclear officials either have cloth ears when it comes to asking such questions, or turn a blind eye to those who remind them of regulatory procedure. When it suited, they would regularly dilute, disperse, and even dispense with protocol. When Udayakumar had asked a NPCIL employee about seeing the EIA report reputed to have been completed by NEERI in 1988 and one in 2004 on the feasibility of four more reactors at Kudankulam that should be in the public domain, he replied, 'but it is too big'.[113] Udayakumar declared, 'but we're educated. We can read it. It is no problem'. Then the bureaucrat said, 'It's in Mumbai. It occupies a whole room and we'll buy you an air ticket to go and see it'. Udayakumar requested, 'Just give it to us here and we will deal with it'. After that exchange, there was 'no word nor plane ticket from the official.

A New Philosophy of Life

India has long been a lively site for informed and innovative grassroots movements for people's democratic demands. Over the last few decades, however, there has been, on the one hand, a hardening of the state, and on the other, the gouging out of the rights of Indian citizens. The notion of the human has been hollowed out from one that had certain rights to a dignified and safe life to one that sees the human either in terms of a hindrance that need be subservient to policy, or as a cost and cog of capital to be arraigned into larger plans for the country's consumer-orientated development. Consumptive rights have taken over constitutional rights. Whereas in the 1980s, those in the political, judicial, and scientific establishment might have taken nuclear concerns seriously and even be converted by them as happened over the nuclear plant near Kothamangalam, by the turn of the century, these anxieties were dismissed out of hand where the right to electricity as the lynchpin for the country's development took precedence over all else.

This state of the 'hard and hollow' is an observation also made by water policy expert Ramaswamy R. Iyer. He states:

> It now seems that the executive and judiciary share a particular understanding of 'development' and subscribe to the proposition that the infliction of injustice and misery on PAPs [project-affected people] must be accepted as the 'cost' of that development.[114]

With the short-circuiting of legislature to a matter of expert policy, punctuality and progress, and the pressure for breakneck development over and beyond protocol, people in peninsular India felt that they were hinterland underdogs for power interests emerging in larger cities to the north. It was precisely the lax implementation of regulations along with India's vast working populations and resources understood as a manageable human cost or capital that attracted multi/transnational nuclear corporations to the subcontinent. Once human and habitat are cast as cheap, they make for a more profitable playing field.

Nevertheless, the post-1990s period also led to a creative widening of the spaces of criticality through connecting with other national grassroots movements, developing new forms of campaigning, and through efforts to entwine nuclear issues in the cultural everyday. While politically handicapped, when it came to working with transnational resources and networks (as we shall further examine in Chapter 9), they linked the threat of the nuclear power plant with miniscule yet essential aspects of their local lifestyles—food and water, for instance—in a bid to reach out to more and more people. It was a tactic that sought to link nuclear developments with the intimate necessities of everyday lives, thereby aiming to countervail accusations of anti-nationalism and take the protest to other levels—from the outer to the inner to then, it was hoped, the stellar.

In 2006, following the popularity of the anti-corruption social activist Anna Hazare in what became known as the 'Indian Spring', *Gandhigiri* became all the rage across the subcontinent.[115] Literally

meaning 'Gandhi force' based on the term *dadagiri* to refer to hood-
lum coercion, *Gandhigiri* owed to the success of the film, *Lagge Raho
Munna Bhai* (*Carry on Munna Brother*, dir. Rajkumar Hirani, 2006).
It was a latter-day attempt at resurrecting Mohandas Karamchand
Gandhi's tactics of peaceful resistance and civil disobedience through
the commodified lens of popular culture. With the power of morals
over muscle, it meant directing the gaze first inwards before change
could be affected outwards. It espoused a life of purity and simplicity
rather than corruption, competition, and conspicuous consumption.
It advocated peaceful action in the face of dishonesty and oppression, not passive contemplation or individual salvation. Although
not directly influencing an anti-nuclear movement, it was a popular
reminder for middle class India of Gandhi's message for the challenges of modern life to add to the official (*sarkari*), overly religious
(*mathwadi*), and social justice (*kujat*) narratives of Gandhism.[116]

The year 2006 also saw the shoots of a local initiative in Nagercoil,
Green Tamils (Pachai Thamizhagam). It was an embryonic and aspirational movement based on a revived Gandhian and green political philosophy that worked from the mind through the body, from the everyday
to the extraordinary, in a slightly different manner. As Udayakumar,
the initiator propounded, 'This would be about how we as individuals can look after ourselves and our immediate surroundings'. It was
another way of asserting self-reliance and control of the desirous and
consuming mind that comes with so much destruction. He suggested
that everyone could form their own groups to meet informally in the
mornings to go for an early morning walk 'when we could focus on one
thing, maybe meditate, and then later have a short discussion about
something that matters to us'. While seemingly solipsistic, it was also
with a view to forge strength through solidarity. He elaborated:

> We can start off at 5 am in the morning when there is less havoc on the
> streets and greet the day with a smile. We don't just want to be seen
> as anti-Kudankulam, but it can be a whole philosophy of life. Start off
> small and eventually it'll take off. Then it'll become too strong for oth-
> ers to oppress.

Tamil *Gandhigiri* was yet another attempt at allying resistance
against the nuclear power plant with the moral high ground of selfless

conduct tied in with a respect for other people and the environment. It was not just about soliciting political support but also developing 'a new way of life, a philosophy of life which starts from the mind'. Udayakumar continued:

> The mind is very powerful. If you think something, it will happen. If there are many people, then there is even more power in the meditation.... Gandhi Day should be a good day to start the movement. He was for peace and he used to go for daily walks to propagate it.

Based on a reflection on historical precedents, to think something was to materialize it. To amplify this thinking among many was to create even more of an impact. Political conduct needed to go beyond meetings and the struggle against obvious ciphers of oppression. It had to stem from the mind. It had to be energized vertically, laterally, as well as orthogonally so as to embolden and enjoin more support even from the 'sideliners'—those who sat on the colloquial fence. In this way, it could be regenerative on the inside while always being tactical on the outside. 'You must be the change you want to see in the world' as goes the now global Gandhian saying. Indeed, in the space of five years, Udayakumar's message drew many in the post-Fukushima Daiichi period that shook the nuclear establishment in India.

When we were walking on one of these early morning walks back in 2006, the sun was waking up. We could see a craggy colossus of a mountain in front of us against a pastel peach sky. Even though tinted by retrospective reflections, it seemed as if the drama of the beautifully intense and chameleon colours bode the rise of a new dawn. Udayakumar greeted it with a smile.

Notes

1. This comment reflects a widespread view. See also Admiral Ramdas and Lalita Ramdas (2012) 'Koodankulam Diary', *Countercurrents*, 9 January https://www.countercurrents.org/ramdas090112.htm, accessed 21 January 2019.
2. On the post-1980s anti-nuclear weapons protests in the USA, see Gusterson, Hugh. *Nuclear Rites: A Weapons Laboratory at the End of the Cold War.* Berkeley: University of California Press, 1998, pp. 164–218.

3. Ruparelia, Sanjay. 'India's New Rights Agenda: Genesis, Promises, Risks'. *Pacific Affairs* 86 (2013): 569–90, p. 569.

4. See Gadgil, Madhav and Ramachandra Guha. *Ecology and Equity: The Use and Abuse of Nature in Contemporary India.* New Delhi: Routledge, 2013; Nixon, Rob. *Slow Violence and the Environmentalism of the Poor.* Cambridge: Harvard University Press, 2011; Srikant, Patibandla. *Koodankulam Anti-Nuclear Movement: A Struggle for Alternative Development?* Working Paper 232, Bangalore: Institute for Social and Economic Change, 2009, p. 3.

5. Subaltern draws upon Antoni Gramsci's description for oppressed communities. Although initially premised on class identities, they have been extended to include a complex of class–caste–ethnic conjunctions as is apparent in the *Subaltern Series* volumes. I concur with Alf Gunvald Nilsen and Srila Roy's summary that subalternity might be best viewed as a category that is relational, intersectional and dynamic. Nilsen, Alf Gunvald and Srila Roy. 'Reconceptualizing Subaltern Politics in Contemporary India'. In *New Subaltern Politics: Reconceptualizing Hegemony and Resistance.* New Delhi: Oxford University Press, 2015, pp. 225–254.

6. Nash, June (ed.). *Social Movements: An Anthropological Reader.* Oxford: Wiley Blackwell, 2004, p. 4.

7. Where the poor have most alarmingly woken up to the dangers of radiation has been in areas around Jadugoda where many have joined political organizations to repel the authorities after being subject to the detrimental effects of uranium mining and tailings from the Uranium Corporation of India Limited. See the documentary, *Buddha Weeps in Jadugoda* (1999), directed by Shriprakash in support of the Jharkhandi Organization Against Radiation (JOAR).

8. Mazzarella, William. 'Internet X-Ray: E-Governance, Transparency, and the Politics of Immediation in India'. *Public Culture* 18, no. 3 (2006): 473–505, p. 485.

9. See Patnaik, Prabhat. 'Neo-liberalism and Democracy'. *Economic and Political Weekly* XlIX, no. 15 (2014): 39–44.

10. See Ho, Ming-sho. 'The Fukushima Effect: Explaining the Resurgence of the Anti-nuclear Movement in Taiwan'. *Environmental Politics* 23, no. 6 (2014): 965–83. With reference to Idinthakarai, see Prabhu, Napthalin. 'Protest Camp as Repertoire for Antinuclear Protest', forthcoming, 2019.

11. Loader, Brian D. and Dan Mercea. *Social Media and Democracy: Innovations in Participatory Politics.* London: Routledge, 2012.

12. See Macadam, Doug, John D. Mcarthy, and Mayer N. Zald (eds). *Comparative Perspectives on Social Movements: Political Opportunities,*

Mobilizing Structures, and Cultural Framings. Cambridge: Cambridge University Press, 1996; Benford, Robert D. and David A. Snow. 'Framing Processes and Social Movements: An Overview and Assessment'. *Annual Review of Sociology* 3, no. 1 (2000): 611–39.

13. Touraine, Alain. *The Post-industrial Society: Tomorrow's Social History: Classes, Conflicts and Culture in the Programmed Society.* New York: Random House, 1971; Offe, Claus. 'New Social Movements: Changing Boundaries of the Political'. *Social Research* 52, no. 4 (1985): 817–68; Laclau, Ernesto and Chantal Mouffe. *Hegemony and Socialist Strategy: Toward a Radical Democratic Politics.* London: Verso, 1985; Melucci, Alberto. *Nomads of the Present: Social Movements and Individual Needs in Contemporary Society.* Philadelphia: Temple University Press, 1989. On an overview, see Edelman, Marc. 'Social Movements: Changing Paradigms and Forms of Politics'. *Annual Review of Anthropology* 30 (2001): 285–317; Della Porta, Donatella and Mario Diani. *Social Movements: An Introduction.* Oxford: Blackwell, 2006.

14. Halfmann, Jost. 'Community and Life-chances: Risk Movements in the United States and Germany'. *Environmental Values* 8, no. 2 (1999): 177–97. Rachel Carson's book *Silent Spring* (first published in 1962, Boston Houghton Mifflin, 2000) and the Club of Rome's report, *The Limits to Growth* (Meadows, H. Donella, Dennis S. Randers, Jorgen Rander and William W. Behrens, first published in 1972, Washington, DC: Potomac Association) were seminal to the growth of environmental movements.

15. See Bettini, Giovanni. 'Climate Barbarians at the Gate? A Critique of Apocalyptic Narratives on Climate Refugees'. *Geoforum* 45 (2013): 63–72; Doulton, Hugh and Katrina Brown. 'Ten Years to Prevent Catastrophe: Discourses of Climate Change and International Development in the UK Press'. *Global Environmental Change* 19, no. 2 (2009): 191–202.

16. Historically, there have been four main approaches to analysing social movements from the 1960s that came together in a more integrated approach in the 1990s: those that focus on collective behaviour, resource mobilization, new social movements, and political processes. See Gibb, Robert. 'Toward an Anthropology of Social Movements'. *Journal des Anthropologue,* Association Francaise des Anthropologues 84–85 (2001): 233–53. See also Escobar, Arturo. 'Culture, Practice and Politics: Anthropology and the Study of Social Movements'. *Critique of Anthropology* 12, no. 4 (1992): 395–432; Tarrow, Sydney. *Power in Movement: Social Movements and Contentious Politics.* Cambridge: Cambridge University Press, 1998; Della Porta, Donatella and Mario Diani. *Social Movements: An Introduction.* Oxford: Blackwell, 2006.

17. Della Porta and Diani, *Social Movements*, pp. 21–2.

18. Pedwell, Carolyn. 'Affective (Self-)Transformation: Empathy, Neoliberalism and International'. *Feminist Theory* 13, no. 2 (2012): 163–79, pp. 164, 165–6.

19. Alex Shankland, Hani Morsi, Naomi Hossain, Katy Oswald, Mariz Tadros, Patta Scott-Villiers and Tessa Leuwin (2011) 'Unruly Politics', Brighton: Institute of Development Studies https://www.ids.ac.uk/idsresearch/unruly-politics, accessed 19 January 2019.

20. Shankland et al, 'Unruly Politics'.

21. G. Ananthpadmanabhan cited in Nair, R. Manoj (2006) 'Greenpeace Arm-twists Taj into Reviewing Power Usage', *Mumbai Mirror*, 7 December, p. 4.

22. Khanna, Akshaye (2012) 'Seeing Citizen Action through an Unruly Len', *Development*, 55(2):162–172, p. 3.

23. Hurd, Madeleine. 'Introduction—Social Movements: Ritual Space and Media, Culture Unbound'. *Journal of Current Cultural Research* 6, no. 2 (2014): 287–303.

24. Holloway, John. *Change the World without Taking Power*, (2002). Available on http://libcom.org/files/John%20Holloway-%20Change%20the%20world%20without%20taking%20power.pdf, accessed 19 January 2019, pp. 20, 58.

25. See Day, Richard. *Gramsci Is Dead: Anarchist Currents in the Newest Social Movements*. London: Pluto Press, 2005; Graeber, David. *Direct Action: An Ethnography*. Oakland, CA: AK Press, 2010; and Maeckelbergh, Marian. *The Will of the Many How the Alterglobalisation Movement Is Changing the Face of Democracy*. London: Pluto Books, 2009.

26. Graeber, David. *Fragments of an Anarchist Anthropology*. Chicago: Prickly Paradigm Press, 2004, p. xvii.

27. Harcourt, B.E. 'Political Disobedience'. *Critical Inquiry* 39, no. 1 (2011): 33–55.

28. See Haksar, Vinit. 'Violence in a Spirit of Love: Gandhi and the Limits of Non-violence'. *Critical Review of International Social and Political Philosophy* 15, no. 3 (2012): 303–24; and Stephan, Maria J. and Erica Chenoweth. 'Why Civil Resistance Works: The Strategic Logic of Nonviolent Conflict'. *International Security* 33, no. 1 (2008): 7–44.

29. Sharp, Gene. 'The Meanings of Nonviolence: A Typology'. *Journal of Conflict Resolution* 3, no. 1 (1959): 41–66, pp. 46–59.

30. Hurd, 'Introduction', p. 288. See James, Jasper. *The Art of Moral Protest: Culture, Biography and Creativity in Social Movements*. Chicago: University of Chicago Press, 1997.

31. See Scott, James. *Seeing Like a State: How Certain Schemes to Improve the Human Condition Have Failed*. New Haven: Yale University Press, 1999, pp. 11–15.

32. Luke, Timothy W. 'Environmentality as Green Governmentality'. In *Discourses of the Environment*, edited by Eric Darier. Malden, MA: Blackwell Publishers, 1999, p. 122. On biopolitics, see Foucault, Michel. *Society Must Be Defended: Lectures at the Collège de France, 1975–1976*. New York: St. Martin's Press, 1997.

33. Todd, Zoe. 'Protecting Life below Water: Tending to Relationality and Expanding Oceanic Consciousness Beyond Coastal Zones'. *American Anthropologist*, 17 October 2017. Available on http://www.americananthropologist.org/2017/10/17/protecting-life-below-water-by-zoe-todd-de-provincializing-development-series/, accessed 19 January 2019.

34. On identity politics, Patnaik differentiates three kinds: 'identity resistance politics', 'identity bargaining politics', and 'identity fascist politics'. 'Neo-liberalism and Democracy', p. 41. Another distinctive movement is the woman-led Gandhiite anti-liquor movement. See Hardiman, David. *Gandhi: In His Time and Ours*. New Delhi: Permanent Black, 2003, pp. 198–238. See also Omvedt, Gail. *Reinventing Revolution: New Social Movements and the Socialist Tradition in India*. ME Sharpe, New York and London, 1993; Sethi, Harsh. 'Groups in a New Politics of Transformation'. *Economic and Political Weekly* 19, no. 7 (1984): 305–16; Sheth, D.L. 'Grass-root Stirrings and the Future of Politics'. *Alternatives* 9, no. 1 (1983): 1–24.

35. Kothari, Rajni. 'The Non-party Political Process'. *Economic and Political Weekly* 19, no. 5 (1984): 216–24.

36. Bornstein, Erica and Aradhana Sharma. 'The Righteous and the Rightful: The Technomoral Politics of NGOs, Social Movements, and the State in India'. *American Ethnologist*, 43, no. 1 (2016): 76–90, pp. 81–2.

37. On the use of non-violence in anti-nuclear movements in USA from the 1970s, see Futrell, Robert and Barbara G. Brents. 'Protest as Terrorism?: The Potential for Violent Anti-Nuclear Activism'. *American Behavioral Scientist* 46, no. 6 (2003): 745–65.

38. Shiva, Vandana and Jayanto Bandopadhyay. *Chipko: India's Civilisational Response to the Forest Crisis*. New Delhi: Indian National Trust for Art and Cultural Heritage, 1986; and Guha, Ramachandra. *The Unquiet Woods: Ecological Change and Peasant Resistance in the Himalaya*. Berkeley: University of California Press, 2000.

39. Ravi Rajan, S. 'Bhopal: Vulnerability, Routinization, and the Chronic Disaster'. In *The Angry Earth: Disaster in Anthropological*

Perspective, edited by Anthony Oliver-Smith and Susanna Hoffman. New York: Routledge,1999; and Fortun, Kim. *Advocacy after Bhopal: Environmentalism, Disaster, New Global Orders*. Chicago: University of Chicago Press, 2001.

40. Fisher, William F. (ed.). *Toward Sustainable Development: Struggling Over India's Narmada River*. Armonk, NY: M.E. Sharpe, 1995; Nilsen, Alf Gunvald. *Dispossession and Resistance in India: The River and the Rage*. New Delhi: Routledge, 2012; Baviskar, Amit. *In the Belly of the River: Tribal Conflicts over Development in the Narmada Valley*. New Delhi: Oxford University Press, 1995; Dreze, Jean, Meera Samson, and Satyajit Singh. *The Dam and the Nation: Displacement and Resettlement in the Narmada Valley*. New Delhi: Oxford University Press, 1997. See also Roy, Arundhati. *The End of Imagination*. Chicago: Haymarket Books, 2005; and Mies, Maria and Vandana Shiva. *Ecofeminism (Critique. Influence. Change)*. Halifax, NS: Fernwood Publications, 1993.

41. Harvey, David. 'The "New" Imperialism: Accumulation by Dispossession'. *Socialist Register* 40: 63–87.

42. 'Construction of Jaitapur Nuclear Plant Expected to begin by Year-end: French Ambassador'. *Business Standard India*, 20 March 2018. Available on https://www.business-standard.com/article/pti-stories/construction-of-jaitapur-nuclear-plant-expected-to-begin-by-year-end-french-ambassador-118032001145_1.html, accessed 25 January 2019.

43. See Vaid, Minnie. *The Ant in the Ear of an Elephant*. New Delhi: Rajpal and Sons, 2016.

44. On the primacy of the nation, see debate with Jameson, Frederic. 'Third-world Literature in the Era of Multinational Capitalism'. *Social Text* 15, no. Autumn (1986): 65–88; and Ahmad, Aijaz. 'Jameson's Rhetoric of Otherness and the "National Allegory"'. In *Theory: Classes, Nations, Literatures*. London: Verso, 1992. The material presented here points to indigenous contestation by way of nation as people and nation as correlate with the state.

45. This is largely due to the paucity of 'anarchists' in the country although this is not to say anti-authoritarian thinking does not configure India's history. See Ramnath, Maia. *Decolonizing Anarchism: An Antiauthoritarian History of India's Liberation Struggle*. Edinburgh: AK Press, 2011.

46. See Weber, Max. 'Politics as a Vocation'. In *Weber's Rationalism and Modern Society*, translated and edited by Tony Waters and Dagmar Waters. New York: Palgrave Books.

47. Compare for instance with the anti-nuclear movement in the USA where participants were prepared to be arrested in the knowledge

that they would be out soon after, as reported by Gusterson, *Nuclear Rites*, pp. 178–80. He also makes the interesting observation that as the majority were white, well-educated middle classes, they shared similarities with those who worked in the nuclear establishment. The main difference was that the former were educated in the humanities and social sciences, the latter in science and technical fields. Gusterson, *Nuclear Rites*, p. 192. This observation is only partially relevant to the case considered here.

48. See also the contrary view by Peter Gelderloos that holds democracy too is allied with (legitimated) violence, a minority view that was articulate in south India and became more evident as the movement got stronger. Gelderloos, Peter. *How Nonviolence Protects the State*. Boston: South End Press, 2007. Anarchist movements, however, may foreground violence as a means to an end but are actively vigilantly against the rise of tyrannical leadership. See Churchill, Ward with Mike Ryan. *Pacifism as Pathology: Reflections on the Role of Armed Struggle in North America*. Winnipeg: Arbeiter Ring Publishing, 1998. On Maoists movements in India's 'Red Corridor', see Shah, Alpa *In the Shadows of the State: Indigenous Politics, Environmentalism and Insurgency in Jharkhand, India*. Durham: Duke University Press, 2010. Shah, Alpa and Judith Pettigrew, eds. (2012) *Windows into a Revolution: Ethnographies of Maoism in India and Nepal*. New Delhi: Social Science Press, 2012.

49. See Hansen, Thomas Blom. 'Governance and State Mythologies in Mumbai'. In *States of Imagination: Ethnographic Explorations of the Postcolonial State*, edited by Thomas Blom Hansen and Finn Stepputat. Durham: Duke University Press.

50. Heller, Patrick. 'Degrees of Democracy: Some Comparative Lessons from India'. *World Politics* 52, no. 4 (2000): 484–519, pp. 487–8. See also Comaroff, John L. and Jean Comaroff. 'Postcolonial Politics and Discourses of Democracy in Southern Africa: An Anthropological Reflection on African Political Modernities'. *Journal of Anthropological Research* 53, no. 2 (1997): 123–46.

51. For details, see Ramana, M.V. *The Power of Promise: Examining Nuclear Energy in India*. New Delhi: Penguin Books, 2013, pp. 85–8.

52. Udayakumar, S.P. (ed.). *The Koodankulam Handbook*. Nagercoil: Transcend South Asia, 2004, p. 35.

53. Gadgil and Guha, *Ecology and Equity*, p. 102.

54. See Benjamin, Walter. 'Critique of Violence'. In *Reflections: Essays, Aphorisms, Autobiographical Writings*, edited by Peter Demetz. New York: Schocken Books, 1986, pp. 277–300.

55. *Indian Express*. 'Koodangulam: A Nuclear Graveyard'. 20 June 1989.

56. Cited in Srikant, Patibandla. *Koodankulam Anti-Nuclear Movement: A Struggle for Alternative Development?* Working Paper 232, Bangalore: Institute for Social and Economic Change, 2009, p. 9.

57. *Indian Express*, 'Koodangulam'.

58. *Indian Express*, 'Koodangulam'.

59. M.P. Varghese. *A Critique of the Nuclear Programme.* New Delhi: Phoenix Publishing House Pvt. Ltd., 2000, p. x.

60. Ramachandran, K. 'The Peringome Antinuclear Struggle—When We Look Back after Two Decades'. 3 April 2012. Available on http://www.dianuke.org/the-peringome-antinuclear-struggle-when-we-look-back-after-two-decades/, accessed 19 January 2019.

61. Ramachandran, 'The Peringome Antinuclear Struggle'.

62. Cited in Varghese, *A Critique of the Nuclear Programme*, p. xi.

63. Varghese, *A Critique of the Nuclear Programme*, p. xi.

64. See Prakash, Gyan. *Another Reason: Science and the Imagination of Modern India.* Princeton: Princeton University Press, 1999.

65. Varghese, *A Critique of the Nuclear Programme*, p. viii.

66. Vaasanthi. *Cut-outs, Caste and Cine Stars: The Word of Tamil Politics.* New Delhi: Random House, 2008.

67. Exceptions include the Marumalarchi Dravida Munnetra Kazhagam (MDMK), and to a lesser extent, the Aam Admi Party (see Chapter 11). Vijay Kumar, S. 'MDMK for Closure of Kudankulam Nuclear Power Plant'. *The Hindu*, 22 March 2014. Available on https://www.thehindu.com/news/national/tamil-nadu/mdmk-for-closure-of-kudankulam-nuclear-power-plant/article5819246.ece, accessed 17 January 2019.

68. Moolakkattu, John S. 'Nonviolent Resistance to Nuclear Power Plants in South India'. *Peace Review* 26, no. 3 (2014): 420–6, p. 425.

69. Muthukrishnan, A. 'Indian Left and the Nuclear Hypocrisy'. *Countercurrents*, 26 March 2012. Available on http://www.countercurrents.org/muthukrishnan260312.htm, accessed 29 January 2019.

70. See Harvey, David. *A Brief History of Neoliberalism.* Oxford: Oxford University Press, 2005; Wacquant, Loïc. *Punishing the Poor: The Neoliberal Government of Social Insecurity.* Durham: Duke University Press, 2009; Brown, Wendy. *Undoing the Demos: Neoliberalism's Stealth Revolution.* Massachusetts: MIT Press, 2015.

71. Iype, George. 'The Russian Connection'. In *Darkness at Noon.* Available on https://www.rediff.com/news/2000/nov/23nuke.htm, accessed 22 October 2019.

72. Ramana, *The Promise of Power*, pp. 86–8.

73. Kakodkar, Anil. 'Energy in India for the Coming Decades'. Proceedings of an International Ministerial Conference on Nuclear Power for the

twenty-first century. Available on https://inis.iaea.org/search/search.
aspx?orig_q=RN:38056572, accessed 29 January 2019.

74. 'Seminar on the Impact of Kudankulam Nuclear Power Plant on
    Pechiparai Dam and Its Health Hazards'. Publicity, Conservation of
    Nature Trust, Nagercoil, 18 August 2001.

75. Dunlap, Alexander and James Fairhead. 'The Militarisation and
    Marketisation of Nature: An Alternative Lens to "Climate-Conflict"'.
    Geopolitics 19, no. 4 (2014): 937–61.

76. National Alliance of Anti-nuclear Movements (NAAM). 'The
    Kanyakumari Declaration: Statement of the National Convention
    on "The Politics of Nuclear Energy and Resistance", 4–6 June 2009,
    Kanyakumari, Tamil Nadu'. Available on http://www.sacw.net/arti-
    cle949.html, accessed 19 January 2019.

77. In south India, they included joining up forces with the River Farmers
    Association, Bhoomi Padukapu Sangam, Najilla Velanmai Vivasayigal
    Sangam or Fertile Land Farmer's Association and Kumari District
    Water Body Protection Federation. See Prabhu, Napthalin 'Antinuclear
    Power Movement in India after Fukushima Disaster: The Case of
    Koodankulam India', forthcoming, pp. 10–11.

78. Chatterjee, Partha (2004) *The Politics of the Governed: Reflections on
    Popular Politics in Most of the World*. New York: Columbia University
    Press, 2004.

79. Cox, Emily, Phil Johnstone, and Andy Stirling. 'Understanding the
    Intensity of UK Policy Commitments to Nuclear Power: The Role
    of Perceived Imperatives to Maintain Military Nuclear Submarine
    Capabilities'. SPRU (Science, Policy Research Unit) Working Paper
    series, p. 2. Available on https://www.sussex.ac.uk/webteam/gate-
    way/file.php?name=2016-16-swps-cox-et-al.pdf&site=25, accessed 19
    January 2019.

80. Cox et al. 'Understanding the Intensity'. See also Chapter 6.

81. Bidwai, Praful and Achin Vanaik. *South Asia on a Short Fuse: Nuclear
    Politics and the Future of Global Disarmament*. New Delhi: Oxford
    University Press, 1999.

82. Gadgil and Guha, *Ecology and Equity*, p. 135. On a critique of large-
    scale alternative energy developments, see Dunlap and Fairhead, 'The
    Militarisation and Marketisation of Nature'.

83. See Udayakumar, S.P. *Presenting the Past: Anxious History and Ancient
    Future in Hindutva India*. Westport, CT: Praegar, 2005; Galtung, Johan
    and S.P. Udayakumar (eds). *More Than a Curriculum: Education for
    Peace and Development (Peace Education)*. Charlotte, NC: Information
    Publishing Age, 2013.

84. Ranciere, Jacques. *The Politics of Aesthetics*. London: Continuum, 2006; and Tolia-Kelly, Divya P. 'Rancière and the Re-distribution of the Sensible: The Artist Rosanna Raymond, Dissensus and Postcolonial Sensibilities within the Spaces of the Museum'. *Progress in Human Geography* 43, no. 1 (2017): 123–40.

85. Ramana, *The Power of Promise*, pp. 279–91.

86. 'Why the Kudankulam Nuclear Power Project near Kanyakumari—A Few Objections' [sic]. Conservation of Nature Trust leaflet, Nagercoil.

87. On feminist recuperation of 'Mother Earth' in social movements, see Mies and Shiva, *Ecofeminism*.

88. Brown, Wendy. 'Finding the Man in the State'. *Feminist Studies* 18, no. 1 (2006): 7–34, p. 188.

89. Roy, Biswajit. 'The Professor and the Politics in Anti-nuclear Crucible'. *The Telegraph*, 19 February 2013. Available on http://www.telegraph-india.com/1130209/jsp/nation/story_16541617.jsp#.V3D-4vkrLIV, accessed 19 January 2019.

90. See Martin, Emily. *The Woman in the Body*. Boston: Beacon Press, 1992.

91. Roy, 'The Professor and the Politics in Anti-nuclear Crucible'.

92. As the protagonist bemoans non-violence in the anti-corruption film, *Gabbar Is Back* (2015, dir. Krish): 'The currency notes only have Gandhiji's face, not his stick. That's why some people have become spoilt'. See Kaur, Raminder. *Atomic Mumbai: Living with the Radiance of a Thousand Suns*. New Delhi: Routledge, 2013.

93. On a catalogue of problems with the VVER-1000 reactors, see Bhawna. 'Nuclear Energy, Development and Indian Democracy: The Study of AntiNuclear Movement in Koodankulam'. *International Research Journal of Management Sociology and Humanity* 7, no. 6 (2016): 219–29, pp. 225–7; Ramana, *The Power of Promise*, p. 87.

94. Udayakumar reports the difficulties of utilizing these avenues against nuclear authorities where the Supreme Court calls upon Sections 3 and 18 of the Atomic Energy Act to consider certain information restricted or prohibited 'in the interest of national security, which was paramount'. Available on https://www.wiseinternational.org/nuclear-monitor/619/india-hazardous-mix-peculiar-act-and-perilous-energy, accessed 19 January 2019.

95. Agarwal, Vidhi. 'Privacy and Data Protection Laws in India'. *International Journal of Liability and Scientific Enquiry* 5, no. 3/4 (2012): 205–12, p. 205.

96. Rakshit, Nirmalendu Bikash. 'Right to Constitutional Remedy: Significance of Article 32'. *Economic and Political Weekly*, 34, no. 34/35 (1999): 2379–81.

97. Goyal, S.L. Court master c/o The Chief Justice, Mr Justice RP Sethi and Mr Justice Arijit Pasayat, writ petition (civil) no 286/2002.

98. Goyal, S.L. Court master c/o The Chief Justice, Mr Justice RP Sethi and Mr Justice Arijit Pasayat, writ petition (civil) no 286/2002.

99. Led by Udayakumar, Power was later replaced by Energy in the name to form PMANE.

100. Available on https://www.iaea.org/about/overview, accessed 19 January 2019.

101. Available on https://www.iaea.org/about, accessed 19 January 2019.

102. See also Srikant, *Koodankulam Anti-Nuclear Movement*, p. 7. On the interpretation of environmental laws in USA, see Masco, Joseph. *The Nuclear Borderlands: The Manhattan Project in Post-Cold War New Mexico.* Princeton: Princeton University Press, 2006, pp. 114–15; and Gusterson, Hugh. *People of the Bomb: Portraits of America's Nuclear Complex.* Minneapolis: University of Minnesota Press, pp. 206–20.

103. See Abraham, Itty. 'The Violence of Postcolonial Spaces: Kudankulam'. In *Violence Studies,* edited by Kalpana Kannabiran. New Delhi: Oxford University Press, pp. 323–4.

104. Ramana, M.V. and Ashwin Kumar (2014) '"One in Infinity": Failing to Learn from Accidents and Implications for Nuclear Safety in India', *Journal of Risk Research,* 17(1): 23–42, p. 27.

105. Nilsen, *Dispossession and Resistance in India,* pp. 130, 137–8.

106. Nilsen, *Dispossession and Resistance in India,* p. 140.

107. Nilsen, *Dispossession and Resistance in India,* p. 130.

108. Bornstein and Sharma, 'The Righteous and the Rightful'.

109. See Thelen, Kathleen. *How Institutions Evolve: The Political Economy of Skills in Germany, Britain, the United States, and Japan.* Cambridge: Cambridge University Press, 2004.

110. Ferguson, James. *The Anti-Politics Machine: Development, De-politicisation and Bureaucratic Power in Lesotho.* Minneapolis: University of Minneapolis Press, 1994.

111. Hecht, Gabrielle. *Being Nuclear: Africans and the Global Nuclear Trade.* Massachusetts: MIT Press, 2012, p. 4.

112. See Gusterson, *Nuclear Rites,* p. 174; Bornstein and Sharma, 'The Righteous and the Rightful', p. 79; Rajak, Dinah. *In Good Company: An Anatomy of Corporate Social Responsibility.* Stanford, CA: Stanford University Press, 2011.

113. Udayakumar, *The Koodankulam Handbook,* p. 42. Independent experts who have analysed the report conclude: 'its assessments on damage are not reliable, and it does not consider the full range of potential

impacts'. Ramana, M.V. and Divya Badami Rao (2008) 'Violating Letter and Spirit: Environmental Clearances for Koodankulam Reactors.

114. Iyer, Ramaswamy R. 'Abandoning the Displaced', *The Hindu*, 10 May 2006, p. 11.

115. See Patnaik, 'Neo-liberalism and Democracy', p. 42.

116. Yadav, Yogendra. 'If Bapu Recedes Again, Who Is to Blame?' *Indian Express*, 26 October 2006. Available on http://archive.indianexpress.com/news/if-bapu-recedes-again-who-is-to-blame-/15434/, accessed 14 January 2019.

4

The World of the In/visibles

Who are we to believe? Government representatives are always keen to use the example of 'there is even radiation from watching television'. Does that mean we should stop watching it? Or that we get radiation every day from the sun. Does that mean we can't get away from it? (Purnima, Kanyakumari District, 2006).

By the turn of the second millennium, concrete markers of an expansive and expensive project backed by the central government were doggedly drilled into the south Indian coastlands. An outlying hinterland had become decisively fused with the arms and armies of the nation's heartland. Accentuated by the international trade circuits of the Nuclear Suppliers Group from 2008, local residents were forced to reconcile with a reconfigured global–national–local nexus in what could be described as 'hinternational'. Although the coastal communities were already attuned to transnational currents, those in this subcontinental peninsular became directly subject to a confluence of powerful decisions emanating from New Delhi, Mumbai, Chennai, and Moscow.[1]

With this seeming loss of local and personal control, came a piquant sense of 'risky times' ahead: a pronounced perception of risks that things *could* go wrong; and even worse, owing to the uncanny convergence between the invisibility of radiation and the opacity of nuclear authorities, things *would* go wrong and nobody ever gets to find out about it until it is too late. Ecological vagaries played well into the vast hands of a secretive state.

Suspicion was not just levelled against central government behind the nuclear project—largely viewed as 'outsiders' or 'northerners' who had no care for south Indians—but also against the Russian government whose secretive conduct and substandard nuclear technology inherited from the former USSR cried foul on so many levels. No matter what their literacy levels, most people had heard of the Chernobyl accident in 1986 and the fact that the mishap was due to the oversight of scientists and engineers with the reactor Unit 4 that the Soviet regime was keen to cover up. Even though strictly speaking the VVER-1000 technology in Kudankulam was a more robust design than the RBMK (graphite moderated) reactors at Chernobyl, the prospect of similar technology run by similarly opaque authorities close to home was all too troubling. When activists raised this issue about the reactors, NPCIL officials argued instead that they were safer versions of those at Chernobyl.[2] When they pointed out that they had not been trialled on coastal terrains, the NPCIL simply switched off.

Indeed, it has been revealed that equipment for the commissioned reactors had not been adequately maintained. Nor, as it was learnt later, were they the ones that were originally planned.[3] Rather, they were provided by Russian companies, ZiO-Podolsk, Informtekh, and Atommash, whose CEOs have been jailed for corruption for the sale of aged and defective Soviet-era equipment.[4]

Against such restless shadows, people would mention that their anxieties had escalated. As one fisherman in Kanyakumari District reiterated on the untested and untrusted technology: 'We are the guinea pigs for it.' A woman expressed her worry: 'We feel like we have lost control of our destinies'. Another local scientist added: 'Worries about cancer have increased. Even if there was a slight lump on the body, people would rush to the doctor'. Generated by the nuclear power plant, radiation ruminations had the inevitable effects of inducing states of anxiety, anger, depression and, with the prospect of their views not mattering, apathy.[5]

Fundamental to much of the discussion on the nuclear plant was a creeping entry into 'the deep time of radiation'.[6] Prior to the late 1980s, radiation was not only an unfamiliar word, but also a strange concept to many residents in the southern peninsula. Locally referred to as *kathirvichu*—literally 'that which emits radiation'—even literate adults were a little uncertain about its characteristics. A mother tried to explain it to her infant son who asked about its effects on the body: 'So does it have a mouth?' 'Yes', she replied, 'It's like an insect and it eats away your cells'.

Whether it be released in the form of natural background radiation, through coastal sand mining for monazite that contains the radioactive thorium, or via radionuclide emissions and contamination from the Kudankulam Nuclear Power Project and related activities, the more conscientious took it upon themselves to find out and circulate as much information as they could through conversation, meetings, and locally produced literature about the nuclear plant, that with the arrival of the worldwide web, was transplanted online. Such aims were integral to the effort to give firmer shape to the imperceptible where the invisible could begin to be visibilized through hint, suggestion, or theory—as the researcher and curator Ele Carpenter elaborates for a more artistic context: '...to bring the fears home, to make them visible not as an external sublime but as a localised uncanny'.[7]

On matters radioactive, however, the NPCIL saw themselves as the ultimate authority in the region. It goes without saying that in their public presentations they would erase or even evade the dangers of ionizing radioactivity. Ambiguous concerns paled next to the plant's multiple boons. As goes the official hymn sheet across the world since the 1940s, atomic development is a wonderful opportunity for the region as well as the nation. Like a stuck record, the reactors would supply cheap power to everyone; the plant would provide jobs and other entrepreneurial opportunities; the project would catalyze the development of the area; it would put Kudankulam on the map, a proud asset to the nation.

Any questions on radiation were filtered through five somewhat contradictory registers. First, there would be denial and dismissal— that radioactivity is a mere fantasy and not a cause for concern. Second, it would be recognized and domesticated like an old friend, or normalized as natural when different kinds of radiation emissions are conflated and held to be benignly present everywhere on the planet. Third, with recourse to both an elite class of nuclear experts and fetishized technologies, officials would emphasize the high levels of management and technological prowess that stemmed any escape of dangerous radioactivity.[8] Fourth, some officials diverged from any negative consequences to stress not just the nuclear industries' but also radioactivity's many virtues by highlighting the mundane benefits of radioactive isotopes in medical science, diagnostic studies, agricultural developments, and the like. Fifth, they often exaggerated and aggrandized the notion of radiation, some even celebrating the fact that ionizing radiation is 'good', because it has led to a variety of mutations and new life forms.[9] Just as in nature, radiation became part of the evolutionary process to do with the 'survival of the fittest'. Although somewhat idiosyncratic, the discourses converged in bulwarking a radiation regime that enabled the promotion of nuclear industries without the need for public admissions of guilt or responsibility. It permitted a display of complete and often, as M.V. Ramana observes, 'absurd confidence' in their technologies, personnel, and procedures.[10] At root are hierarchical dualisms that see science as outside of and in control of the vagaries of nature, and where the mind is figuratively detached from the body as the ultimate arbiter of knowledge.[11]

As the reactors grew in height, so spread the crosswise rhizomes of vernacular knowledge in the region. The (quasi-)scientific radiation regime that was deployed to colonize reality was countered by its unravelling. Nuclear critics challenged most if not all of what they saw as distortions that beset this discursive mesh of denial, domestication, manageability, digression, and aggrandizement. They tried to iron out, pull apart, and weigh down the authorities' lofty radiation regime with a certain gravitas. They were particularly influential in pointing to the limits of technological managerialism with recourse to a lifetime experience of bureaucratic incompetence and intransigence.[12] Calling upon several studies and imagery from those marred by the release of radioactivity around the Chernobyl disaster, the Hiroshima and Nagasaki bombings, and the uranium tailings in Jadugoda in north eastern India, they held up the sanctity of the body against the prospect of genetic mutations, cancer, and, infertility, as well as reminded people of water scarcity and the threat of displacement from their homes.[13] As the anthropologist, Hugh Gusterson observes for the antinuclear weapons movement in the

USA, a 'public culture of fear [that] had profoundly subversive potential' was activated.[14]

Development with its grid of governmentality brought with it, not the fruits of modernity in terms of 'goods'. Rather, as the sociologist Ulrich Beck maintains, it introduced more 'bads':[15] the fear of contamination and accident alongside a keen sense of repression evidenced in nuclear authorities' indifference about local grievances, and the increased policing and militarization of the region in the name of national development and security. Boons were inverted as burdens. Uncertainty began to be radicalized.

In this chapter, we focus on the kinds of knowledge, ignorance, and uncertainties that circulated in the region. In the next chapter, we profile three people from different backgrounds, showing how people dealt with palpable risks in conflicted social contexts. Together, the chapters demonstrate that the ground was tilled across the region for support to an anti-nuclear movement, one that flourished in the post-Fukushima disaster era from 2011, even though individuals prior to that period may not have actively resisted the nuclear plant.

Knowledge, Ignorance, and Uncertainty

Wealth and status do not always underpin peoples' 'social risk positions', Beck maintains.[16] Rather, it is knowledge that is more critical in giving shape to an uncertainty as a risk.[17] The anthropologist Åsa Boholm adds: 'The concept of risk can be understood as a framing device which conceptually translates uncertainty from being an open-ended field of unpredicted possibilities into a bounded set of possible consequences'.[18] Knowledge framed risks away from rumination into a set of navigable coordinates. This knowledge is not simply about the realm of expertise with its respective hierarchy of credentials. Knowledge may be imbricated in power complexes, but it is also multi-layered, situated, reflexive, contested, as well as subject to much alterations depending upon who is speaking and when. This phenomenon is less about a regulated discourse of knowledge but the more dynamic and changeable *knowing*, or the contingencies of being 'in the know'. Awareness of radiation differed according to which place one visited, which person one talked to, and at what point one talked to them. Against the grids of 'knowledge-power', there were also more organic, fluid understandings that were mobilized through discussions on issues of concern.[19] As distinct from vertically positioned specialists and intellectuals, the political philosopher Antonio Gramsci termed those who are rooted and conscientized members of the working classes as 'organic intellectuals'.[20] To similar ends, there was a discernible growth of organic intellectuals on nuclear matters in and around Kudankulam. They rose from fishing and agricultural labouring backgrounds to challenge official aspersions of their 'irrational fears' and 'ignorance'.[21]

Ironically, ignorance is not only derided but also actively (re)produced. An agnatological perspective identifies the political investments and opportunities in creating and sustaining public ignorance.[22] On a related note, the epidemiologist David Michaels highlights the vested interests in 'manufactured uncertainty' with respect to the tobacco industries.[23] Similarly, while nuclear authorities were quick to label critics as marred by 'ignorance', they also invested plenty of energy and funds in ensuring that ignorance and doubt proliferated with a medley of concoctions. Clarity of communication and purpose was not a priority for the nuclear authorities, even while they might prize rational discussion in the Habermasian vein among themselves.[24] While they set out to 'educate' the public, the discursive production of ignorance was essential to create uncertainty among the populace, circumvent their duty to be transparent and accountable, and to undermine people's anti-nuclear views. Circulating contradictory discourses around radiation simply added to the manufacture of ignorance and uncertainty.

Yet others asserted that nuclear officials were themselves wilfully ignorant for misleading others with their 'tall claims, half truths and outright lies', to use the words of the activist K. Ramachandran.[25] Like knowledge, ignorance too was multi-layered, situational, and perspectival steeped in contested hierarchies. A striking example of this phenomenon is if we look sidelong to IREL planted on the monazite sands of Manavalakuruchi along the Malabar coast. IREL officials and members of the panchayat (village council) typically state that there are no incidents of cancer in the village. Flying against the face of lived experience, this representation is no doubt one of the outcomes of closely liaising with the mineral mining company based on the coast. Enquiry off people who live in the village tells one otherwise, as there are plenty of cancer afflictions, particularly among those who worked for IREL, as we shall see in Chapter 7. However, this was a difficult enquiry to ascertain statistically, for there were several people who did not want to talk about cancer. This was not necessarily because such disclosures were not welcome by IREL officials who had a phenomenal amount of power in the village, but because they could indict the self and/or family. A virulent social stigma was attached to the disease, principally because it could jeopardize the marriage prospects of related youth.[26] Even if living in a toxic environment, cultural repertoires of reticence to do with fear or shame associated with the harms of nuclear industries meant that cancer in the family was not always acknowledged. Aside from its stochastic nature, understanding and transmitting knowledge about radioactive realms and their somatic effects was made an even more difficult and unpredictable endeavour. Information rarely travelled in linear and clear conduits—it may fissure, mutate, be suppressed, become obscure, and be channelled again through 'Indian whispers' where chimerical traces take on other ambiguous avatars.

Enchantment

Many publicity drives were held by the nuclear authorities and 'training' given to key workers who might reproduce the radiation regime as with teachers, civil servants, community and social workers, and other people of standing in the neighbourhood. The nuclear authorities would deftly engage such people to their own advantage, and intensify their measures around times of crisis as transpired in 2011 after the Fukushima Daiichi nuclear disaster. Viewing it more as a shortcoming in public relations rather than a nuclear crisis, a 'propaganda blitzkrieg' was unleashed on the Kudankulam reactors' safety features including padayatras—foot pilgrimages by persons of standing made famous by Mohandas Karamchand Gandhi to connect with ordinary people. Of course, these were considered wrong-footed by activists who mobilized Gandhian strategies for social justice.

Official padayatras came with the stepping up of more lectures, public meetings, seminars, report and booklet dissemination, regular e-newsletters, films, and awareness campaigns in schools and colleges to raise technical awareness over the safety aspects of the nuclear plant.[27] In their 'civilising nuclear mission', local leaders were invited to well-catered seminars and symposia where state officials talked about how building the nuclear power plant was 'good for the nation and jobs', and that this was an important message to relay to young people and students in their classes so as it becomes inscribed in their futures.

One teacher recalled an official's comments in a 2006 seminar: 'There are a few problems with the power plant, but they had told people not to worry ... the government is taking care of these matters'. She continued: 'Radioactive waste will be well-sealed and deposited deep in the sea bed, they say.... It won't affect the sea life so there is nothing to worry about'. 'They won't let the waste come out', relayed another person with more conviction. Officials would regularly tell their 'trainees' not to not listen to other voices—the voices of those who they decried as 'anti-developmentalists who are only out to cause trouble—trouble-makers'.

Backed by a litany of institutional experts, a sanitized version of their ambitions and achievements is reproduced and distributed in their publicity material, many of which they would use to inform their public presentations in the region. In their Tamil-language brochure translated as *NPCIL: How to Produce Healthy Power*, the first page explained the science of hydroelectric dams, coal power stations, and wind farms with the aid of pictures and diagrams. The second asks: 'Why atomic power?' The response: because it '(i) produces no environmental pollution; (ii) radiation is minimal—natural radiation is much more; and (iii) it creates no acid rain and no air pollution'. The next page goes on to note the problems with other sources of power with cavalier dismissals beginning with thermal power plants:

7 million tons of gas pollutes the air with SO_2, CO_2, and CO [sulphur dioxide, carbon dioxide and carbon monoxide] every year. There is also a lot of coal wastage in the form of sulphur: 150–250,000 tons. However, with the nuclear power plant, only 25 tons of uranium per year is required to produce 1,000 megawatts. It doesn't pollute the air. Whereas 300 million tons of coal are needed to produce the same amount of power, here just 25 tons is enough. In India there are not enough coal deposits. But we have enough thorium for power production in nuclear power plants. With hydroelectric power, only certain areas can be dammed to produce power. A nuclear power plant can

be put anywhere. With wind farms, we need a lot of land. But the
NPP [Nuclear Power Plant] does not require so much land. With wind
farms, power is only produced in windy season. With the NPP, you can
have power all year round.

Once again, the prevalence of thorium is propagated as the golden
ticket for India's nuclear ascent to higher orbits. It is as if merely
pointing to the abundance of this magical material in the subcontinent is enough to enhance and enchant its role in India's electricity
portfolio. Moreover, when faced with the visible pollution of fossil fuel,
Indian nuclear plants were seen as not just climatologically clean, but
also visibly clean. It is a message that they seek to inculcate in the
youngest of minds. A NPCIL animation puts it across with saturated
sentimentalism to juvenile music: beaming smiles on cooling towers that stand next to 'clean, safe and green' reactors amidst verdant
foliage and wild life with the captions, 'less place is enough', 'frugal
electricity option' and '24hrs of electricity production' [sic].[28] All other
alternative sources of energy are belittled as unreliable, insufficient,
and/or inconvenient as they require too much land, suggesting that
they could cause even more disruption to peoples' lives.

The next page in the aforementioned booklet underlines the
NPCIL's custodianship of an arcane science, by detailing the science
of the atom's structure and how energy is produced. It then provides
a discussion on the high levels of safety and security comparable to
international standards. The reader is informed that:

> There is an active safety system and passive safety system.... Everything
> is meticulously planned.... And there are automatic fire safety mea-
> sures and an anti-fire unit at the plant. The construction is high-tech
> and well-protected.

Meticulously planned, high-tech, and well-protected is the relentless
message. The leaflet then details the reactor designs, before normal-
izing radiation, this time described as 'exposed particles'. Sources of
radiation including cosmic radiation, natural radiation on the Malabar
coast, radiation from aircrafts, X-rays, radio and television are listed,
summarizing that '84% of radiation is natural. Only 12% is artificial,
of which 0.04% is from nuclear discharge'. The final page lists the
amount of radiation for different sources in millirads per year. The

list, apparently sourced from the IAEA includes emissions from a concrete house, brick house, wooden house, space radiation, radiation in the air, food and water, in another paean to their normalization.

Such content on 'healthy power' is typical of brochures and presentations by nuclear representatives.[29] Ramana concludes: 'The aim, evidently, was to impress the reader with euphoric descriptions and complicated jargon that would be incomprehensible to most'.[30] Even to those who have an understanding of the intricacies of nuclear science and technology, the plethora of unexplained facts and unaccounted for numbers only fogs the picture so that a proper assessment becomes impossible.

The banalization of danger signals the danger of banalization. Making something appear normal, safe, and even delightful when in fact it can threaten one's health is seen by critics as part of a treacherous conspiracy to keep people in the dark. As with the 'greenwashing' of toxic industries, twisted so as to appear beneficial to humans and the environment, here we have a warped complex of radiation banalization, rationalization, and/or glorification.[31] When, for instance, emissions from cosmic rays and television sets are cited in the same breath as sources of alpha, beta, and gamma radioactivity, and safe and dangerous radiation is conflated as if they were one and the same thing, we might deem these representations as a case of 'radia-washing'.

To similar ends, hegemonic views on radioactivity had long been disseminated through leaflets and public relations exercises in the region by kindred industries as with the IREL plant in Manavalakuruchi. They add another strain to the representational regime—'radiation reduction'. IREL publicity for the Manavalakuruchi plant claims that they actually reduce radiation in the area: 'As the tailings used for refilling are free from monazite, it has resulted in reducing the radiation level of these areas'. Some people concurred with the official line that the IREL claim that they are taking away radiation from the area, and that the company was 'doing a good job' for 'taking away the monazite' and thus decreasing radioactivity in the area. Thus, sand mining is actually beneficial for the village in that black sands were being removed. Otherwise, it placed their region on the national map, some people even led to believe that sand mining for atomic minerals in Manavalakuruchi played a part in the manufacture of the 'nuclear bombs' that were tested in the deserts of Rajasthan in 1998. Such workers saw it as a national duty to support the IREL in another way: if they did not mine the rich mineral sands, then it would go to waste and could even be carried away by ocean currents down to Sri Lanka. The enchantment of jobs and national interest swept all other concerns to the sides.[32]

Singing from the same sheet as the nuclear plant, the health and safety of IREL employees was also subject to a techno-managerial discourse:

> A separate wing under the control of Health Physics Division of BARC, Mumbai is looking after the Health Physics and Industrial hygiene aspects. Persons working in the plant are provided with TLD [Thermo-luminescent Dosimeter] badges in order to monitor and control the radiation dose received by them. The dose received by individuals is limited well within the tolerance limits fixed by the Atomic Energy Regulatory Board (AERB) and International Commission for Radiological Protection (ICRP).

The TLD badges, a combination of lithium fluoride crystals in the form of either a metal foil or aluminium planchet, record the level of radiation received by the wearer. Any fogging of the photographic film in the badge can then be analysed at the Bhabha Atomic Research Centre (BARC) in Trombay, 1,300 km to the north.[33]

Unsurprisingly, practice never quite fits the ideal. Information is cherry-picked to make everything seem like a bowl of cherries. As one retired Nagercoil-based scientist clarified about the TLD badges:

> The question is: is it properly monitored? Is it done monthly like it should be done? This is the area which is most likely to suffer due to prioritisation of other work.... Even in the X-ray labs not all workers wear these badges. The girls, the nurses are paid very low. Even though they are well qualified they are not given badges. They know that this is wrong but they value their jobs over everything else. They asked doctor not to tell anyone about this as they feared losing their jobs.

While employees knew that black sands on the beach, referred to as *kathirvichu manal*, contain monazite with thorium that is radioactive, many of them subscribed to the IREL view that the risks were exaggerated. As it were, they were constantly monitoring radiation levels with a dosage badge that workers wear. However, when I queried a former IREL worker about what the badge does, he was not able to explain, saying that 'they tell us to wear it for our safety'. One worker thought that the badge 'catches radioactive waves'. Others thought that it provides 'full proofing from radiation'. Few knew that it was no more than a monitoring device that were at the time supposed to be regularly sent to BARC every month to check radiation levels.

On the nuclear power plant, a lack of transparency between official and project-affected person was compounded by the fact that reactor technology was not fully understood by all. One man perceived a nuclear plant as 'a power plant that uses *anushakti* [literally, atom power] Something great. Something powerful'. On the related subject of radioactive waste, several residents talked about as if it was rubbish. They sometimes referred to it with the Hindi expression for waste, *kachra*, that could spoil sea life, associating the contamination to material residues as a consequence of 'invaders from the north'. Conversely, others preferred to think of radioactive waste as a 'limited problem'. One young man expressed an anxious hope by comparing radioactive elements as something that could be 'filtered' out of a nuclear reactor, a procedure that is followed by Indian householders with respect to tap water up and down the country. Explaining the procedures surrounding water coolant in the nuclear reactor at Kudankulam, he stated:

> First, it goes through the first black building, then it is processed so as it's more pure and then it goes into second building until eventually by the time it gets to fourth building it is fine to release it into the sea.

No matter how incongruous, this was indeed the kind of domesticated views that nuclear officials were content to live with.

Safe as Houses?

The topic of 'safe radiation levels' is another point of contention. Historically, in the Indian nuclear industry, Radiation Protection Rules were advanced in September 1971; they were primarily intended for industrial and medical installations at the behest of the Department of

Atomic Energy.[34] The formation of the International Atomic Energy Agency in 1974 led to the preparation of codes and guides to ensure safety in the design, siting, and operation of nuclear power plants. They were codified in nuclear safety standards programme that Indian nuclear experts draw from for their own establishments. After the inauguration of India's Atomic Energy Regulatory Board in 1983, two apex committees were set up to advice on generic safety issues to do with Nuclear Safety and the other on Radiation Protection. Retired atomic scientists C.V. Sundaram, L.V. Krishnan, and T.S. Iyengar declare:

> SARCOP [Safety Review Committee for Operating Plants] called for a thorough review of radiation exposures to plant workers. Cases of exposures in excess of the individual limits laid down are quite rare. The safer philosophy in nuclear installations, unlike the practices in conventional chemical plants, requires however that exposures should be as far below the prescribed limit as possible.[35]

Highlighting that overexposure is 'quite rare' in nuclear power plants, the authors go on to chart a trajectory of identifiable problems, measures implemented and remedies sought. Oftentimes, officials profess to have confident control over the vast world of the invisibles:

> The effects of radiation are better understood than is the case with many chemical pollutants. Severe exposures are quite uncommon. At low levels, for the most part, exposures cause no perceptible effects. Any effect that appears might do so after a long delay. There are other agents, which cause similar effects, and these cloud the analysis.[36]

It is not so much radiation exposure that is the problem, they argue, but 'other agents' that can 'cloud the analysis'. Radiation comes off as a relatively knowable and benign product of nature. In echoes of the above discussion, they add that people's perception is not based on an evaluation of the evidence, but rather an irrational fear:

> The most difficult factor is the fear of radiation, which is widely prevalent. This makes it difficult for the regulatory body to convince the general public of the adequacy of safety measures taken. Capability to detect radiation present in the environment in very low levels, too low

to be of any significance also seems to be a disadvantage. Reports of such presence are known to contribute to undue perception of fear.[37]

Officials repeatedly state that any danger with this product of nature is overstated.[38] Even if it is present as a potential danger, it is too low to be of any significance. Next to such proposals, it is local populations who are irrational, who need to be 'educated' to the benefits and wonders of nuclear technological prowess—educated so as they continue to be confidently ignorant.

Such public representations are at odds with literature that my interlocutors came across regarding radioactivity training for personnel in the nuclear industries. From notes for a vocational course on radiation protection available at the Department of Atomic Energy, it is evident that nuclear authorities acknowledge the dangers of radioactivity and the need for stringent measures to ensure safety of both the plant and employees.

When compared with such data, it becomes starkly evident that nuclear authorities have a different mode of address for their contracted trainees and employees as compared to the literature that they make available to the wider public. For staff, they make no bones about the dangers of radiation and the need for proper protective measures. Reiterating received wisdom on the subject, they point to different kinds of radiation—alpha radiation that consists of positively charged heavy helium nuclei and cannot travel very far but can be hazardous when in the body. Beta or fast electron particles have a very small mass, travel further and can be obstructed by a thin sheet of aluminium. Gamma radiation at the extreme end of the electromagnetic spectrum can penetrate most if not all matter and can have innumerable damaging effects on life forms.[39] It is these forms of dangerous radiation that are accounted for under the term, radioactivity, for their mutating and thus hazardous consequences.

The document on radiation protection continues by outlining two types of effects on the human body, (*a*) stochastic effects that have no clear threshold and can cause genetic problems and cancer at any level of exposure; and (*b*) deterministic effects that have a clear threshold below which ailments like skin burn and cataracts do not occur. The notes explain that when a person has an uptake of 2 MBq/l of tritium (with reference to their work at the Madras Atomic Power Station

[MAPS], in Kalpakkam), then the person is under the 'CAUTION category'. If this goes up to 4 MBq/l, then they are under 'RESTRICTION', where the person is not allowed to work in a radioactive area until lower levels are registered on the body. 1 Bq or Becquerel is the rate of radioactive decay equivalent to one atomic disintegration per second.

In a procedural manner, the document outlines four methods of controlling external exposure: (a) time, (b) distance, (c) shielding as with the use of aluminium, concrete, and lead, and (d) decay due to the inherent half-life of unstable atoms. Contamination can either be fixed—that is, it cannot be washed away; or it is loose—that is, loosely attached to the surface, and can be removed but might still be inhaled, ingested, or absorbed through the skin. It is also stated that the inhalation of radioactivity can be prevented by wearing respirators, such as orinasal respirators, air-supplied respirators, self-contained breathing apparatus, along with the protection of rubber gloves, gum boots, and plastic suits.[40]

When it comes to safety regulations and procedures, there is not only an effort to contract figures to do with safe levels of radioactivity, but also a double-discourse going on among officials depending upon the intended audience. One encapsulates themes to do with denial, domestication, the prowess of manageability, digression and/ or aggrandizement for the public as we have already outlined. The other continues the theme of technoscientific managerialism but also encapsulates discourses of procedure, precaution, and even care for those deemed as part of their 'extended nuclear family'. All purport to be objective but are framed by subjective interests determined by the intentions and interrelations of social actors. For the general public, sleek and domesticated presentations serve the slippery purpose of glossing over the reality of radioactivity as we understand it. In their paternalistic manner, nuclear authorities deem the public as only suited to diluted and dispersed logics, echoing somewhat their attitude to radioactive waste . For nuclear workers, the reality of dangerous radioactivity is spelled out with scrupulous, black and white details, all the while stressing safe levels, protection and methodical procedures in conducting any work in a nuclear installation.

Activists contested the infallibility of *safe* radiation levels. They doubted any figures offered by authorities: after all, a higher threshold would allow for more industrial work and less public panic.[41] In the process, objective knowledge steeped in the language of science and technology as publicly presented by nuclear authorities began to be dismantled as negligent knowledge. As the scientist and feminist Donna Haraway might argue, it is another instance of 'irresponsible knowledge' for inadequately acknowledging its partial and biased agenda.[42] We could add that it is not only irresponsible for acknowledging its subjectively biased agenda, but also dangerous when it comes to its operational neglect and bias against those outside of their silo community.

Disenchantment

Clearly, nuclear authorities presented a smooth, sterile and anodyne vision of a brave new atomic world.[43] While they become the arbiters of specialist knowledge, they also try to exact social control. However, people had become adept at piercing their performative propaganda. A war of words and visions ensued, behind which was fathomless incredulity. While nuclear technocrats could not understand why local communities were not able to see the benefits of nuclear power, the latter could not believe how bureaucrats and scientists could believe what they preach. Can it all be '100% safe' as officials were wont to say? 'Can everything be solved with a graph, equation or measurement?' 'How can they not see the social and environmental effects of nuclear technologies?' We could add: is this a very clever way to either convince or confuse people as to the value of nuclear power where harsh truths are obfuscated? Or can the content be seen as blatant lies to promote vested interests by nuclear authorities that, along with the Indian Space Research Organisation (ISRO), remain one of the 'spoilt darlings of the Indian state' in the words of one activist?

One man inverted official claims to authoritative knowledge. He recounted with a sense of disbelief an incident when an employee at the Kudankulam Nuclear Power Plant said to him, 'If there is an earthquake, the safest place would be inside the reactor!' Another person relayed how a nuclear official waxed lyrical that he has even been 'inside the reactor and my family has had no health problem'. Presumably this was before the 'inside' of the reactor went critical. Their confidence appeared beyond absurd. The man continued: ' ... they may as well come from a super-race of aliens! But these aliens are not very intelligent.' Or if they can be deemed to be intelligent, it is on a different vector, at odds with the acumen of the populations they seek to subjugate.[44]

Through a litany of hazy screens, nuclear industries are able to expand their operations into regions dominated by subaltern populations, exploiting internal hierarchies and nuclear ignorance as much as they entrench new divisions and discourses. In addition, many of the communities provide a pool of casual and cheap labour for the nuclear industries to use. There is a devil-may-care approach when it comes to subcontracting them. At the IREL site in Manavalakuruchi, for instance, fishermen manually heap up sand on the beaches so as manned lorries can come and pick it up to take into the factory compound. These casual workers do not have recourse to TLD badges or any kind of monitoring scheme. The division of labour safety was patently obvious, as one of them confirmed: 'There are no check-ups for ordinary labourers or nearby residents. There are only there for contracted workers' and even here, as noted above, there was little in terms of a systematic implementation or understanding of practice.

Indian nuclear authorities take full advantage of the marginality and/or illiteracy of the poor. Casual workers are not accounted for in any IREL nor indeed NPCIL books. Sub-contracted through middlemen, not only are they unmonitored for radiation doses, but they are also not eligible for safety standards and company health services should they become ill.[45] Workers unfortunate to contract a life-threatening disease have little recourse, whether it is to pay for costly treatment themselves or to go to underfunded government hospitals where those with more advanced afflictions are simply left to die on rope-bare charpoys.

Clearly, nuclear-scapes depend a great deal upon nuclear scapegoats—the availability of cheap and unofficial labour who are on the peripheries of health and safety concerns and regulations.[46] It is with the 'dispensability' of the labouring classes that uranium has been extracted, monazite has been mined, nuclear complexes have been constructed, and their buildings continue to be maintained and cleaned.[47] These workers are not sufficiently resourced with training and protection against exposure to radioactivity. Such a labour force is conceived to be at the bottom of the pile of nuclear necropolitics. They remain the 'dog's bodies' to neo-Brahmanic hierarchies entrenched in the nuclear establishment as was argued in Chapter 2.

With a trail of broken promises left by nuclear officials, and caught between a swirl of information from governmental and activist sources, people were not sure what to do or say about radiation and its impact on health and ecology. One teacher, for instance, was earlier grateful for the expansion of nuclear industries in the region, but later changed her mind. Her father had found relatively well-paid work in the IREL until he was dismissed due to health reasons. He had contracted cancer. She was sorely disappointed by the company's measures to help former employees. There were insufficient health provisions for his ailment even if he had been working full-time for the company. This is despite the fact that IREL maintain that they have a rigorous welfare policy including a 'full-fledged Dispensary with two medical officers and adequate supporting'[sic]. In their printed publicity material, they boast that:

> Dispensary works around the clock.... Periodical medical examinations of the employees are also carried out. Employees enjoy many amenities like House Building Advance, Hire Purchase Scheme, Conveyance allowance etc. A heavily subsidised Canteen wholesome food to the employees at nominal rates.

However, employee experience was far from wholesome. The daughter of the former IREL employee began to cut through the churn of their public relations. Born out of her experience of inadequate protection and provisions for her ailing father, her scepticism extended to other nuclear authorities including the Kudankulam Nuclear Power Plant, a development that she also saw as the harbinger of more problems.

The teacher's narrative of disenchantment is yet another to add to the growing number of people on top of activists, environmentalists, and fishing and farming communities who began to question and resist the nuclear plant, forging new spaces of criticality. They are the ones who initially may have supported nuclear developments through the lure of jobs, contracts, and new opportunities brought to the region. However, over time, they gradually turned against the plant seeing only capsuled communities behind shrouds of deception.

W/holes

The reactors were less generators of power as they were of uncertainty, anxiety, and division. A combination of nuclear state opacity and the arcane nature of the effects of radiation meant that people's lifeworlds were clouded by imperceptibles.[48] In recurrent spins on propaganda, while authorities organized public education campaigns, they also benefited from ensuring that a haze of ignorance persisted. Local residents gained a certain degree of knowability where full knowability remained outside the horizon of possibilities—ideas at the edges of perception in a physical and political world of in/visibles.[49] Unknowns might become vaguely known that might then become more known. However, even here there were shadows of the unknown or the uncertain, particularly when other perspectives were thrown into the picture. Knowledge and ignorance became relational entities as much as they became highly charged politically. Such curves of learning and ignorance compelled a seismic reorientation about how things are, ontology, and the way we know about such things, epistemology.[50]

In this political ecology, the way we understand radiation (epistemology) becomes a principle point with which to be (ontology) and experience life with all its perceptible orders and disorders. In a space of criticality, neither ontology nor epistemology has primacy: the ontological basis of being and its epistemological framings in terms of coming to know and understand perceivable risks to life creates a spiralling double-helix phenomenon. In other words, awareness and explanations of radioactivity became a compelling frame with which to understand (potential) disease and disorder, and disease and disorder became increasingly explained with respect to radioactivity and the nuclear industries in the region. One young man claimed, for instance, that when he visited the sand mining beach in Manavalakuruchi, his hair on his arms stood on end, a corporeal awareness stimulated by his knowledge about high levels of radioactivity on this part of the Malabar coast. A pregnant woman avoided walking on black sands altogether understanding it to be a source of radiation and potentially hazardous for her growing baby. One student whose family had recently migrated to Nagercoil, reasoned retrospectively:

My mother was teaching in a military college in Assam. Before that
we were in Punjab. We have been here for eight years. My mother has
developed breast cancer in the last few years. We feel it is because of the
high radiation in this area.

This feeling was yet another sharp expression of an onto-episte-
mological conjunction, one that began to influence my thoughts and
actions too in terms of where I would go and what I would imbibe.
Such emergences underline how obscure phenomena are understood
and made coherent through a reciprocal worlding even while some
elements remain at the edges of perception and knowability.[51] In this
world of in/visibles, the line between visible and invisible, reality and
representation, and the body and the environment becomes diffuse.[52]

Despite views to the contrary, even scientists and bureaucrats
cannot be said to have the upper hand outside of such onto-episte-
mologies.[53] They too are deeply influenced by particular narratives
of reality. These are largely invested with authoritative interests to
do with ease and order. It is an onto-epistemology that is hinged on
Janus discourses—one door facing inwards with methodical signs to
do with procedure and precaution; the other outwards reliant upon
swings of denial, domestication, manageability, digression, and
aggrandizement. It is only against this conviction of belief rooted in a
mind–body dualism that an educated official's suggestion to go inside
a reactor as a safe haven from an earthquake without a hint of satire
can make any modicum of sense. The radiation regime is taken to
the extreme and applied to the reactor: in other words, radiation does
not exist in there, and even if it does, it is harmless, manageable, or
even advantageous. This overbreadth is not merely about propagan-
dist presentations, but another vindication of the mind as 'abstract,
disembodied and disembedded from popular social contexts' in the
words of the anthropologist Arturo Escobar.[54] The bodiless mind then
comes across as a mindless body. Joseph Masco describes it as a case
of the 'invulnerable scientific body', one that would even be available
to rebuild a city after a nuclear blast.[55] In a parody of a self-obsessed
scientist, defensive logic can easily descend into the ludicrous when
the body bizarrely disappears in a place best named Absurdistan.

In the aftermath of the 2004 tsunami, questions began to be raised
about whether high tidal waves could do reactor damage, as water

was reported to have flooded the sea water pump houses for the two MAPS commercial reactors and the excavated pit for the prototype fast breeder reactor under construction at the Indira Gandhi Centre for Atomic Research in Kalpakkam.[56] Even though the Sri Lankan landmass and the Gulf of Mannar around Kudankulam acted like a buffer zone against the high waves, the tsunami still destroyed buildings and killed people along this southern shoreline. A human and social crisis in the wake of the tsunami waves was, nevertheless, dismissed by nuclear officials as if nothing has or will happen. In an edited volume, S.P. Udayakumar states:

> Although the Minister of State for Science and Technology of the Government of India, Mr Kapil Sibal, went on record saying that the government would not have constructed the Kalpakkam plant on the sea if there was the danger of the tsunami, the Dept of Atomic Energy [DAE] is persisting with the Kudankulam plant. Neither the DAE nor the GOI [Government of India] has come up with any concrete statement on the impacts of the tsunami on the Kalpakkam and Kudankulam plants.[57]

A medical activist noted at a 2006 meeting in Nagercoil:

> When the tsunami hit Kalpakkam, people died including from the DAE. But DAE hospital was kept under lock and key. They did not want to report the deaths. From a friend who works there, there were about 200 people who died including those who were in their residences. The [mainstream] media reports it was 4!

A feat of mankind, however grand, could not stand up to the phenomenal force of the ocean, but nuclear officials begged to differ. These were people who were content to rewrite the realities of yesterday's weather. As we shall revisit in Chapter 8 with respect to the disaster at Fukushima in Japan, representatives reverted to their comfort blanket of denial and manageability. They were like the mild-mannered janitors to an unyielding netherworld. Nothing could or indeed should get in the way of their combined nuclear commitments. Mankind had after all cracked the atom, the source of all power that fuelled a megalomaniac ambition for more power and control over nature, and for the nation among other nations. Holding these atomic magi in their mesmerized hands, people's queries and confrontations were merely a nuisance that could be slapped aside. After all, self-interests in their employment and welfare were premium, and India had to keep up with global modernity and develop, come what may. This, they proffered, is the route for national greatness. Radiation remains a mere afterthought.

Notes

1. See Ferguson, James and Akhil Gupta. 'Spatializing States: Toward an Ethnography of Neoliberal Governmentality'. *American Ethnologist* 29, no. 4 (2002): 981–1002.
2. '"Kudankulam Power Plant Is More Advanced than Chernobyl Nuclear Power Plant"—R.S. Sundar'. Available on https://www.youtube.com/watch?v=4PfB9hcIQKw, accessed 10 December 2017.
3. Padmanabhan, V.T., R. Ramesh, V. Pugazhendi, K. Sahadevan, Raminder Kaur, Christopher Busby, M. Sabir, and Joseph Makkolil. 'Counterfeit/Obsolete Equipment and Nuclear Safety Issues of VVER-1000 Reactors at Kudankulam, India'. *Nuclear and Atomic Physics*, 2013. Available on http://vixra.org/abs/1306.0062, accessed 17 January 2019.
4. Padmanabhan et al., 'Counterfeit/Obsolete Equipment'.
5. There are several theories of rumination in psychology, the act of obsessive reflection and rehearsal of something that distresses. The way I invoke the term is in a more anthropological sense to investigate collective discourses regarding the nuclear power plant and radioactivity evident in people's behaviour, words and actions. See Smith, Jeanette M. and Lauren B. Alloy. 'A Roadmap to Rumination: A Review of the Conceptualization of this Multifaceted Construct'. *Clinical Psychology Review* 29, no. 2 (2008): 116–28.
6. Carpenter, Ele. 'The Smoke of Nuclear Modernity Drifts through the Anthropocene'. In *Power in the Land*, edited by Helen Grove-White. Wales: X-10, pp. 17–29, p. 29.
7. Carpenter, 'The Smoke of Nuclear Modernity', p. 18.
8. See Beck, Ulrich. *Risk Society: Towards a New Modernity*. Newbury Park: Sage, 1992, p. 29.
9. Manu Joseph reports on scientists in the Bhabha Atomic Research Centre (BARC) in Trombay: 'Radiation, that much-maligned thing in the outside world is somewhat liked in BARC. "Without radiation, species will not mutate", a senior scientist whispers'. Joseph, Manu. 'Spot the Indian Nuke Scientist'. *Sunday Times of India*, 12 March 2006, p. 13.

10. Ramana, *The Power of Promise*, p. 216.

11. See Csordas, Thomas. 'Introduction: The Body as Representation and Being-in-the-World'. In *Embodiment and Experience: The Existential Ground of Culture and Self*, edited by Thomas Csordas. Cambridge: Cambridge University Press, pp. 1–24.

12. On the relevance of this point to Indonesia, see Amir, Sulfikar. 'Challenging Nuclear: Antinuclear Movements in Postauthoritarian Indonesia'. *East Asian Science, Technology and Society: An International Journal*. Available on https://pdfs.semanticscholar.org/b633/0706fd657b d50fe994194b303baca1ed9b2e.pdf

13. See Udayakumar, S.P. *Disaster Management*. Nagercoil: Transcend South Asia, 2005.

14. Gusterson, Hugh. *Nuclear Rites: A Weapons Laboratory at the End of the Cold War*. Berkeley: University of California Press, 1998, p. 203.

15. Beck, *Risk Society*.

16. Beck, *Risk Society*.

17. With his more discursive approach, Michel Foucault elaborates on how knowledge is normalized through discourse that is perpetuated by institutions and people in order to legitimate a particular truth and regulate practice. However, as Donna Haraway adds, this may well be the case but all knowledge is 'situated' dependent on individual positionality and orientation. Foucault, Michel. *Power/Knowledge: Selected Interviews and Other Writings*, edited by Colin Gordon. New York: Pantheon; Haraway, Donna. 'Situated Knowledges: The Science Question in Feminism and the Privilege of Partial Perspective'. *Feminist Studies* 14, no. 3 (1988): 575–99. On their relevance to nuclear weapons scientists and the antinuclear movement in USA, see Gusterson, *Nuclear Rites*.

18. Boholm, Åsa. 'The Cultural Nature of Risk: Can there be an Anthropology of Uncertainty?' *Ethnos: Journal of Anthropology* 68, no. 2 (2003): 159–78, p. 167.

19. Knowledge-power is a messier variation of Foucault's oft-cited proposal that subjects are created as one of the discontinuous effects of power-knowledge. Ethnography demonstrates that while determinable processes produce knowledge orders, there is always ambiguity and contestation. It is in this light that I have inverted the terms of reference. See Chapter 7, and; Foucault, *Power Knowledge*, p. 28.

20. Gramsci, Antonio. *Selections from the Prison Notebooks*. New York: International Publishers, (1917).

21. On the distinction between the 'expert rationalism' of nuclear weapons scientists and the 'emotionalism' of those that critique them, see Gusterson, *Nuclear Rites*, p. 204. In a bid to avoid accusations of

anti-development, Indian anti-nuclear campaigners too sought to make a claim on rational expertise.

22. Proctor, Robert N. and Londa Schiebinger (eds). *Agnatology: The Making and Unmaking of Ignorance*. Stanford: Stanford University Press, 2000, p. 1.

23. Michaels, David. 'Manufactured Uncertainty: Contested Science and the Protection of the Public's Health and Environment'. In *Agnatology: The Making and Unmaking of Ignorance*, edited by Robert N. Proctor and Londa Schiebinger. Stanford: Stanford University Press, 2000, p. 22.

24. See Habermas, Jürgen. *Theory of Communicative Action, Volume One: Reason and the Rationalization of Society*, translated by Thomas A. McCarthy. Boston, Massachusetts: Beacon Press, 1984.

25. Ramachandran, 'The Peringome Antinuclear Struggle'.

26. On its stigmatizations in the global north, see Sontag, Susan. *Illness as Metaphor*. London: Allen Lane, 1979.

27. Subramanian, T.S. 'Full Steam Ahead'. *Frontline*, 20 April 2012, pp. 114–21; 'Kudankulam Update: Arrests of Peaceful Protesters Under the Sedition Law in Tamil Nadu, India'. Available on https://indian2006. wordpress.com/2012/03/20/koodankulam-update-arrests-of-peaceful-protesters-under-the-sedition-law-in-tamil-nadu-india/, accessed 17 January 2019.

28. 'NPCIL 24 Hrs English Film'. Available on https://www.youtube.com/watch?v=pNeG1tU6eOk, accessed 17 January 2019.

29. One short film asserts that nuclear energy is also economical and does not 'case any pollution in the environment'. Without which there would been no 'golden future' [sic]. 'NPCIL need Nuclear Power English Film'. Available on https://www.youtube.com/watch?v=3zk5h2xgvgI. Another underlines the NPCIL's 'important role in making of our country', 'NPCIL We too English Film'.Available on https://www.youtube.com/watch?v=xKE5rDodEeg, accessed 17 January 2019.

30. Ramana, *The Power of Promise*, pp. 126–7.

31. See Greer, Jed and Kenny Bruno. *Greenwash: Reality Behind Corporate Environmentalism*. Lanham, MD: Apex Press, 1997. With the claim to reduce carbon emissions, Alexander Dunlap calls it 'green violence'. Dunlap, Alexander. 'Counterinsurgency for Wind Energy: The Bíi Hioxo Wind Park in Juchitán, Mexico'. *The Journal of Peasant Studies* 45, no. 3 (2018): 630–52.

32. Such enchantment practices are combined with IREL corporate social responsibility measures in terms of providing employment, contributing substantially towards fishermen's' welfare schemes, and extending aid to local schools for building class rooms, compound walls, providing

furniture, and the like. In addition, they conduct medical camps for the benefit of nearby village residents and promote sports activities throughout the district.

33. Sundaram, C.V., L.V. Krishnan, and T.S. Iyengar. *Atomic Energy in India: 50 Years.* Government of India: Department of Atomic Energy, 1998, p. 138.

34. *Radiation Protection Rules, 1971.* Available on http://www.hp.gov.in/dhsrhp/Radiation_Protection_Rules_1971.pdf accessed 10 October 2019. On an earlier history on radiation levels, see Hecht, Gabrielle. *Being Nuclear: Africans and the Global Nuclear Trade.* Massachusetts: MIT Press, 2012, pp. 183–212.

35. Sundaram et al., *Atomic Energy in India,* pp. 169–70.

36. Sundaram et al., *Atomic Energy in India,* p. 171.

37. Sundaram et al., *Atomic Energy in India.*

38. On the deployment of rationality as a parallel discourse to legitimate nuclear policy, see Wynne, Brian. 'Risk and Environment as Legitimatory Discourses of Technology: Reflexivity Inside out?' *Current Sociology* 50, no. 3 (2002): 459–77.

39. There is the additional kind of neutron-based radioactivity where free neutrons are released as a result of nuclear fission or fusion that then react with other atomic nuclei that could emit radiation.

40. The notes for nuclear plant staff designate Radioactive Zoning for which there are four zones of contamination control:

   1. Zone 1—Turbine building, dosimeter TLD Issue Room, TLD storage area, etc.
   2. Maintenance shops, clothing crib, washing area.
   3. Shift Health Physics Office, High Bay, Service Area, etc.
   4. Reactor Buildings, spent fuel bay, and decontamination centre.

   At the barrier of every zone, personal monitoring instruments are provided. After describing the functioning of the Dosimeter, Radiological Work procedures, Emergency Procedures, the notes end with a section entitled 'Safety standards in our nuclear programme' where it assures that 'The radiation exposures to DAE workers are generally well within the permissible limits. This is the result of effective use of radiation protection procedures at the nuclear installations'.

41. See Wynne, *Rationality and Ritual,* p. 153.

42. Haraway, 'Situated Knowledges', p. 58.

43. See Hecht, Gabrielle. *The Radiance of France: Nuclear Power and National Identity.* Massachusetts: MIT Press, 2009.

44. On the 'interplay of distorting mirrors' between nuclear plant managers and the public, see Zonabend, Francoise. *The Nuclear Peninsular.* Cambridge: Cambridge University Press, 1993, p. 5.

45. In her study on African countries that has some parallels here, Hecht notes:

> Since the 1970s, the International Commission for Radiological Protection (ICRP) has promoted the exposure philosophy of ALARA: As Low As Reasonably Achievable. 'As low as' reflects the rough consensus that all radiation exposure has some health effect: 'reasonably achievable' represents a concession to economic and political imperatives (and power). Buried deep in the ICRP's philosophy is the assumption that human lives have different values ... this philosophy has been interpreted as legitimation for spending less to protect workers in poor nations who have remained invisible to experts.

Hecht, *Being Nuclear,* p.44.

46. On -scapes, see Arjun Appadurai who describes them as 'deeply perspectival constructs' that do not look the same from 'every angle of vision'. With what appear as uneven fields of consent, negotiation and contestation, Appadurai outlines five -scapes along which these perspectival constructs are arraigned: ethno-, finance-, media-, techno- and ideo-scapes. According to his rationale, anything to do with nuclear issues would largely fall under technoscapes. However, the nuclear industries straddle a whole range of terrains—ecological, material, demographic, and discursive from the potentials and constraints of the enterprise to discursive strategies to promote and resist the plans, down to the impact it may have on peoples' lives in terms of their thoughts, conduct, livelihoods, and health. Appadurai, Arjun. 'Disjuncture and Difference in the Global Cultural Economy'. *Theory, Culture Society* 7, nos 2–3 (1990): 295–310, p. 296. See also Pitkanen, Laura and Matthew Farish 'Nuclear Landscapes', *Progress in Human Geography* 42, no. 6 (2018): 862–80.

47. On implications for racial hierarchies, see Hecht's account on uranium mining in Africa, *Being Nuclear,* pp. 213–318.

48. See Harman, Graham. 'I am also of the Opinion that Materialism must be Destroyed'. *Environment and Planning D: Society and Space* 28, no. 5 (2010): 772–90.

49. See Barad, Karen. *Meeting the Universe Halfway: Quantum Physics and the Entanglement of Matter and Meaning.* Durham: Duke University Press, 2007.

50. See the debate: Carrithers, Michael, Matei Candea, Karen Sykes, and Martin Holbraad. 'Ontology Is Just another World for Culture: Motion

Tabled at the 2008 Meeting of the Group for Debates in Anthropological Theory, University of Manchester', ed. Soumhya Venkatesan, *Critique of Anthropology* 30, no. 2 (2010): 179–85.

51. Heidegger, Martin. *Being and Time* (originally published 1927), trans. Joan Stambaugh, revised by Dennis J. Schmidt. Albany: State University of New York Press, 2010. On 'worlding', see Zhan, Mei. *Other Worldly: Making Chinese Medicine through Transnational Frames.* Durham: Duke University Press, 2009.

52. See Bateson, Gregory. *Steps to an Ecology of Mind: Collected Essays in Anthropology, Psychiatry, Evolution and Epistemology.* Chicago: Chicago University Press, 1972; and the literature on embodiment where, as Thomas Csordas has argued, the body is 'the existential ground of culture'. Csordas, Thomas. 'Embodiment as a Paradigm for Culture'. *Ethos* 18, no. 1 (1990): 5–47, p. 5.

53. On a related argument about 'civic epistemology', see Jasanoff (2005). *Designs on Nature: Science and Democracy in Europe and the United States.* Princeton: Princeton University Press, 2005. On plural epistemologies with regard to nuclear politics in Indonesia, see Amir, 'Challenging Nuclear: Antinuclear Movements in Postauthoritarian Indonesia'.

54. Although the analysis was originally applied to conventional academics, it is equally relevant to nuclear scientists. Escobar, Arturo. 'Culture, Practice and Politics: Anthropology and the Study of Social Movements'. *Critique of Anthropology* 12, no. 4 (1992): 395–432, p. 419.

55. Masco, Joseph. *The Nuclear Borderlands: The Manhattan Project in Post-Cold War New Mexico.* Princeton: Princeton University Press, 2006, p. 45.

56. Ahmad, S.N. 'Press Releases from Nuclear Power Corporation on the Impact of Tsunami that struck Kalpakkam on 26 December 2004'. Available on http://www.dae.gov.in/press/tsunpcil.htm; Jayaraman, K.S. 'India's Nuclear Debate Hots Up after Tsunami Floods Reactor'. *Nature* 433, no. 7027 (2005): 675, accessed 1 June 2006.

57. Udayakumar, S.P. 'Nuclear Reactors and the Tsunami'. In *Tsunami and Its Impact,* edited by Dr R.S. Lal Mohan. Nagercoil: Conservation of Nature Trust, 2005, p. 164.

5

Full Lives

'Isn't it a risk to live here?'

'Life is a risk' (conversation with Daniel, Kanyakumari District, 2006).

The construction and commissioning of nuclear reactors threatens to make proximal lives insignificant even while it promises to animate them with plentiful supplies of electricity and their spin-off benefits. How did people in the region navigate this limbo status? In this chapter, social and political dynamics are fleshed out with respect to the densely textured intimacies of specific people. Nuclear risks did not emanate from the solar plexus of the reactor alone, but in a circuitous fashion, were rerouted through mundane practice—revisited in terms of changes and challenges to peoples' health, diets, homes, livelihoods, the expense of living, the future of their children, marriage prospects, and their specific worldviews.

The first part of the chapter is a prequel that throws some light on how and where opinion, resilience, and/or resistance against a nuclear power plant might emerge. The second part provides a finer focus on three people, and how their 'social risk positions' are nuanced and anchored in wider socio-political contexts that go beyond just nuclear industrial and environmental concerns.[1] The narratives are but a slice of meshed conjunctions to do with risk and resilience in their everyday lives.

One person is Josef who had just turned 20 when I met him, a student and fisherman, who in the long run, wanted to train to become a priest. Second is Savitri who was in her mid-twenties, a teacher from a farming background. Third comes Rajesh who was in his forties, a writer and family man with infant children living in a town close to the nuclear plant. The three life stories were selected to highlight variable locations (coastal, agricultural, urban), religio-ethnicity (Christian, Hindu, and related castes complexes), lifecycle (celibate, single on the marriage market, married with children), and gender positionality. The material presented in this chapter is from fieldwork in 2006, when the two reactors had not been commissioned, and five years before the Fukushima Daiichi disaster that propelled the anti-Kudankulam plant struggle into other stratospheres. Altogether, they demonstrate how resistance was fermenting indigenously and not at the behest of outsiders such as NGOs and foreign funders or agencies as nuclear state officials were wont to dismiss.

Countering Risks

Between the raw edges of anger and apathy, resistance and compliance, were other zones of orientation. People developed a variety of coping mechanisms to enable them to counter risks to do with a nuclear power plant. The drive for the acquisition of knowledge about nuclear power and radiation has already been recounted in the previous chapter and will be revisited at length in Chapter 7. Effectively, people took it upon themselves to try and become experts on nuclear issues. This also had the effect of wresting control from authorities and countering accusations of them as unreasonable. Fully acknowledging that the granularity of real life can never be succinctly or fully captured in words or phrases alone, other orientations might be heuristically outlined in terms of adaptation, diversion, diversification, solace, and satirical release. These themes are viewed in terms of a multi-leafed palimpsest, rather than as part of linear chains of actions and reactions.

Adaptation was with regards to altering routine behaviour. The circulation and interpretation of information on nuclear industries and how they might be related to diseases such as cancer variously affected essential aspects of life: people's thoughts, conduct, diets, and politics. Conversation had corporeal consequences—the more the subject of radioactivity was raised, the more physically disconcerted people became as if a phantom was encircling their bodies. Radiation awareness began to be internalized such that daily practice began to gradually change with the understanding that perhaps they could contribute to preventative measures and remedies from any malady. People might take recourse to a proliferating range of health options to boost their immunity systems. These included allopathic, homeopathic, Ayurvedic, and the local tradition of siddha forms of healing. According to a medic I talked to in 2006 in Kanyakumari District, there had been about 10 per cent reduction in cancer due to people taking preventative measures over the previous decade. Nowadays, local people are well tutored due to the work of activists, NGOs, schools, health workers, parish priests, and self-help groups.

They combined such health practices with (plans for) changes to their diet, eating high-fibre foods seen to be 'anti-cancerous' or at least effective in boosting the body's immunity system. These included cabbage and beetroot that were known to provide antioxidants and possibly, as one woman put it, 'positive immunity against cancer'. Some people had resolved even to stop or cut down on eating their staple diet of fish once the nuclear reactors became critical, stating that they would eat more 'vegetables and meat'. However, whether inland or coastal, fish was a daily delicacy and to abstain was a prospect that many did not relish.

Diversion from the perspective of local residents was when they might adopt tunnel vision and/or deferral strategies. In some cases, rerouting radiation risks might even lead them to being tolerated: for instance, the threat of higher radionuclides posed by a government-backed enterprise might be viewed as less directly intimidating when compared to the prospect of reprisal should people try to protest against the plant. Fears of disruption to their lives, harassment by police, and possible victimization could seem like greater dangers than the threats of living with radioactivity. So risks become scaled from the immediate to the distant future with more gravity given to the former. Some people seemed to be caught in a spiralling bind. In the privacy of their own homes, they may complain about the sand mining plant and the nuclear power plant, but believed that little could be done against the might of the authorities. So, they did not raise the topic in public.

Diversification led on from diversion as people's thoughts about moving further away from the plant and the effort to educate their young, so that they could develop transferable skills and move out of the region, took on more urgency. As had been the case since the introduction of trawlers in the 1970s, fishing was not a guaranteed source of livelihood for the future of coastal communities.[2] Young people went to college to train in other skills, such as in engineering, IT, working on a commercial ship, or even leaving the world of work and family altogether and becoming a Catholic priest. Those who did not have the educational credentials entertained ideas about going abroad to the Gulf countries to work. Even though they valued their coastal heritage, traditional vocations were no longer dependable. The nuclear plant only pronounced this decline. Born out of forced disruption, diversification was the only viable way forward.

Solace was with regards to taking comfort in spiritual faith in order to reconcile with problems and overcome any anxieties. Whatever their religious background, many people sought comfort in prayer and ritual to help them overcome their difficulties, with an understanding that 'it is God's will' and 'we live by God's grace'. More specifically, there were a number of elements that Christians took from their faith in terms of the environment. With regards to their resolve to resist, they would talk about 'godliness is next to cleanliness—so no pollution' and 'we need to be kind to the earth and not exploit it'. With regards to their resilience, they highlighted that 'God is your protector', 'there is strength in suffering', and a belief in humanity and the power of the oppressed collective, conveyed by the need for 'justice in the world' and a resolute conviction that ultimately, 'the meek shall inherit the earth'.

Most people regularly worshipped in a temple, church, or mosque, and periodically conducted penance. At key periods in their life, Hindu children's heads were shaved as 'a sacrifice to God' so as they would look after them. One person explained that:

> It's like offering the head to god, a sign of loyalty and humbleness. People do it when they go on pilgrimage to Tirupati, Thiruchendur or Suchindram. Also if there is a death in the family and the parents want

children to remain healthy, they say to God, keep our children healthy and we will devote our lives to you. It is like a sacrifice. They sacrifice the beauty of children's hair so as they may be protected in the future.

Alongside periodic tonsuring were a scattering of large round black marks that perennially decorated young children's faces and bodies to avert the 'evil eye'. This notion of malignance also spread to the nuclear power plant. By association, the nuclear plant and officials who supported it became the demonic. For some Christians, the apocalyptic scenes in the Book of Revelation in the New Testament became a way with which to imbue atomic energy as an evil force outside of the control of humanity. From Psalm 23, the 'shadow of the valley of death' alluded to the abysmal anxiety, uncertainty, and trouble that a nuclear plant created in their lives which they had to navigate with God's help.[3]

Numerous images of St Michael defeating the demon hung inside or outside coastal Christian homes, providing another inspired expression of protection from, and victory over, nuclear nightmares (Figure 5.1). As one fisherwoman put it: 'St Michael goes ahead of all parades—before the parish saint and even Jesus. He is the warrior saint. He fights evil such as the police. He clears the way'. In the spirit of concordance, believers in the Hindu god, Vinayaka, added that their strong faith in this 'remover of obstacles' would also vanquish nuclear structures and hurdles.[4]

A minority of local residents were sceptical of religion, however:

It is just for desperate people. Like when you see an elephant for the first time. First you wonder what it is. Then you begin to fear it. Lastly, you start to worship it. This is there for people who believe God can cure diseases.

From a fatalist understanding of phenomena came a more modernist appreciation that stressed less destiny and more environmental and man-made cause. One did not altogether replace the other, however, as Ulrich Beck may argue in his focus on the rise of risk discourse in modernity.[5] Instead, they tussled with each other, born out of an interlocking of hybrid worldviews that resonated most with people's lives.[6] What may be described as pre-modern reasons where risks were viewed as the workings of fate or divine will or karma, loosely translated as destiny, continued and jostled with more modern

Figure 5.1 External Household Shrine of St Michael Defeating a Dragon, Idinthakarai, 2012.
Source: Raminder Kaur.

ideas where risks were firmly placed with new developments in their area as with the science of a nuclear reactor, radiation readings, and their environmental and somatic associations. As we shall see in Chapter 10, recourse to supernatural practice and explanations need not amount to mystical behaviour, but might even be construed as a rational approach to a situation in which people felt they had minimal say.[7] Indeed, when the authorities offered little space for dialogue and recompense, a religious figure like Mother (Matha) Mary might even be seen as more accessible and just. She was the harbinger of hope and strength for many of the Roman Catholic fishing communities.

Resigning oneself to the plant was not necessarily a token of defeat. Nor was it ever complete. Some saw it rather like a black comedy, where satire was king.[8] This strain was evident in much of the political theatre developed in the region as much as it is evident elsewhere. Gabrielle Hecht describes such interventions as 'counter-spectacles' to the 'technological spectacles' of the aggrandized 'radiance of France'.[9] S.P. Udayakumar's play, *Anushakthi Amma* (*Atomic Mother*), sums up this irreverent streak in peninsular India—wildly absurd were it not striking some painful truths. Centred on the desolate lives of women and children among self-important leaders, the play makes a mockery of the nuclear lobby's vested claim to develop the country when in fact they are held to destroy the very heart of its ecological, moral, and social fabric. Udayakumar set one of the scenes in an air-conditioned conference hall at the KNPBP, the apocryphal Kalankulam Nuclear Power and Bomb Project:

> Sitting at the center of the dais is Dr. D.A.E. Das, who is the Supreme Commander of the Nuclear Industry of India. Seated on his right is the President of India and to his left is the Prime Minister of India with the dreaded nuclear button that, on a small push, can obliterate Pakistan and the whole of South Asia. At the back of D.A.E. Das are the Three Holinesses, Paramhans Powermaniacharya, Janab al-Muslim all-I-care, and Father Ignorantacious, followed by the Supreme Justice Honorable Mr. Blindeyer, the National Editors Guild Chief Mr. Bigliar, and the Government of India's Principal Secretary Mr. Bootlicker. Every time Dr. D.A.E. Das releases some radioactive gas, his accompaniments take a deep breath and go into a state of delirious trance. If and when they seem to be coming to their senses, Dr. D.A.E. Das farts again and sends them to their 'developed' stage.

D.A.E. is of course an allusion to the Department of Atomic Energy who are supported by a close-knit circle of power-wielders including 'investors, stockholders, Generals, scientists, engineers, geologists, seismologists, doctors and most importantly, people from the AERB, Atomic Energy's Reliable Backers', or in other words, the supposedly impartial Atomic Energy Regulatory Board.

As goes its character, satire is reliant upon humour, irony, exaggeration, and ridicule to corrode dominant logics.[10] This attitude of ingratitude spread to virtually all aspects of life in the shadow of nuclear reactors. One woman asked in her discussion on counteracting the effects of ionizing radiation on the body, for instance: 'What is more important—God or cabbage?' In another instance of its proliferating poiesis, a five-year-old boy asked his parents for a mobile phone, who disapproved saying that 'there's radiation coming out of it'. The boy reasoned with the extension of an adaptive logic: 'But I can use the phone. Then I can eat lots of cabbage afterwards.' The suggestion rang on deaf ears.

Josef

Josef lived in a coastal village of mainly Christian low-caste Paravar and Chambars. Its central landmark was an attractive pink and white church, built in 2000 on the beach with help from a German missionary. Next to the church was an outhouse where the fishermen would sit, repairing their nets, talking, drinking, sleeping. Josef's compact, one-storied four-roomed house was situated about a 100 m away directly on the beach. Sometimes during the monsoon, the waves would come up to within 10 ft of the gate to his house (Figure 5.2). It was in a stunning location, but with the enduring memory of the 2004 tsunami, the sea took on the aura of dreadful necessity: dreadful because of the horrors it had brought to people's lives in recent years; a necessity because it was central to their lives and livelihoods.

In the tidal rhythms of melody and brutality, the sea was the basis of an elemental and existential life. It was their mother, Amma or Kadalamma (Sea Mother) who both nourishes and admonishes them. Referring to it as 'our Great Mother', Josef explained that the tsunami was like a moral backlash, the forces of nature against the conceit of humanity that brutalized it.[11] He added, 'Man is not stronger

Figure 5.2   Looking Out onto the Sea from a Fishing Household, Kanyakumari District, 2006.
Source: Raminder Kaur.

than nature', alluding to the threat that nuclear authorities posed to its sanctity. 'When she gets angry, it takes it out on the people'. The depths of the ocean sustained a profound 'moral meteorology' that influenced his views on the plant.[12] The reactors were not respectful to nature. Neither could they withstand the forces of nature that inevitably could add to its destructive potential.

Fortunately, for Josef and his village, the tsunami waves were broken by the high sand dunes, a rubble mound groyne that projected into the sea, and the northeastward direction of the beach on which they were located. As a consequence, there were no tsunami-related deaths in his village. While the force of the tsunami damaged boats and catamarans (*kattumaram*) that were left further down the beach, it did not destroy any of their homes. Neighbouring villages were not so lucky, however. A coastal village that could be seen to the north of this beach did not have natural or man-made barriers and its residents were not spared the ferocity of the high waves. While the sand dune in front of his house had been totally washed away, Josef held an abiding belief that the tsunami would not happen again for 'another fifty years. Scientists said that'. His faith in the divine blended with a modicum of faith in expert science even if this sentiment did not extend to scientists from the nuclear authorities. They were wilfully worlds apart.

The fact that Josef's family and community were 'saved from the tsunami' re-enhanced his religious beliefs. Images of Jesus (holding the sacred heart), St Anthony and Tamil lines of religious verse next to scenic pictures decorated virtually all the walls of his home. He translated the text on one of the posters: 'It says "don't fear. I am with you"—it is in the gospel but also God has saved us. Here, nobody died from the tsunami.' He elaborated that God would also save them from the nuclear power plant 'should there be any problem'. He saw its construction as part of a larger social predicament, emphasizing that 'God is the light—spiritual light is more important than material light. People have a lot of material light, and want more and more. But they need more spiritual light': one reason why he was determined to follow the path of priesthood himself.

There was also a framed poem about footprints in the sand on the wall—a Christian souvenir that is widely available but took on particular poignancy here due to the centrality of the beach to the lives of the community:

One night I had a dream
I was walking along the beach with God
And across the skies flashed scenes from my life
In each scene I noticed two sets of footprints in the sand
And to my surprise
I noticed that many times along the path of my life
There was only one set of footprints
And I noticed that it was at the lowest and saddest times in my life
I asked God about it
'God, you said once I decided to follow you
You would walk with me all the way
But I notice that during the most troublesome times in my life
There is only one set of footprints
I don't understand why you left my side
When I needed you most'
God said, 'My precious child, I never left you during your time of trial
Where you see only one set of footprints,
I was carrying you.'

Such words of solace and support kept the coastal community upbeat especially when seeing, in the media and in real life, the shattering damage of the tsunami in neighbouring villages. It was as if God had lifted Josef and his family from dangers on the beach. It was this abiding faith that also made him endure all the changes and problems that life threw at him, his relatives, and community.

A VCD (video CD) made by his friend's uncle from found footage, showed the tsunami wave smashing into the two rocks and beaches off Kanyakumari town. We all watched it together in his front room with his family—mother, father, elder sister, cousins, and auntie. The film was received with dread, sympathy, reprieve, and anger. It was a retrospective fear that sent a chill down the spine. Seeing a big wave rising and racing towards the Kanyakumari rocks where people were still stranded, resurrected their memories of the thousands who died. Images of dead babies on the beaches understandably evoked most compassion. Relief registered when they looked round and felt blessed that they had been saved from the worst. Disgust was levelled at the high court judge who was transported first off one of the rocks by boat, while the other people were left behind. They were also offended on seeing the police on the beach: 'If we fishermen can't save the people, what can the police do?'[13]

They had little confidence in the political and judicial systems, viewing them as thoroughly corrupt. They were peopled with individuals who had minimal understanding of the lives of the fishing community, and who supported the development of damaging industries. The police were deemed as no more than the brutal arms of a draconian state. This distrust of authorities extended to contempt for the NPCIL, 'outsiders' who treated the sea as if it was a ditch in which to throw all their rubbish. The officials demonstrated a careless cruelty that hurt their oceanic consciousness.

Nevertheless, Josef believed that fishermen would be able to detect any changes in the sea due to the nuclear plant, almost intuitively: 'We can see it in the sea as we know it very well. If there's any changes, we will notice it', he asserted. He also vindicated a prevalent idea that the fishing community is strong and able to withstand health hazards 'better than city people'.

After NGOs replaced boats that were damaged by the tsunami, fishermen started their regular routine of going out to sea at 2 am to return around 8 am. Then they would work on the nets, eat around

1 pm, sleep for a little, and then go out again for a few hours to catch fish in the late afternoon. Their timetable was geared towards the morning and evening fish markets. When Josef was not busy with his college work, he too would go out with his father and cousins. On an average day, they may make about 600 rupees after selling their catch to share between the three of them.

Fresh fish was both a source of income and a staple part of their diet: 'We cannot live without fish curry and rice'. But now the nuclear power plant jeopardized such daily relishes. Josef considered how they could vary their diet—eating more vegetables with high antioxidants, with the understanding that they minimized the risk of cancer. Whereas he would have liked to eat more meat, it was much more expensive. '50 rupees for chicken compared to 10 rupees for half a dozen fish', he rued.

He was conscious that big changes were imminent. Local communities had fought against the nuclear plant, but the government was adamant on pressing ahead with the plans. At the time, Josef seemed resigned to the prospect of the plant in their midst. Village residents knew that one reactor would go critical the following year, then reputed to be December 2007 but later postponed. However, they did not know what to do about it: 'The plant has already grown, we resisted it at first but did not win'. The protests had tapered off by 2002 after the crushing gavel of the Supreme Court justice verdict that we learnt about in Chapter 3. This was to add further injury to police harassment of fishing communities who stood up against the plant.

Residents in Josef's village, while still against the plant, had to be both fatalistic and flexible—to be ready for what they had then thought would be the inevitable. Other residents had begun to think of strategies that would mean that they are not as reliant on fishing in the area as before. Some fishermen he knew chose to go away to the Gulf countries to fish in trawlers. It was not a desirable proposition for they loved their home village, but the income was more attractive and the catch more reliable. Others had migrated to cities like Mumbai to pursue careers on commercial ships. Younger people were encouraged to improve their educational qualifications by the village parishioner. Therefore, the majority of the older children were going to college in Nagercoil. Josef along with his village community were wary that fishing may well be a thing of the past: an even more uncertain future lay ahead with the prospect of a nuclear power plant heating the sea due to the release of water coolant and contaminating the environment with radionuclide emissions.

Josef confided that when his family got enough money together, they would move further inland, but due to the love he harboured for his village, only as far as to the other side of the road. Distrust ran so deep that fishing communities feared that should they move too far from their homes, they could be disconnected from the beaches by state and/or corporate ventures. First off, however, his family had to marry his sister off who had graduated that year with a B.Ed. The burden of a dowry was a heavy one. He complained about its manipulative role in defining gender relations:

> A man can be very cynical about dowry. He will get further education just so as he can get more dowry money. They say the boy has a degree so it costs and we need to ask more money for it. Sometimes it could be degree at any old college. Then they do no work. For instance, a man who does mechanical engineering cannot get any work here. So he does nothing.

It was as if dowry was treated as reparation for the likely unemployment or underemployment of young men in the region. By helping with marriage arrangements for his sister, Josef saw first-hand how unfair society was to girls and women: 'Both sides should give equal amounts which will help the couple out in their new life. Usually, they ask for five lakh rupees [500,000] and then they can also ask for lots of gold', a figure that has since rocketed. Household goods such as a refrigerator or a television may also be brought into the bargain. There were the additional costs of clothes, hospitality, and food for the wedding borne by the bride's family. Higher expectations came with rising expenses, and with fishing livelihoods threatened, financial burdens were commonplace.

Ordinarily, Josef rather than his father would go and check out a family if any of them advanced with a marriage proposal for his sister. Then if it worked out and dowry issues were settled, there would be a brief meeting between the boy's and girl's families. The next time the couple would see each other would be at the marriage ceremony. The institution of arranged marriage was considered the best to secure a stable future for Josef's sister. He was not too concerned about his own future, for well versed in the Catholic tradition, he knew that his mission was to 'dedicate his life to the Lord and his people', not to a particular woman.

For Josef, family, nature, and God overruled everything. Against this, the nuclear power plant was a gross aberration and eyesore from so many different angles. His resilience was borne out of a communal netting of knowledge acquisition, religious solace, reluctant adaptation, diversion from the nuclear issues at hand, and the compulsion of diversification. While he himself may not feel directly vulnerable to risks, he certainly felt it through intimate associations. With the experience of the tsunami, he reflected on people's mortality and inferiority next to God and nature. In the process of helping to arrange his sister's marriage, he gained a gender-conscious understanding of social risk positions. Through his life experience as part of a coastal Christian community, he saw the nuclear power plant as yet another transgression by 'inlanders'. They were encroaching on their lifeworlds, further depriving them of their livelihoods and threatening their lifestyles down to the very food that they ingest. There was an inveterate distrust of figures of authorities who misunderstood, misrepresented and/or mistreated them. The resilience that fishing communities had acquired through their marginal status strengthened their resolve to do something about the nuclear plant, whether it be directly or indirectly, but after more than a decade of anti-nuclear resistance, quite what this might entail at the time was not clear.

Savitri

In her mid-twenties, Savitri was the regular breadwinner in her Nadar farming household. Her father was no longer working due to ill health. Savitri's mother looked after their fields—a small holding of coconut, mango, neem, amla, and cashew nut trees from which they earned extra income. Her biggest worry at the time was 'money problem'. This mattered at the time seeing as her parents were looking for a man for her to marry, and they would need to raise a dowry for a marriage. Saving was proving to be very difficult. The age to get married averaged around 27 for men, and 23 for women, which in the larger context of Tamil Nadu is considered relatively late.[14] The later ages were due to the emphasis put on further education and/or because the girls' side needed more time to raise a substantial dowry from relatively paltry incomes. As noted above, higher education did not lead to enlightened views on dowry practices, however. Paradoxically, it had reinforced the custom due to the rising costs of education that the boy's family was keen to use as a bargaining counter to recuperate costs. Savitri observed the negotiable scale of charges in the marriage market:

> If they were professionals, like a doctor, he could ask for crores [10,000,000]. An advocate [lawyer] can ask for more than 50 lakh rupees [500,000]. They tend to marry other professionals like themselves.

Others such as a police officer may ask for ten lakhs [100,000] which could be on top of gold and other consumables. Nonetheless, there was some manoeuvrability. If the girl was considered exceptionally pretty, dowry could be decreased. Or if there was land that came with the woman, it might also be reduced. If the couple were in love, then they may not ask for a dowry at all.

Because of the need to find 'a suitable boy', restrictions were placed on Savitri's comings and goings. Her mother did not want anyone to talk about her in a slanderous manner at this critical stage in her life. Even when Savitri wanted to go for a walk in the village, someone had to accompany her. However, she seemed content with her lot. She did not really want freedom: 'Freedom for what?' She was happy with the idea of marriage. In fact, it was love that she thought was too risky:

> You may fall in love with a man, and he may leave you. Then your heart is broken. Even when I was at college all my friends were following the path of love. They were pushing me as well, but I said I only want to get married to someone. I don't want just love on its own. If the guy is good, you can fall in love with him after marriage.

On one occasion as we walked near her home, Savitri pointed out a person who she described as 'a mad woman who became mad after the tenth standard' when she was aged 16. The person in question was an unkempt and temperamental woman, who would sit on the side of the road, practically left to vegetate. Savitri informed me that she was only 34 years old but looked a lot older. She explained:

People say it's because she wasn't allowed to marry the man she wanted to. Her brother looks after her. Sometimes she freaks out. It is said that if she hears a child cry, she goes out and beats the mother. Then she starts throwing things, like stones, at people.

However much she tried to sympathize, the woman's life was a dreadful testimony of how love could go wrong and turn someone mad.

Come what may, arranged marriage promised her security and stability. Even though emotionally attached to her village, being a woman was a socially sanctioned passport out of the region as they are expected to leave their natal homes after marriage. While worried about radionuclide emissions from the nuclear power plant, however, she did not express too much concern about background radiation as she lived a few miles inland:

That is only [high] in Manavalakuruchi and there's a village near Kudankulam. This village is fine. It has everything. We have water, a beautiful climate. I really love my village. After marriage I would love to live here. But people will talk. It's not the done thing for me because I'm a woman—to stay in the same village as my birth.

In Savitri's opinion, whoever she married had to be a Nadar whether he be Hindu or Christian. Indeed, there were several customs that she as a Hindu shared with her Christian neighbours such as going on a pilgrimage to the Vellankani shrine where the Holy Mother Mary is venerated, and the practice of burial for the deceased in her village rather than the standard of cremation for Hindus. Caste was more important than religion when coming to consider a potential husband, and he could not be a coastal Christian: 'They are of Paravar Caste. SC [Scheduled Caste]. And we're BC [Backward Caste], Nadar—we're above them. Also they have a reputation for drinking and fighting'. She too had absorbed and come to accept hegemonic views about coastal communities.

There was one arterial road junction in particular where fights were known to flare up. The men who gathered there were renowned to be 'eve-teasers', and to be seen in their company was not good for a woman's reputation, Savitri admonished. Once she expressed shock that a Hindu friend of hers went out with fishermen on a boat when a female tourist visited. In her mind, she generalized that to associate with fishermen was not a good idea. It would pose grave risks to her reputation and her marriage prospects.

Health concerns about radiation were understood but not immediately felt. When the mosquito-borne Chikungunya epidemic emerged in 2006, it was this that was of paramount concern rather than the rogue rays of radiation. Still, her observations illuminated rising health expenditure and how people were prepared to sacrifice their health for money:

> There are many diseases here because people here don't invest enough on their own health and proper care and medicine. They are more interested in saving money. They are thinking about the future, about the future of their children especially if they have girls to marry off.

Concerns about health were tied in with economic factors in the fact that illness could stop her from earning if she had to take time off from a job where there was no paid sick leave. This extended to health concerns regarding dangerous radioactivity, a topic that she was learning more and more about.

She had attended several seminars on the nuclear plant organized both by officials as well as activists. The plant had created a state of curiosity and uncertainty that she was keen to dispel. The more she learnt, the more she began to see through the sanctioned narratives about nuclear power. From one seminar organized by local activists, she recalled the BBC docudrama, *Chernobyl Nuclear Disaster* (2006, dir. Nick Murphy), which was showcased on World Environment Day. A scene of a recently married couple and an entourage of children oblivious to the dangers of the reactor was particularly upsetting. It was akin to a future that she too was entertaining for herself. The parallels with the Russians in south India added to her concerns:

> We can see the reactions of an accident at a plant more [in the film]. They are also Russian. Kudankulam is with the Russians, so it's worrying having these reactors near our house. Accidents happen all the time. It can happen here...There are so many similarities with India [and Russia], the way they respect their scientists. And the way they keep things secret about nuclear plants.

The docudrama was shown alongside the documentary, *Buddha Weeps in Jadugoda* (1999, dir. Shriprakash), on the environmental damage of uranium mines and effects of tailings on nearby communities who lived in the north-eastern state of Jharkhand. She saw for herself disturbing images of Indians with carcinogenic mutations. Savitri stated: 'I had heard about Jadugoda but never seen anything. Images are more powerful than words and they can really hit home.' Savitri was also very moved by Sumitra Raghuvaran's PowerPoint presentation on congenital abnormalities at the event. For over a decade, Raghuvaran had done various surveys on complications during pregnancies in the Chavara (Kerala) and Manavalakuruchi (Kanyakumari District) areas where IREL were sand mining the beaches for minerals that include the radioactive, alpha-emitter, thorium in the monazite. She had noticed a high rate of miscarriages, abortions, and physical and intellectual disabilities in these high-radiation zones.

Prospects for children's health were paramount for Savitri. She knew of someone who had lost her baby recently. He was born premature and died a few days later. After the seminar, Savitri reflected:

> She had a baby that died after six days. Since then she has got pregnant again. But after three months, she had to have an abortion. The baby was 'imbecile' [intellectually challenged]. Now she has gone to live with her mother. People say it happened because she worries a lot about her family, her husband in the Gulf and he smokes a lot. But maybe it was due to the background radiation.

Even though she could not pinpoint radioactivity as the definitive cause, it remained a point of reference for an experience that she could share. Her awareness of people with intellectual disabilities was also sharpened as a result of the seminar that she had earlier attributed to fate.

There tended to be three lines of thought regarding mental retardation (MR) and mentally ill people among the wider populace. One was the dismissive attitude: to describe them as colloquially, having 'a screw loose'. Such was the reaction from many who had little time to view the intellectually disabled with any emphatic understanding. Another reaction was to fear people with MR, seeing them as wild and dangerous who could possibly turn psychotic, or perhaps even those who could spread disease. A third attitude prevalent among those who might be related to people with MR, was a combination of embarrassment and shame, sometimes denying the problem or even their existence in order to uphold their family's status.

Savitri was most worried about 'sleepwalking' into the third dimension, compounded by thoughts about the kind of family that she might marry into–another unknown which she would step into at some point in her life. There was a degree of paranoia about families marrying into those with a history of disabilities, the fear being that MR could be genetically transmitted to the next generation.[15] She elaborated:

> There is one family who has a MR child. She just lies in bed all day. It's been like that for the last twenty years. They're hoping that she'd die. Others lock them away, hide them or abandon them.

We had heard about a married woman who had four sisters, one of whom was intellectually disabled. Her in-laws were not informed about this sister. When they went to stay at her parental house, they thought she was a servant as this is what they had been told. In such circumstances, the person might remain unmarried or be betrothed to 'outsiders' from the village or town—that is, people who would not be partial to local knowledge.

Savitri had also heard about a woman who was led into thinking that a prospective husband did not need to work because he was very rich. Encounters between the couple were brief, the dowry was paid and the marriage held. Later, the woman realized that the man was intellectually disabled. Albeit a mild case of MR, the family had deceived her into thinking he was 'sane of mind'. The woman's life became extremely difficult, for she had to not only look after the house, but also became the main bread earner. Divorce was little of an option as in a conservative society, separation was looked upon disapprovingly. Even though the newly married woman had studied to the twelfth standard (at the age of 18), she had to do all kinds of jobs to make ends meet, and effectively became a domestic servant. However sympathetic others were, she along with others deemed that this was her lot, her karma, and she had to make the most of her circumstances. Karma then became a necessary burden, a fatalistic risk that a woman in a conservative family could do little about. Such social complexes were made worse by the fear that there could be a rise of intellectual as well as physical disabilities due to the increase of dangerous radiation in their surroundings.

When compared to the unknowns of the marriage market, the Kudankulam Nuclear Power Plant seemed more 'knowable'. Savitri felt that she could at least try and do something about the plant for herself. Even something as basic as teaching young children about what it was and its associated dangers to physical and genetic health was a worthy pursuit. Therefore, while Savitri was tied to social expectations of a young unmarried woman, the nuclear plant afforded her an opportunity to become more politicized. For such endeavours, she could operate outside conservative spaces for women.[16] She reflected:

> Nobody wants it [the plant]. People's power can be very strong if we all get together. Christians, Hindus, Muslims. I do not want the plant to open. It will affect the fish. And I love fish. I eat it every day. I cannot live without eating it.

'What will you do if the reactor starts?' I asked. 'I don't know', she seemed dismayed, 'I will make my mind up then.'

For Savitri, resilience to challenges in life was born out of enmeshed strategies to do with knowledge acquisition and diversion from the immediate risks of living near a nuclear power plant when money and marriage were more of a premium. Risk perception depended upon where she was in her lifecycle, where the unpredictable tracts of marriage was on top of the cards, and for that she needed money for a suitable dowry that could go some way towards her having a comfortable life. Even if a sufficient dowry was collected, the arranged marriage itself was a great step into the dark, but still considered less risky than a romantic relationship in terms of longevity. Walking behind her husband holding for the most part the marriage tether from his shoulders was a necessary risk to embark upon in her life. As she stated: ' ... man domination will always be there'. But marriage was also a possible route to have more agency through the prospect of motherhood, as well as escape a monumental nuclear-scape if her husband lived away from the region bearing in mind her disquiet about how radioactivity could affect genetic pools and the health of children.

Gendered and caste expectations played a palpable part in her activities and perceptions of her present and future. Simultaneously, the construction of the nuclear plant forged new pathways for the attendance of seminars and the dissemination of what she had learnt. She expressed unity with fishing communities when it came to resisting the plant, as if the prospect of greater radioactivity brought the community together. But she disassociated herself from them as much as possible when it came to everyday life for fear of social aspersions on her character. While constrained by social convention, anti-nuclear politics offered her a wider platform for public engagement. She believed that, despite people's backgrounds, people would come together to resist this project which she deemed ' ... terrible. We do not need it. Nor do we want it'. Fresh possibilities were presented for her in the public arena, ones that seemed to carve out a new space for women's participation in public fora, a subject that we return to in Chapter 8.

Rajesh

Rajesh was a writer of Nadar background living in a town about 30 km from the nuclear plant. Although he too suffered financial strife in a place that was getting increasingly expensive to live in with an average salary, his main anxieties arose out of the consequences of living near a nuclear power plant. He felt that he had led a happy and healthy life in his ancestral land and, at the very least, owed it to his children to have the same opportunity. Yet he was constantly reminded of their vulnerability to increased levels of dangerous radiation in the region that made this prospect highly unlikely: 'They are very young, growing and vulnerable to increased radiation. They may get cancer'. Apprehension was writ large on his face. The visceral memory of his grandparents' deaths due to cancer spurred him: 'It's a horrible death. I sat through it in their last days.... It was so painful, so degrading to humanity'.

If the nuclear plans could not be halted, he had to resign himself to sending his children away for their education:

We cannot escape from the NPP [nuclear power plant] if it is to be built.... Our hope is that the reactor does not go critical when planned. Maybe that'd be delayed by a few years. And then the next one would take at least five more years. And by that time, I'll send my kids off to study somewhere else.... In fact people are thinking like that. If they can't beat the plant, they'll think about getting their degrees and finding work elsewhere.

According to the Atomic Energy Regulatory Board's code of practice, high-density population towns with a population of more than 10,000 people were not to be within 30 km vicinity of a nuclear plant.[17] The higher density of his town was within this aerial range. He felt bitter about what he thought was the imposition and twisting of arbitrary rules. Fearing the increase of health hazards, he wondered whether his family should move further away. He asked in anger, 'Where do we move to? This is my home for generations and generations'. And then more forcefully, 'Why should I move?' It was like he was forced to be a refugee in his own home. He was extremely disillusioned by the nuclear authorities and their sleight of hand that enforced prospects of displacement from his homeland. He felt like a victim to their excessive and aggrandized ambitions:

> They believe that the greater good is for the maximum number of people. Some people have to suffer for the greater good and this happens to be us. People who suffer may not be a risk to national security but still they are held as suspect.

His anxiety turned into resentment that then turned into cynicism. He highlighted how people come to the peninsular to scatter cremation ashes into the sea and how they too were like the 'living dead':

> Even people in Kerala are now concerned [about the nuclear plant]. They have heard about eight reactors and now are phoning me up asking what we can do. It's also been heard that Thermal Power Corporation also want to start up two more reactors so that would make a total of ten! They're all coming here because the site evaluation's been done, the density of people is low, they say. They can't find other suitable locations as they're dumping all the new reactors here. People from over the world come to Kanyakumari to die. They say it's the edge of India and the end of their life.

With the lack of reliable information from nuclear authorities, people were stuck in a tensile web of rumours. Rajesh wanted to raise more awareness about the implications of living near a nuclear power plant. However, it was compounded by people's inability to commit themselves to a greater collective cause even if it was for their ultimate good. He elaborated:

> The problem is that most people only think about themselves. They have a short vision, short-term. Not long-term. This is from the level of politicians to the people. There are quick bucks to sell land to factories, for SEZ [Special Economic Zones] areas, but they're not interested in agricultural development any more.

A lack of foresight, intellectual culture and collective conscientization were the main hurdles:

> The ironic thing is that this society is very futuristic. It thinks about the future. There are about five astrological charts done per family. And these are also drawn up for key life events. But the future here is only focused on self. About making more money. About education. Jobs. Education here is how to make more money. There is no strong intellectual culture here. In the best college in this area I met some research scholars and I asked them—about forty of them—how many of them read the daily newspaper. Only two women put their hands up. And one of the papers mentioned was a Tamil tabloid. The other a daily Tamil newspaper. But the Tamil news does not cover as much news. It's not as global.

Limited opportunities and/or imagination meant that people were locked into relatively preordained lives—education, jobs, finances, and not to forget marriage and children. Everything else seemed peripheral. Even though himself a Hindu, he had much respect for Christians, fully aware that great strength lay latent amongst the fishing community: 'Fishermen are more environmentally conscious than farmers—they have to rely upon fish and the elements much more'. The fact that they had to live with and daily navigate one of the most formidable forces of nature, the ocean, became the grounds for an organic environmentalism:

> There are about one and a half lakh [150,000] of fishermen in this district. What will they do when the reactors are running and they can't find live fish easily? Would people continue to eat fish? I doubt it.

He was keen to utilize his talents to spread awareness on the dangers of the plant and radioactivity. His interest in disseminating information about nuclear power was a threat to the authorities in terms of its wildfire potential to affect large numbers of disaffected people. As with activists that preceded him from the late 1980s, he felt that the Criminal Investigation Department (CID) would have begun 'a file' on him in order to intimidate him into compliance with nuclear plans. A sense of paranoia had settled in with increased surveillance of his activities. Nonetheless, he did not let it beat him into defeat, only that he had to be more cautious about his available options to raise awareness:

> The police and CID can accuse you of anything. You will have little comeback. Defence matters are taken very seriously here. Anything nuclear comes in as defence even if it's for civilian purposes. If you're seen as a spy, it is a very serious charge.

In September 2006, the then President A.P.J. Kalam was the talk of the town and newspapers when he visited the Kudankulam Nuclear Power Plant. A couple of days before, Rajesh received a phone call from someone who said he was calling from Mumbai and that he had heard about his anti-nuclear views. He asked: 'What are you now up to?' Rajesh responded by saying that he was not doing much at the moment but 'there may be a few things in the near future'. Then he hesitated to consider whether the phone call was from a CID officer tracking him to check what his movements were so that he did not cause a security concern for the president. After that experience, Rajesh did not say much to anyone he did not know.

Intelligence departments could collate a file on anyone they deemed suspicious. These persons could simply be spreading news and information of which the authorities did not approve. Nevertheless, Intelligence officers in the local police office were also peopled by men who knew the individuals personally. If not close, they were at least known to them and the possibility that Rajesh could be 'a sleeper terrorist' was far off the mark. Surveillance was more out of duty to their office and following instructions in a chain of command from distant headquarters in Chennai and New Delhi. On one occasion in the early 2000s, the daughter of one anti-nuclear activist was to marry a police inspector. He told everyone about how they went to the police station (Intelligence Section) and a very large file was brought out on him. 'See how much we have on your father', said the officer. However, when the activist went to read the contents of the file, he found nothing in it that was 'actionable', just a long list of fairly benign and non-criminal activities. Even though security personnel followed him around, they had nothing substantial on him. Like a hovering wasp, Intelligence was both an irritating nuisance and a source of stirring anxiety. This was a phenomenon that was to dramatically change after the Fukushima Daiichi nuclear disaster in 2011 as we shall see in Chapter 10, when even the 'nothing actionable' became overly actioned.

Rajesh saw the presidential missile-man's visit as yet another nail in the nuclear coffin. But the big joke of the day that provided a round of merriment against claims of nuclear manageability was that, even after Kalam had stated that the scientists, technologists, and staff 'will excel in operational performance in Kudankulam Power Plants [sic]', a man fixing the air conditioning unit above Kalam at the nuclear plant fell through the false ceiling and landed a few feet behind the president.[18] On hearing about it, Rajesh quipped: 'And this is in a plant where they say safety is very high!' A critical disposition to the nuclear plant disposed him to satirical interpretations of iconic figures, no matter what their standing in the nation.

For Rajesh, resilience was borne out of endeavours for knowledge acquisition and the effort to disseminate it far and wide, diversification particularly when considering the fate of his children, and satirical release. His fear for the future resided in his experience of the past when his grandparents died of the terribly slow and clawing spread of cancer. His overall stance was a refusal to adapt but an ambition for change with the hope to alter not just nuclear plans but also the docile public who did not rise up against such onslaughts. Perceptions of risks were strongly triggered by the nuclear power plant into which other aspects of his life became cordoned—a sign of a most worrying triple-bind of state repression, environmental destruction, and mortal dangers.

Risky Times

Risks are a phenomenally complex and changing category with varying degrees of potency and traction in peoples' lives. They were contingent on knowledge/ignorance, time, place, life experience, life-cycle, economics, environment, and the orientation and idiosyncrasies of individuals. What remained constant were anxieties that centre on an individual–family–community nexus through which radiation burdens—biomedical, ecological, economic, socio-psychological, and political—were rerouted. Livelihoods, marriage, and the health of children, for instance, were of a premium, and for young women and their families in particular the rising costs of a dowry and the character of the new family were a massive concern. Oftentimes, these more socio-cultural concerns reared their heads over the towering risks presented by a nuclear reactor in their midst.

For Josef, risks were rooted in a liminal space characterized by both the marginalization and resilience of fishing communities. Their coastal exceptionality powered them onwards in the conviction of their alternative worldviews and lifestyles. He sought to provide some kind of religious, social and quasi-scientific salve to people's problems including those do with the expansion of nuclear industries near their homes. For Savitri, they were directly to do with her gendered and caste identity, dowry prevarications, the unknowns of love and marriage, and having children into which health and radiation concerns played a piquant part. Awareness campaigns against the plant offered her scope for political interventions and a degree of unity across social divisions where custom and convention did not. For Rajesh, risks to do with family and environment were very much shoehorned into the nuclear power plant as if radionuclides could penetrate every pore of his body. Raising awareness about the potential hazards was paramount even if it meant contending with the threat of state surveillance and intimidation. As a herald of radiation burdens, the plant was just too overbearing to be cast aside for the sake of future generations.

Notes

1. Beck, Ulrich. *Risk Society: Towards a New Modernity,* translated by Mark Ritter. London: Sage, 1992. On how cultural context affects perceptions of risk, see Douglas, Mary and Aaron Wildavsky. *Risk and Culture.* Berkeley: University of California Press, 1982.

2. Subramanian, Ajantha. *Shorelines: Space and Rights in South India*. Stanford: Stanford University Press, 2009.

3. On the invocation of the apocalypse among Catholics against nuclear power stations in France, see Hecht, Gabrielle. *The Radiance of France: Nuclear Power and National Identity after World War II*. Cambridge, MA: MIT, 1998, p. 228.

4. See Kaur, Raminder. *Performative Politics and the Cultures of Hinduism: Public Uses of Religion in Western India*. London: Anthem Press, 2005; Fuller, Chris J. 'The Vinayaka Chaturthi Festival and Hindutva in Tamil Nadu'. *Economic and Political Weekly* 36, no. 19 (2001): 1607–09, 1611–16.

5. Beck, *Risk Society*.

6. This combination of discourse recalls Walter Mignolo's proposal on the borders of ontologies where multiple worldviews might be appreciated while undermining monolithic views of reality. Mignolo, Walter. *Local Histories/Global Designs: Coloniality, Subaltern Knowledges and Border Thinking*. Princeton: Princeton University Press, 2010.

7. See Stambach, Amy. 'The Rationality Debate Revisited'. *Reviews in Anthropology* 28, no. 4 (2000): 341–51.

8. On a satirical look at the Indian nuclear industries, see the documentary, *Nuclear Hallucinations* (2016, dir. Fathima Nizaruddin).

9. Hecht, *The Radiance of France*, pp. 201, 205.

10. See Kaur, Raminder. 'Atomic Schizophrenia: Indian Reception of the Atom Bomb Attacks in Japan, 1945'. *Cultural Critique* 84, no. Spring (2013): 70–100.

11. On humanity's domination of nature, see Deckard, Sharae. *Paradise Discourse, Imperialism and Globalisation: Exploiting Eden*. London: Routledge, 2009.

12. Burman, Anders. 'The Political Ontology of Climate Change: Moral Meteorology, Climate Justice, and the Coloniality of Reality in the Bolivian Andes'. *Journal of Political Ecology*, 24: 921.

13. On how the state ignored fishermen caught out at sea on Cyclone Ockhi in 2017, see the film, *Orutharum Varela* (*No-one Came*, dir. Divya Bharati 2018). Available on https://www.youtube.com/watch?v=LIyOF5BTEJ4, accessed 10 July 2018.

14. M. Gopalakrishnan (ed.). *Gazetteers of India, Tamil Nadu State, Kanyakumari District*. Madras: Government of Tamil Nadu, 1995, p. 1070.

15. Such stigma also applies to people with physical disabilities and deformities. See James Staples who discusses stigmas against leprosy sufferers in south India. Staples, James. 'Personhood and Masculinity in a South Indian Leprosy Colony'. *Contributions to Indian Sociology* 39, no. 2 (2005): 279–305.

16. On how crises and upheavals can lead to new spaces for women's agency, see Menon, Ritu and Kamla Bhasin. *Borders and Boundaries. Women in India's Partition.* New Delhi: Kali for Women, 2007.

17. On criteria for site selection, see Ramana, M.V. *The Power of Promise: Examining Nuclear Energy in India.* New Delhi: Penguin Books, 2013, pp. 44–6.

18. Abdul Kalam, Dr A.P.J. 'Speeches: Interaction with the Scientists of Kudankulam Nuclear Power Project, Kudankulam, Tirunelveli, Tamil Nadu'. 2006. Available on http://www.abdulkalam.nic.in/SP220906-5. html, accessed 12 July 2018.

6

The Plot Thickens

In the 1940s, scientists moved towards nuclear science which was in the phase of development in the western world. Once out of the bottle, the genie cannot be put back in. That is how the nuclear science plot proliferated. (Manik, Nagercoil, 2006)

Their science policies are anti-people. And human security is more important than national security. (Anthony, Nagercoil, 2006)

In an unprecedented move in the region, a public hearing with the Nuclear Power Corporation of India Limited (NPCIL) was scheduled to be held in the Tirunelveli District Collector's office in Tamil Nadu on 6 October 2006. This did not happen for the first two reactors that were agreed in 2001, the NPCIL claiming that they had environmental clearance in 1989—that is, before the Environmental Impact Assessment Notification Act came into force in 1994 as we have already seen in Chapter 3. Still, the later public hearing was planned with little publicity in order to swiftly pass the construction of four additional reactors at the Kudankulam Nuclear Power Plant. Word quickly got round about the public hearing and in the space of a day, about 600–700 people turned up, much to the NPCIL's surprise. Usually operating behind the screen of heavy securitization, this officially sanctioned public forum offered a rare glimpse of nuclear representatives arm-in-arm with the all-seeing panopticon in the form of police and intelligence agencies coming to face local communities in the region, and revealed their foiled ambitions to convince the public of the advantages of constructing more reactors.

As Wendy Brown contends, the state is fundamentally contradictory: 'the paradox that what we call the state is at once an incoherent, multifaceted ensemble of power relations and an apparent vehicle if not agent of massive domination'.[1] It is dispersed in terms of its effects on everyday lives, and it is cohesive when experienced as a mighty monolith allied with a military–corporate–industrial complex. On this occasion, attendants at the public hearing felt they were going to encounter the agents of domination directly. This was as one person put it to me at the time, 'a once in a lifetime opportunity where the people will meet the nuke people'.

The public hearing provided an exemplary occasion with which to consider the clash of epistemologies between the nuclear state and local residents, and therefore an opportune moment to consider the sparks that circulate in a space of criticality. Attendants at the hearing entered into an intensive force field. Both sides were aware of the hurdles of democratic protocol but engaged with them as a matter of cause, entertaining aspirations for the nation that differed markedly from each other. On the one hand, nuclear authorities wanted to be seen to be following due procedure with a token nod to procedures of public consultation. They nurtured paternalistic ideas about developments that were tied into a national security defence agenda. For them, the public hearing was akin to a cat-and-mouse game where their feline operations were with a view to tick a mandatory box while controlling proceedings so as the 'hearing' was only selectively heard and 'consultations' went according to plan. On the other, social actors from different backgrounds with varied political orientations nurtured ideas of democratic development that put people and the environment before nuclear power. Fishing and farming communities, environmentalists and activists hoped for genuine consultation, justice, and recompense from a fear of contamination, bodily deformations, and loss of resources and livelihoods due to the nuclear plant. For them, the event was not just a matter of protocol but one of an overdue and urgent entitlement.[2]

In this chapter, we consider what underpins the clash of epistemologies with first, a focus on the sovereignty of the nuclear state that is embedded in a colonial carcass and the supremacy of techno-science and law.[3] We then cast a lens on the preparations, processes, and the aftermath of the public hearing, noting some of the direct, creative, and nuanced ways a range of people negotiated with and challenged the nuclear state.

The Nuclear State of Exception

Techno-scientific developments have long been corralled into state enterprises. Historically, the science–state collusion was sealed in the early twentieth century driven by national goals to do with strategy and self-reliance. Military and surveillance industries began with the creation of ordnance factories to produce offence and defence goods. Physicists in particular were incorporated into the World War II effort when it was realized that 'science in the service of the state could be of significant mutual benefit', as Itty Abraham proposes.[4] Atomic energy and the race for the nuclear bomb became 'the epitome of the science-state relationship', playing a seminal role in the rise of the USA in the post-war period in which national security developed as a model under which state and civilians should conduct their behaviour.[5] Abraham elaborates, this period encompassed:

> ... a totalising condition of civilian militarisation beyond simply border defence or even inter-state war, the indistinguishability of war and peace in relation to the practices of state security institutions and the panoply of legal instruments that support their activities, the militarisation of information and the enormous growth in intelligence agencies, related to which can be observed ever increasing degrees of state surveillance, deeply dependent on technology ...[6]

The technologies of war were reassigned as perpetuators of peace. Informed by the Cold War context of superpower antagonism and calls for the defence of its populations, national security became deeply entwined with nuclear prowess, a 'nuclear nationalism'.[7] In comparison with *discipline* that is centripetal in that it concentrates and encloses, *security* is centrifugal in that it constantly widens its circuits to the point that it merges with the atmosphere and appears as a neutral and undeniable good for the population.[8] So goes the controlling dynamics of nuclear nationalism in the ever-expanding hands of national security.

Technologies of mass destruction become the state's overt signature of power and influence. Law comes to operate in the service of the state to legitimate their special status in the interests of national security where privileged status is attributed to the atomic departments and their personnel.[9] The state then becomes indistinguishable from law, and nuclear authorities become compulsory custodians of a tyrannous technology presented as nationally and socially beneficial. Indeed, as the writer Robert Jungk elaborates, authoritarian exceptionality applies to all nation-states with nuclear industries, not merely to those who have declared nuclear weapons.[10]

Gabrielle Hecht refers to this 'essential nuclear difference—manifested in political claims, technological systems, cultural forms, institutional infrastructures, and scientific knowledge—[as] nuclear exceptionalism'.[11] At another level, this nuclear exceptionalism is, paradoxically, also part of a process of nuclear banalization. Over the years, the extraordinary as it pertains to the possession of nuclear weapons and the associated development of atomic industries to provide modern goods and services such as electricity has invariably become the ordinary.[12] The phenomenon is another instance of a 'crisis ordinary' where the exceptional becomes a part of mundane norms as goes the philosophical argument by Lauren Berlant.[13] When the crises of the nuclear state are made ordinary, threats to its legitimacy can be decried as not only aberrant but also abhorrent.

To appreciate the development of the state in India, models developed from the Greco-Roman paradigm need to be modified in order to account for the exercise of colonialism and indigenous articulations of power that continue to hold sway in the subcontinent. While Indian nationalists adopted the lingua franca of modern sovereignty, its wholesale incorporation only applied to an elite sector of the populace.[14] So it could be argued that whereas the European state developed mainly from autochthonous developments and was ruled by a process of including local powers and civic subjects into its expanding and disciplinary frame, the colonial state came from without and continues to function through a process of exclusion where it is only the elites who remain its paramount citizen–subjects. Others remain outside of its primary frame of address aside from occasions such as electoral periods where their vote becomes of intermittent importance.[15]

From this perspective, the modern state in India has always been the exceptional, a structure to which national security issues such as nuclear weapons and power plants adds another concretizing layer of exceptionality. The particularities of the postcolonial state owe to the exclusionary mechanisms of colonial government, an apparatus that has carried over into the contemporary era. As the historian David Hardiman underlines: 'India had inherited an autocratic system of government from the British, and very little of the repressive apparatus was dismantled'.[16] Instead, more liberal and progressive models were woven on to this inherited structure.

When it comes to focusing on the nuclear issue in India, on the one hand, there has been a hardening of the state as a consequence of security and scientific industries converging such that atomic science took on the role of a national fetish.[17] Since at least the early twentieth century, Indian nationalists had felt left out of the tracks of modernity for being denied an equal part in availing of the aggrandized glories of techno-science under a colonial regime. Modernist politicians such as Jawaharlal Nehru went on to monopolize the postcolonial state and decreed that this lag had to be addressed and several development and funding programmes were initiated to play 'catch up with the west'.[18] M.V. Ramana adds 'that science, through its association with "freedom and enlightenment, power and progress," has contributed in a major way to the Indian state's effort at legitimising itself'.[19] At a time of sublime associations attached to nuclear science from the 1940s, Indian statesmen even imagined a means with which to skip a beat and come into universal time of global parity, hence the inordinate importance attached to atomic science in postcolonial contexts such as in India.[20]

On the other hand, there have been discrepant inscriptions on the populace. As Partha Chatterjee notes, the marginalized 'are only tenuously and even then ambiguously and contextually, rights–bearing citizens in the sense imagined in the constitution'.[21] The 'overdeveloped state'—to draw upon the work by the sociologist and activist Hamza Alavi—is attendant with what could be described as an 'underdeveloped civic space' (although this is not to pose civic space in the West as the normative standard).[22] Large swathes of the populace in India are not decreed citizens in the sense of participating in what could be called 'civil society', and more, as Chatterjee describes, an arena of 'political society' that encompasses subaltern populations.[23] For the overdeveloped Indian state, even the questioning of the nuclear authorities can become a suspect act such that those from 'civil society' can fast be demoted as we will see in Chapter 10.

The state monopoly of nuclear matters is compounded by the scarcity of independent expertise outside of the DAE and kindred organizations based in India. Nuclear scientists, for reasons of funding and job security, straddle the administrative domains of state departments as well as the commercial exploitation of companies such as the NPCIL responsible for the construction of nuclear power plants and energy distribution. Those peopling this nuclear brotherhood exchange information and personnel amongst themselves relatively freely but not outside their securitized walls. As has already been noted in Chapter 3, even the supposedly independent AERB with the objective to regulate the health and safety of nuclear installations is peopled by those known to the DAE. Indeed, the political scientist George Perkovich observes that 'there are no means within India's institutional structure to provide independent scientifically expert checks and balances on the nuclear and defense establishment'.[24] Nuclear activity is entirely state-run or, at the very least with recent moves to open up to select parts of the private sector, a state-endorsed enterprise. Nuclear authorities have become a pumped-up sanctuary of state apparatus receiving special status as provider for and defender of the nation, and endowed with ample funds with which to conduct their daily affairs. This prerogative of the nuclear state to do with national security is shored off from other state departments even though the latter may be utilized to their cause as and when they see fit. Describing it as the 'Indian nukedom', S.P. Udayakumar, observes how 'it defines national security, science policy, energy paradigms, and the future vision of the country'.[25] To challenge it was a mammoth undertaking.

The Days Before

Seeing the nuclear state as an intractable no-go area, residents around the Kudankulam Nuclear Power Plant felt that they had no authority to whom they could effectively address their objections about more nuclear reactors in their vicinity. If a person had any complaints, they may go through the District Collector or the state's Pollution Control Board, knowing full well that this approach would be futile because these comparatively low-level officials could not proceed far with anything that contravened the nuclear state. Local authorities and regional political patrons were mere minions to the might of the nuclear state emanating from India's megalopolis.

Residents idealized that their views may matter yet were also conscious that the powers of the nuclear state knew no ends: they were far removed, impenetrable, and authoritarian. However, for the first time in the region, on 1 October, a photocopy of an announcement of a public hearing on the Koodankulam Nuclear Power Plant was circulated by an NGO to a person based in Nagercoil who I have called Anthony. The notice in the English language *Economic Times* dated 25 August 2006 was placed by the Tamil Nadu Pollution Control Board for 'the proposed expansion activity for Reactors III, IV, V, and VI of capacity 1×1000 MW each in the same premises of Kudankulam Nuclear Power Project (as per the Environmental Impact Assessment Notification Act, 1994)'. It announced that:

Suggestions, Views, Comments and objections of the Public are invited within 30 days from the date of Publication of this Notice by the District Environmental Engineer, Tamil Nadu Pollution Control Board Tirunelveli. All persons including Bonafide Residents, Environmental Groups and Others, located around plant sites/sites of displacement/ sites likely to be affected can participate in the Public Hearing and they can make Oral/Written suggestions to District Environmental Engineer, Tamil Nadu Pollution Control Board, Tirunelveli on the above subject [sic].

The notice spawned a flurry of activities. Despite the lack of transparency of the nuclear state, the prospect of a public hearing wrenched open an aperture so as people's opinions might be heard and registered by the relevant authorities.

Public hearings have become a vital tool to speak out about development projects impacting people's lives in irreversible ways, yet this belief exists in a context of chronic distrust. The public hearing was something people wanted, yet they were also wary of how the state could channel the proceedings so as it favoured their directives. It offered a peculiar vista of both hope and doubt.

That very day, Anthony quickly called round to double-check whether a public hearing was really going to happen as one had not been organized in the region before despite their earlier appeals. Should this be the case, he wanted to ensure that as many people as possible would get to know about it, and plan how he and his associates could get together to draft a petition.

He phoned up the Tamil newspaper, *Dinamalar*, and managed to have an article printed on the very first page a day before the hearing. Announcing what appears to be a 'call for arms', the report stated:

> A secret public hearing will be conducted tomorrow at the District Collector's office in Tirunelveli.... An inadequate organisation, NEERI [National Environmental Engineering Research Institute], has issued the EIA [Environmental Impact Assessment] report where it is stated that the required water will be brought in from Pechiparai Dam by underground pipelines. The Kudankulam project will need 7.5 lakhs cu metres per year. The Pechiparai Dam water is not even enough for the farmers.
>
> For these reasons the public hearing must also be conducted at the Kanyakumari Collector's office. As the public oppose the plant, the nuclear authorities have shifted the public hearing to Tirunelveli.... We will oppose the plan to take the water tomorrow at the public hearing.[26]

The article was a rallying cry for those in the Kanyakumari District likely to be affected by water shortages to get together with fishing communities among others to oppose the nuclear plant.

The day before the hearing, 5 October, turned out to be a very busy day in Nagercoil. Phone calls were made, meetings were held, and tactics were discussed. Word spread like wildfire in the three adjoining districts likely to be most affected by the plant—Kanyakumari, Tirunelveli, and Thoothukudi.

Krishna, a medical expert, had come down to the town of Tirunelveli the night before the hearing and a group of people were to meet him at his hotel. He had been exhaustively studying the EIA report to produce a petition. To gain access to the report was itself a momentous feat for the nuclear authorities had until then tried to keep it out of public circulation. He was to be leading a group of six including Sunil and Manick, all friends from different walks of life united by the common purpose of resisting the nuclear power plant.

At the hotel where Krishna was staying, a letter of appeal was carefully worded. It was decided that the petition should be based on 'hard facts' and focus on discrepancies in the EIA report itself so as to demonstrate their more lucid grasp of scientific arguments. The group decided to base their appeal by reworking the available evidence so as to move the terms of debate away from presenting an outright challenge to nuclear nationalism to instead foregrounding their rights as Indian citizens to the essentials of life such as water. They stressed, first, that there was no proper assessment of the Pechiparai Dam in Kanyakumari District. Krishna concluded that this would risk the livelihood and crop of numerous farmers in the district. Periodically, there is a water shortage, and more water being siphoned off for a large-scale development could potentially lead to widespread social unrest. Second, it was known from the nuclear reactors in Kalpakkam that the sea heats up by 10 degrees after the release of water coolant, 3 degrees more than the permitted amount. It was estimated by the group that if and when the Kudankulam plant has six reactors, it will use 83 per cent more water, so the sea would heat up even more as more water coolant would be released. This would have an extreme impact on the ocean's biodiversity as well as affect fishing practices such that boats would have to go further out to sea for their daily catch.

Krishna recited while Manick wrote the appeal down after group consensus. The letter came to a total of six pages and five copies were made to hand out to the nuclear representatives at the hearing. One of the objectors thought that this campaign should be made into a transnational challenge by involving countries like Sri Lanka whose sea waters were also likely to be affected by any plant emissions. The others agreed but felt that they had to focus on domestic issues first. This is largely because they did not want to be accused of colluding with foreign forces. They were in a double bind: to make a group like this stronger, it had to go beyond the artificial oceanic boundaries of the nation. However, if that was to happen they would be subject to allegations of anti-nationalism and colluding with foreign donors—a theme that we will return in Chapters 9 and 10. The idea to go transnational was duly dropped. Still, Krishna was grateful for the opportunity of a public hearing and announced, 'Let us play the dice'. When the grape juice came at dinner time, we imagined it as wine, and drank to the occasion.

The Public Hearing

The hearing was in a large rectangular hall. In the middle was positioned a large 6-foot-tall black granite statue of Veerapandiya Kattabomman, a local rebel king who fought against the British East India Company, and was eventually captured and hung in 1799. His symbolic significance was not lost on the people who attended the hearing, seeing the nuclear state as yet another colonizing intruder on their beloved land. About a hundred long tables with red chairs lined the room. On top of the tables were microphones. On the podium in front was a table with more microphones, surrounded by two PowerPoint presentations welcoming everyone:

> *Kudankulam Nuclear Power Plant*
> *Units 3, 4, 5 and 6*
> *WELCOMES*
> *Public Hearing Committee Members,*
> *Invitees and Members of Public*

The room was already packed to overflowing. More than half the attendants were women, sari-clad, barefoot or wearing flip-flops with some carrying their babies. They tended to gather in the back of the hall, with many leaning against the back wall as there was insufficient seating. Others sat at the tables, admiring the new Formica furniture, relaxing by putting their feet up on the tables. The Catholic priest, 'Father', from this village turned up besmocked on his motorcycle. In the centre of the room in front of the podium appeared to be more NPCIL workers, one of whom was the Associate Director. In front of the staff were stacks of NPCIL publicity material, the likes of which we read about in Chapter 4.

Media reporters were already in the room, one of them bearing the logo of Sun News. However, there were also others with cameras shooting both still and moving footage. They could have been private users but more likely, plain-clothes officers from the Criminal

Investigation Department (CID) or the Intelligence Bureau (IB), a hunch that proved to be true as later, at a moment of high commotion later, one of the police officers grabbed a video camera from one of them, and stood on the podium filming those who were shouting the loudest—that is, the rabble-rousers, the ring leaders, the trouble-makers. The officers were taking pictures without even looking to see where they were pointing just to make sure everybody's faces were recorded for surveillance.

Even though the authorities were hoping that nobody would get to know about the public hearing, thereby limiting the number of attendees, by the time of the proceedings, an impressive number of people had made it to the venue. There was already quite a lot of noise in the room, but this was exceeded by a commotion outside as another white Ambassador car pulled up and out walked four men led by a short man with a gold headband and cummerbund around his portly waist, the District Collector's peon. Fashionably late by fifteen minutes, the peon led the group to the front of the hall, straight up to the table on the podium. The men sat at the podium included the District Collector, the Environmental Pollution Engineer from the Tamil Nadu District Pollution Control Board, and two other personnel from the top ranks of the NPCIL.

The NPCIL representatives spoke in English with an interpreter translating into Tamil. One began by saying: 'We the Commission want to start with a presentation'. In what by now had become familiar anodyne displays, the first PowerPoint slide showed a chart about the uses of nuclear energy. In the centre was a circle with nuclear energy, and around it other circles with labels of water, fluid fuel substitute, electricity, industry, technology, national security, and a couple of others outside of my view. As soon as one man began to talk, a woman from the back of the room began to start shouting in Tamil, 'Go, go, we don't want it'. Another added, 'We didn't come here to hear you talk. This is a public hearing. You should hear us talk'. There was exasperation and impatience amongst the audience with such Public Relations talk. Other women joined in, and one of the men began chanting slogans in a mixture of Tamil and English. '*Vendam, vendam Kudankulam vendam*' (No, no, Kudankulam, no') they exhorted. It was impromptu and not even expected by the cheerleaders themselves who picked up the momentum as more people joined in. '*NPCIL thirumbi poo'* (NPCIL go back'). Even Russia got a look-in—*'Russia thirumbi poo'*. Attendants cheered the slogans on with more and more inventive sayings created on the spot. One chant triggered off another and soon the whole room was full of an antiphonal chorus. *'Vendam, vendam. Cancer vendam'* ('We don't want cancer'); *'Vendam, vendam, khadirvichu vendam'* ('No to radiation'); *'Edukathe edukathe Pechiparai thaneerai edukathe'* ('Don't take Pechiparai water'); and most poignantly, *'Kollathe, kollathe, kuzhandhaikalai kollathe'* ('Don't kill our children'). A release of anger and frustration was energized by the elation of collective call and response.

The men on the podium were almost dumbstruck. They tried to calm the attendants, but there was no stopping the public. The authorities seemed a little panicked by their number and jubilant intensity. Following the lead of the press and Intelligence officers taking photographs, some of the people stood on tables to get a better view. Other men around them, presumably plainclothes police men or men from the IB started looking at each other, walked out, and made a few phone calls. Feeling swamped, they called for more police, and later two uniformed police officers came marching in. They were looking towards the left where most of the people from Nagercoil were chanting, including Anthony and a representative from the Kanyakumari Fishworkers' Union, the ones who were most energetic in leading the slogans. Anthony had come to the hearing separately in a minibus with his own coterie of fishermen and farmers.

Arms were held aloft in the air, as protesters were enraged at the prospects of their lives being irreversibly affected by the nuclear power plant (Figure 6.1). Some people in the audience were smiling, enjoying this brief moment of conquering what they thought of as their formidable enemy. Considering the short notice, they seemed pleased that there was a big turnout at the hearing. Amongst the charge of emotional energy, people were advancing to the table with their petitions, either written or stated. Some attendants protested that the EIA report stated that water from the Pechiparai Dam would be used for the plant. The NPCIL representatives argued that it was not the case, that it would not be used, and that the EIA report is outdated. 'So why are you using an outdated report for a public hearing?' they asked angrily.[27] Another complainant went to the Collector, furious that they had not organized a hearing in Kanyakumari, the district

Figure 6.1   Public Hearing for Further Reactors at the Kudankulam Nuclear Power Plant, Tirunelveli, 2006.
Source: Raminder Kaur.

which was most likely to be affected due to the water issue? He almost scolded the nuclear representatives, holding aloft an article that they had produced, adding: 'How can you say nuclear energy gives you 100% protection? How can man create anything that's 100% perfect?'

The most vocal women were from the village next to the V.V. Minerals plant, Kuttankuzhi, a few miles to the west of the Kudankulam Nuclear Power Plant. For the last decade, V.V. Minerals, locally known as the 'sand mafia' had been sand mining for minerals such as garnet and ilmenite in the monazite-rich sands, that is also a container of the alpha emitter thorium. Health problems were already manifesting themselves amongst the nearby villages. The residents, mainly fishing communities, observed that now they were getting 'children with no arms or hands and other physical handicaps', something that they had never seen there before. One woman shouted that all the children in her family were born disabled. They had never had this happened to them before the sand mining industry began. She grabbed her marriage chain (*thali kodi* or *mangalsutra*) and hammered her chest with her hands: 'You are killing my family!'

The Member of Legislative Assembly for Radhapuram, M. Appavu (then with the Dravida Munnetra Kazhagam), got up from the audience and spoke to try and pacify the people. He was essentially supportive of the nuclear plant, the immediate suspicion being that he had been paid off to do so. However, people jeered at him, one shouting, 'You never say anything in parliament. All you do is stand up and sit down. Even I can do that! What use are you?' Appavu's efforts to calm people came to nothing and he sat down.

The sharp-nosed CID and IB officers, although a tad perturbed, walked confidently about trying to amass as much information as possible on anyone they marked as the rabble-rousers. Even though in plain clothing, their stocky height made the men stand out from the crowd. They made occasional sidelong comments to police officers. Together, they could come down on the audience like a ton of bricks at any moment. There was another khaki-clad police officer looking in through a window from a side room but he just watched. They seemed well-instructed not to create a stir, for that would be to generate more media attention for the protesters. Furthermore, as there were many women and children in the room, it would not bode well to barge in there in order to contain the crowd. Later, after several phone calls, another police car pulled up outside with four officers followed by two large riot vans. Eventually, there were about eighteen uniformed officers to add to the plainclothes officers who surveyed and recorded but did not make any attempts at arrest. Instead, the riot officers nonchalantly stood outside the door repeatedly banging their five-foot batons on the ground in anticipatory agitation.

Meanwhile, in a hopeless bid to try and impress, NPCIL booklets began to be distributed amongst the attendants. One was a glossy colour booklet, another a black and white one, both in Tamil. The first booklet proclaimed the NPCIL Mission: 'to develop nuclear power technology; to produce nuclear power in a safe, environmentally benign and economical manner and to provide expert assistance to the power and allied sector'. The colour booklet showed the workings of a reactor and how it would provide power to the people. It included images of other nuclear power plants claiming that those like the Kakrapar, Narora, and Kaiga atomic power stations had won the Ministry of Power's Gold Shield for Outstanding Performance, and the plant at

Tarapur was awarded the Silver Shield and a Certificate even though the plants have been subject to accidents and oversights.[28]

Initially, there was a bit of a scramble for the booklets. However, later after flicking through them, some people had torn them up in disgust and left the shreds scattered on the floor. This was not the kind of power they seeked. They could see the leaflets for what they portended: self-aggrandized publicity for nuclear temples of modernity in which they had little part to play other than as sacrifice.[29]

Eventually, the District Collector was compelled to bring the hearing to a close and promise that further hearings would be held in all three neighbouring districts with adequate publicity in the locally available Tamil newspapers as per peoples' requests. Then along with the NPCIL, the four men on the podium ventured towards their awaiting car. After the hearing, people hung about outside, talking and networking (Figure 6.2). An active contingent arranged to organize another meeting a fortnight later in Tirunelveli as it was a central location for all three districts.[30] The public then began to disperse with a nervous satisfaction that a small but important battle was won.

Figure 6.2   Outside the Tirunelveli District Collector's Office, Tamil Nadu, 2006.
Source: Raminder Kaur.

The riot police also left in their vans, one with a fur-covered plastic monkey feverishly swinging away on the windscreen.

The CID and IB officers remained and milled about outside. They knew what their objectives were and wasted no time in trying to gather more information from the lingering crowds. They were asking people for names of people and villages. There was another group of people, a couple of whom were wearing the signature NPCIL uniform of a beige shirt and brown trousers. A few were speaking in Hindi to each other. They registered the mention of Nagercoil and some names of leaders with a familiar chord of recognition. They were eager to find out who else had attended the hearing, and most importantly, who had organized the protest.

One tall moustachioed man approached Manick and asked for his name. He replied by saying that he was a press representative, and did not offer any further information. The man let it go and Manick careered away in an auto-rickshaw. Slightly dismayed, Manick later commented:

> I wanted the police to arrest someone. If they had arrested someone, then there would be more news coverage and there would be more anger and a stronger movement could be generated.

More publicity, preferably sympathetic, was the oxygen for their struggle against the nuclear plant.

Through the grapevine, it was learnt that the NPCIL had an impromptu meeting shortly after the public hearing for a debriefing. The officials were surprised at the public reaction. They had thought that protest had died down a few years ago when the first reactors were inaugurated in 2002. They did not know where this resurgence had come from and were knocked off their stride. They asked: 'Who funded this protest?' They complained to the District Collector saying: 'There were people from the CIA there!' referring to the US Central Intelligence Agency. As had become almost predictable practice, they blamed public protest on external infiltration. They could not accept that it could have arisen as a result of genuine grievances on the part of the attendants.

Sunil, who related this account to me, quipped: 'When it comes to dealing with the Americans for the Indo-US deal, it's OK. When something like this happens then they blame it on US espionage'. He continued to recount what else had happened:

After the meeting the guys from NPCIL came along and started talking with us in a pally way. Why do you have to do all this? What's your problem? This man has been working in nuclear power plants for fifteen years and *he's* well, his children are perfectly healthy. He has been to the *core* of the reactor. I have been working in them for 35 years and nothing's happened to me. Why can't you be democratic and fair?

To which Sunil asked: 'Are you democratic and fair?' Their edgy smile answered that question. They asked Sunil which organization he was from? He replied that he was from no organization. On top of instigators and infiltrators, the NPCIL officials allocated blame on organizations. It was further testimony to their distance to local lives, and their lack of understanding of the nature of right-to-lives movements.

Then the NPCIL associate director advanced towards the small group. Sunil noted that suddenly everyone became very obsequious. He observed: 'Yes sir, no sir. Their whole body language had changed.' The NPCIL official was seething and could not look Sunil in the eye. Obviously, he had to answer to his colleagues about why the public hearing did not go as planned. Still he shook hands with Sunil in a civil manner, and asked for his card. Sunil lied that he did not have one. By that time his friend had left him and Sunil was alone surrounded by a crowd of 'nuke people'. He felt uncomfortable and decided to make his excuses and leave. 'To hang around with these guys would not be a good idea,' he explained, and asked an auto-rickshaw driver to take a route that did not involve him going past this political hub of toxicity.

Clearly, the nuclear authorities were humiliated at the hearing: 'They were caught with their pants down', as Sunil put it. The public hearing was organized by the NPCIL with the support of local officials, but played into the hands of communities as the event provided a boost to awareness and networking in the south Indian region. While the authorities tried frantically to gather as much information as possible about the organizers and attendants, the participants exchanged experiences, information, and telephone numbers in order to increase the momentum of their anti-nuclear machinery. It was a significant achievement, and however tenuous, appreciated for the fact that 'noise was created' over and against the nuclear plant, noise that was widely circulated in the local and statewide media the next day.

The Days After

The following morning, Anthony reflected about related issues and what the next step may be. He discussed how the Ground Water (Regulation and Control of Development and Management) Act (2005) stipulates that you cannot sell water dug out from a well. However, this did not stop landlords from selling water to the nuclear power plant officials:

> But this is what they are doing in [a town close to the Kudankulam plant] selling tank loads of water for construction of the plant. If this was stopped, they would not be able to build anymore. They may have to turn to the dam. It could become a big problem and the army may be bought in. We could come under national security. This place could become like Assam [a north eastern border state overrun by the Indian military].

He talked about how peaceful this area was otherwise, and how easily it could turn into a bed of state oppression and heavy-handedness, on the one hand, and violence and disorder, on the other. Anthony added:

> Even places like Chattisgarh, Jadugoda, Assam [where authorities, police and/or the (para)military have placed repressive and punishing measures on peoples' lives] were peaceful places at first. It's due to government policies on people's lives, which leads to a volatile situation. But they [the state] fail to recognise this, and the media goes along with it, reporting these protesters as violent and anti-national. They do not look at how the problem was created.

He seemed to portend the post-2011 scenario when the (para) military descended on the region to subdue the intensifying protests. Nevertheless, Anthony was cautiously pleased about the proceedings at the District Collector's office and mentioned that the authorities will now have to organize another one, but most likely in a different fashion. He had heard that, by comparison, a public hearing for the prototype fast breeder reactor at Kalpakkam was relatively subdued.[31] Even so, their views were overturned as if the difference between sanctioned and unruly politics did not seem to matter:[32]

> Everyone gave their petition, the Collector passed them on to the government, but they still went ahead with the project and did nothing about those displaced. Fishermen were left to rot. No compensation. They don't care about the fishermen. As far as they're concerned, they don't exist. As it is, they don't eat fish so think they're dispensable.

Pointing to a neo-Brahmanic conspiracy against fish-loving communities, he stressed that whatever their powers, he did not fear them: 'I'm not worried about that [being interrogated and intimidated by the IB]. We're not being anti-national. Just anti-nuclear', he replied confidently: 'It's our right! Who will talk for the common man?' Even though they might lose the political fight, he saw that their moral authority lay with local people as rightful custodians of the nation. The question then became who represents the nation when the terms of the debate are so skewered against the people in favour of the nuclear state.

Anthony also recalled how the 'ladies were really opposing the plant. They were talking about babies being born without legs, arms and were very emotional'. He wanted to learn more about these women from Kuttankuzhi, as well as contact the Catholic Father and help spread news around other villages. It was felt that, however dispersed and disorganized, the fishing community could come together and present a strong mass against the power plant. Others could join them. To build upon the momentum gained was imperative, but as he confided, whatever they planned, they had to tread carefully in view of V.V. Minerals supporters in the area.

Later that day, a CID agent had phoned up Anthony to say that they had written a report about him the previous day. Anthony was a little ruffled but laughed it off, as he mentioned that he had several friends in the CID—local familiarity and chords of connection with officials warded off any paranoia about a draconian administration. On this occasion, as Achille Mbembe observes, it is as if 'the logic of "conviviality", on the dynamics of domesticity and familiarity ... inscribe the dominant and the dominated within the same episteme'.[33]

Meanwhile, the CID had rang up Sunil as well to tell him that they had written a report on him. By comparison, he demonstrated a logic of cautionary conviviality:

> It was like a polite warning.... When a meeting or campaign is successful, people get very jubilant. But it's also cause for concern. Because it means Intelligence steps up its surveillance and intimidation. They want to work out what your next move is and stop it at every stage.... What they were hoping to do is that hardly anybody would come, they would write down the complaints, send it to the government, done and dusted and then they would tell everyone that they did the public hearing and everything is fine and go ahead with the next four reactors.

There was a subtle hint of fear in people's discussions and a sense of feeling slightly cornered. The invisible force of Intelligence hovered around like a dangerous dose of radiation. The fact that the 'people' had gotten the better of 'them'—that is, the nuclear state—could almost be a Pyrrhic victory. He continued:

> People don't have any faith in the political or judicial system. We can only rely on people's protest but even that can fail people. It is a double bind—if it fails, then it's of no value. But if it's too successful, they will come along and crush it ... see what happened to the Narmada issue. It was such a strong movement and it got nowhere.

The people's movement, Narmada Bachao Andolan, was a phenomenal campaign to ensure people displaced in the construction of the dam in Gujarat, Madhya Pradesh, and Maharashtra were properly rehabilitated and compensated. Sunil also noted how terrorism ultimately worked in the government's favour: 'In the name of fighting terrorism, they are becoming more and more anti-democratic'. As with other countries across the globe, India was no stranger to attacks of terror. It led to the dilution of democratic laws and procedures along with the strengthening of surveillance and policing, that fist-in-hand, can be thrust at anyone who posed a threat, terrorist or otherwise. Directly and indirectly, terrorism begets state terrorism between which peaceful movements stand to get crushed.

When reflecting overall on the public hearing, Anthony summarized: 'The results were good', and, 'It was a good meeting. It showed that there was very strong opposition. It showed a way. We are very happy. It instilled confidence'. For Sunil, however, the view was less optimistic: 'There are many more miles to go before we sleep', and added, 'They did it to appear democratic. Or maybe they wanted to identify trouble-makers and pacify or intimidate people'. Sunil's reflections recall Giorgio Agamben's point about the double-edged sword of rights discourse—both an aid in that it furthers a cause, and a hindrance in that it remains trapped in the logics of a polity that favours the status quo. However, this dual dynamic was also accompanied by a state of constructive ambiguity or ready-and-waiting potentiality.[34] The nuclear power plant had led to a situation that Sunil was not prepared to leave unchallenged even if this meant not doing anything until the opportunity permits.

Equivocal Victories

In neo-liberal India where development projects are greeted with golden handshakes with corporate powers, bureaucrats, and politicians alike, the task to resist on grounds of environmental contamination and disruption and harm to local lives appears of Himalayan proportions. The nuclear state is the Everest among them all. It is one of the extreme expressions of a virulent and secretive science–state collusion in the professed interests of national development, sovereignty and security. However, with the first ever notice of a public hearing on the Kudankulam plant came what appeared to be a small window of opportunity with which people entertained thoughts that perhaps the nuclear authorities would listen to them after all.

In the face of peoples' appeals, the NPCIL was lost for words for an afternoon at least. However, the public were always in a precarious place. Jubilation is short-lived when it comes with a sinking feeling that this taste of triumph can only ever be transient. Everyone was fully aware of the powers of the nuclear state, and the way they could twist the outcomes and arms of other authorities to their advantage. They could dismiss the hearing as simply one attended by 'unruly masses'. They could try and organize another hearing in a controlled environment where attendees were either invited or selected beforehand in their public orchestrations. They could identify the ringleaders at this public hearing and further intimidate them. Or they could take the problems up and try and do something about it with pacificatory antidotes—'warfare as welfare'—that is, by constructing a couple of schools and hospitals, provide 'community service' for nearby villages and offer other incentives, another feature of the benev(i)olent state so as it can be left alone to continue with its plans.[35]

At the hearing, attendants expressed their opinions against the prospect of ecological hazards, social injustice, and loss of livelihoods, constantly moving their arguments sideways and onwards. Their presence at the hearing cannot be taken simplistically, and there is more to elaborate in subsequent chapters. What is clear is that there were several kinds of actions and reactions in this space of criticality. Fishing women who almost scolded the officials subscribed to the umbrella of direct confrontation. The activities of attendants such as Anthony and Krishna suggest the significance of 'permanent provocation' where they are subject to the capillaries of disciplinary power but also become its productive re-creators, to draw upon the work the political philosopher Michel Foucault.[36] Accusations of 'anti-national' were reworked to form pro-nation in terms of working for the health of its people and environment. Those akin to Sunil's mood, in comparison, subscribed more to the tune of what Agamben has called 'the greatness—and also the abyss—of human potentiality', this being evident in Sunil's interests in attending the hearing and critiquing the operations of the nuclear state, but preferring to refrain from demonstrable action and remain in the background whilst other future possibilities were worked out—an excess that did not register in display at the time.[37] Indeed, in view of colonial continuities and the machinations of state surveillance, the appearance of non-activity was a distinct tactical advantage. These are just a few of several interventions that require constant and creative mitoses in the encounter with the nuclear state, more of which we will see in Chapter 8.

As it transpired, the NPCIL did not in fact fulfil its promise to have another hearing with sufficient notice and publicity in all districts likely to be affected by the nuclear project. Instead, a couple of months later after a number of postponements, a hearing was held, supposedly public, in the heavily guarded confines of the nuclear power plant township, Anuvijay, for which passes were strictly needed for entry. Even though local communities protested, the nuclear state went ahead with the 'public hearing' to vindicate the victory of the atom. The officials recorded, logged and filed the hearing, and this time, left a 'paper trail' on its mandatory execution of a public hearing—quite literally, another act of control especially against the not so literate.[38] In January 2007, it was announced in the press that the construction of four more reactors would go ahead.[39] Still, the hearing and announcement was seen as controversial and further announcements were made for another public hearing that were postponed due to unrest in the region.[40] One more hearing was held in June 2007 for the clearance of, this time, two more reactors. Thousands attended from the three districts with about 800 admitted into the hall among a phalanx of heavy security that outnumbered the number of attendants. As Ramana recounts of his attendance, there were '1200 policemen in riot gear, tight security cordons and armoured personnel carriers'.[41] After the hearing was abruptly ended by an official with a statement that all the petitions have been 'answered', the public disbanded. The minutes of the public hearing obtained through a subsequent appeal through the Right to Information Act, acknowledged only two of the multiple petitions submitted on the day.[42]

As in several other cases of large-scale developments, the public hearing was perceived as a fig leaf for democracy, a mere chimera of civility. S.P. Udayakumar's comments in his play, *Anushakthi Amma (Atomic Mother*, 2004), could not be further from the truth:

AERB: .... And, of course, you held a Public Hearing where people could express their concerns after reading the EIA?

Dr. D.A.E. Das: Er ... hmm ... we ... ah ... spoke to gullible farmers about bringing water to their villages, gave innocent fishermen computers to check the prices of the fish they will never be able to catch, brainwashed college students about non-existent job opportunities at the plant, they all accepted what we said ... nobody objected.

In addition to 'nothing happened' was another negative reality—'nobody objected'. Behind the satire, what remains at issue are the strong-arm measures employed by the state to ram development projects down the throats of people. Even though on paper at least, they seek to proceed with the matter in a diplomatic manner, and citizens, particularly project-affected people, have a say in any large-scale development project, in reality, diversion, dishonesty, surveillance, and intimidating policing are used to ensure decisions are made in favour of the nuclear state and those it favours. This only fuelled the fire. Despite the fact that those opposing the plant have been subjected to penalty and punishment, intimidation and surveillance, judiciary dismissals and extra-judiciary threats, people's power went on to grow.

Notes

1. Brown, Wendy. 'Finding the Man in the State'. *Feminist Studies* 8, no. 1 (1992): 7–34, p. 12.
2. On other analyses of public hearings, see Hugh Gusterson who describes one held in the USA as 'rituals of assent'—events in which laboratory scientists 'paternalistically displayed their superior knowledge in the expectation that the local population would then assent to their plans'. Brian Wynne makes a similar observation with respect to the rationalist discourse of judicial procedures in the Windscale Inquiry in Britain, a public hearing about the future of nuclear fuel reprocessing where the terms of the debate along with routinized rituals were pitched against a genuine dialogue with members of the public. Dipesh Chakrabarty elaborates on a public hearing more broadly as a 'ritual of humiliating officialdom', a characteristic trait of protest in India's postcolonial democracy. My account provides yet another perspective: a processual and contextual examination as to the articulation of power relations in a space of criticality that goes beyond the orchestrated assumptions of a ritual. Gusterson, Hugh. *People of the Bomb: Portraits of America's Nuclear Complex*. Minneapolis: University of Minnesota Press, 2004, p. 213; Wynne, Brian. *Rationality and Ritual: Participation and Exclusion in Nuclear Decision-Making*. Abingdon, Oxon: EarthScan; Chakrabarty, Dipesh. '"In the Name of Politics": Democracy and the Power of the Multitude in India'. *Public Culture* 19, no. 1 (2007): 35–57, p. 52. A more theoretical argument is developed in Kaur, Raminder. 'Sovereignty without Hegemony, the Nuclear State and a "Secret Public Hearing" in India'. *Theory, Culture and Society* 20, no. 3 (2013): 3–28.
3. See Jungk, Robert. *The Nuclear State*, trans. Eric Mosbacher. London: John Calder, 1979.
4. Abraham, Itty. *The Making of the Indian Atomic Bomb: Science, Secrecy and the Postcolonial State*. London: Zed Books, 1998, p. 42.
5. Abraham, *The Making of the Indian Atomic Bomb*, p. 56.

6. Abraham, *The Making of the Indian Atomic Bomb*, p. 13.
7. Mian, Zia 'Nuclear Nationalism'. *Nuclear Age Peace Foundation*,1999. Available on https://www.wagingpeace.org/nuclear-nationalism/, accessed 9 October 2019. See also Bidwai, Praful and Achin Vanaik. *South Asia on a Short Fuse: Nuclear Politics and the Future of Global Disarmament.* New Delhi: Oxford University Press, 1999.
8. Foucault, Michel. *Madness and Civilization: A History of Insanity in the Age of Reason.* New York: Vintage Books, 1998, pp. 44–6.
9. See Serres, Michel. *Conversations on Science, Culture, and Time, Michel Serres with Bruno Latour*, trans. Roxanne Lapidus. Ann Arbour: University of Michigan Press, 1995, p. 137.
10. Jungk, *The Nuclear State*, p. vii.
11. Hecht, Gabrielle. *Being Nuclear: Africans and the Global Nuclear Trade.* Massachusetts: MIT Press, 2012, p. 4.
12. Jungk, *The Nuclear State*, p. 135.
13. Berlant, Lauren. *Cruel Optimism.* Durham: Duke University Press, 2011, p. 9.
14. See Chatterjee, Partha. *The Politics of the Governed: Reflections on Popular Politics in Most of the World.* New York: Columbia University Press, 2004.
15. See Banerjee, Mukulika. *Why India Votes? Exploring the Political in South Asia.* New Delhi: Routledge, 2014.
16. Hardiman, David. *Gandhi: In His Time and Ours.* New Delhi: Permanent Black, 2003, p. 198.
17. Abraham, *The Making of the Indian Atomic Bomb*.
18. Gupta, Akhil. *Postcolonial Developments: Agriculture in the Making of Modern India.* Durham: Duke University Press, 1998.
19. Ramana, M.V. 'La Trahison des Clercs: Scientists and India's Nuclear Bomb'. In *Prisoners of the Nuclear Dream*, edited by M.V. Ramana and C. Rammanohar Reddy. New Delhi Orient Longman, 2003, p. 211.
20. Abraham, *The Making of the Indian Atomic Bomb*, p. 29.
21. Chatterjee, *The Politics of the Governed*, p. 38.
22. Alavi, Hamza. 'The Structure of Peripheral Capitalism'. In *Introduction to the Sociology of 'Developing' Societies'*, edited by Hamza Alavi and Teodor Shanin. New York: Monthly Review Press, 1982.
23. Alavi, Hamza. 'State and Class under Peripheral Capitalism'.
24. Perkovich, George. *India's Nuclear Bomb: The Impact on Global Proliferation.* Berkeley: University of California Press, 2001, p. 9.
25. Udayakumar, S.P. 'India's Department of Atomic Energy: Fifty Years of Profligacy'. *South Asians Against Nukes*, 16 July 2004. Available on www.s-asians-against-nukes.org/PMANE/spuk16jul2004, accessed 12 October 2019.

26. 'Enquiry', *Dinamalar*, 5 October 2006, p. 1.

27. Indeed, it is contended that the NPCIL 'constructed the KKNPP campus in violation of the terms and conditions laid down by the AERB [Atomic Energy Regulatory Board]'. Padmanabhan, V.T., R. Ramesh, and V. Pugazhendi. 'Water Balance Sheet of Kudankulam Nuclear Power Plant (KKNPP)'. *DiaNuke*. Available on http://www.dianuke.org/water-balance-sheet-of-koodankulam-nuclear-power-plants-kknpp/, accessed 21 January 2019.

28. See Udayakumar, S.P. (ed.). *The Koodankulam Handbook*. Nagercoil: Transcend South Asia, 2004. On an overview of the lack of adequate safety review, see Ramana, M.V. and Ashwin Kumar. '"One in Infinity": Failing to Learn from Accidents and Implications for Nuclear Safety in India'. *Journal of Risk Research* 17, no. 1 (2014): 23–42.

29. See Kaur, Raminder. 'The Many Lives of Nuclear Monuments in India'. *South Asian Studies* 29, no. 1 (2013b): 131–46. On cathedral- or chateaux-like citadels of modernity that project the 'radiance of France', see Hecht, Gabrielle. *The Radiance of France: Nuclear Power and National Identity*. Massachusetts: MIT Press, 2009. On national sacrifice, see Abraham, Itty. 'Geopolitics and Biopolitics in India's High Natural Background Radiation Zone'. *Science, Technology and Society* 17, no. 1 (2012): 105–22.

30. Such strategies led to the formation of the Peoples Right Protection Movement that was to merge with People's Movement against Nuclear Energy in 2011. See Prabhu, Napthalin. 'Protest Camp as Repertoire for Antinuclear Protest', forthcoming, 2019.

31. Ramana, M.V. *The Power of Promise: Examining Nuclear Energy in India*. New Delhi: Penguin Books, 2013, p. 60.

32. Shankland, Alex, Hani Morsi, Naomi Hossain, Katy Oswald, Mariz Tadros, Patta Scott-Villiers, and Tessa Leuwin. 'Unruly Politics'. Brighton: Institute of Development Studies, 2011. Available on https://www.ids.ac.uk/idsresearch/unruly-politics, accessed 21 November 2016.

33. Mbembe, Achille. 'Provisional Notes on the Postcolony'. *Africa: Journal of the International African Institute* 62, no. 1 (1992): 3–37, p. 9.

34. See Agamben, Giorgio. *Potentialities: Collected Essays in Philosophy*. Stanford: Stanford University Press, 1999, p. 181. This is a different proposal to Craig Jeffrey's observations on lower middle class youth who are 'people in wait'...'to acquire new skills, fashion new cultural styles and mobilize politically'. Jeffrey, Craig. *Timepass: Youth, Class and the Politics of Waiting in India*. Stanford: Stanford University Press, 2010, pp. 2, 4.

35. See Dunlap, Alexander and James Fairhead. 'The Militarisation and Marketisation of Nature: An Alternative Lens to "Climate-Conflict"'. *Geopolitics* 19, no. 4 (2014): 937–61.

36. Foucault, Michel. 'The Subject and Power'. *Critical Inquiry* 8, no. 4 (1982): 777–95.

37. Agamben, *Potentialities*, p. 181.

38. See Sharma, Aradhana and Akhil Gupta. 'Introduction: Rethinking Theories of the State in an Age of Globalization'. *Anthropology of the State: A Reader,* edited by Aradhana Sharma and Akhil Gupta. Oxford: Blackwell, 2006, pp. 13–14.

39. Baruah, Amit. '4 More Reactors for Koodankulam'. *The Hindu,* 26 January 2007. Available on http://www.hindu.com/2007/01/26/stories/2007012616170100.htm, accessed 21 November 2016.

40. Srikant, Patibandla. *Koodankulam Anti-Nuclear Movement: A Struggle for Alternative Development?* Working Paper 232, Bangalore: Institute for Social and Economic Change, 2009, pp. 9–10.

41. Ramana, *The Power of Promise*, p. 91.

42. Ramana, M.V. and Divya Badami Rao. 'Violating Letter and Spirit: Environmental Clearances for Koodankulam Reactors'. *Economic and Political Weekly* 43, no. 51 (2008): 14–18.

7

Discipline and Deviance

Still it's a devious knave.... Too wily to be tethered to a solid noun, the conundrums of cancer match its craftiness. (Lochlann S. Jain, 2013)[1]

Capricious elements to do with the ecology, the body, and state machinations enveloped peninsular India. While large amounts of research funds are committed to nuclear research, only the minimal amount is channelled to studying the industries' effects on public health and the environment. As the physicists M.V. Ramana and Surendra Gadekar maintain, ' ... the nuclear establishment has a tremendous influence over universities ... who may be starved of research funds' should they want to commit time and energy to doing such research.[2] More widely, vested interests lie in quashing any health anxieties, especially around cancer afflictions, even within the confines of nuclear industries and townships.[3] Common hearsay is that there is a high incident of health problems among staff and families of nuclear power plants, a view that is upheld by studies from other regions as with the analysis of over 400,000 workers in 15 countries by a group led by the radiation epidemiologist Elizabeth Cardis. They conclude: 'an excess risk of cancer exists, albeit small, even at the low doses and dose rates typically received by nuclear workers in this study'.[4] However, among nuclear state elites, such acknowledgements, where they exist, remain for in-house articulation with the promise of a comprehensive health system that will cater to any health concerns. As we saw in Chapter 4, officials are fully aware of the dangers and the risks that could lie ahead, but they present and even come to believe that they are all fully governable and remediable. It can and does behave well, '99.9 per cent' even '100 per cent' of the time even, as one IREL official maintained in Manavalakuruchi. The benefits of personal and family security, along with more electricity, development, possibly more bombs, and political clout for the country overturned all other considerations about possible failures in the health of the techno-human system.

It is a similar case elsewhere. Lochlann S. Jain reports on the USA:

> The most powerful culture-makers—industry, government, medicine—have been slow to recognise, let alone advocate for research into the connections between environmental toxins and the disease. If fear results in part from the unknown, then keeping things unknowable contributes to the circulation of an unattributable anxiety.[5]

Any anxiety about radiation had to be kept as precisely that—an unattributable anxiety. This concern could neither be rendered reasonable nor bulwarked by research and reports that could elevate the apprehension to anything approaching grounds for evidence. As goes the argument in Chapter 4, ignorance had to be maintained if not actively encouraged.

Keeping things unknowable may be the modus operandi for the nuclear authorities but it was not good enough for many of my interlocutors. The need to navigate and master the unknown was compelling. The drive to sanction alternative perspectives was a pressing one. The effort to produce a reliable evidence base to challenge the assumptions of state-backed authorities regarding public health and the environment was premium. They wanted to collate data for themselves.

Such motivations led to the emergence and engagement of 'citizen scientists' committed to finding out more about the impact of the nuclear plant. This was evident with the PMANE expert committee and their report on the safety, feasibility, and alternatives to the Kudankulam Nuclear Power Plant completed in 2011.[6] Here, independent scientists among other professionals compiled a dossier of scientific, environmental, and statistical facts and figures on the nuclear power plant and its surroundings so as it could become an accessible and alternative source of information to the inadequate and piecemeal information supplied by government-appointed committees and organizations.

Preceding such initiatives was an 'indigenous survey' on health and environmental radiation, a statistical methodology that was pursued by local residents in 2006. They sought to resist toxic industrial developments through a knowledge strategy that emphasized systematic procedure rather than cultural interpretations or impressions of health problems in a region of high background radiation.[7] The strategy signalled a move to embolden the space of criticality with supportive facts. They wanted to represent anti-nuclear campaigners as reasonable and rational, not child-like, emotional, and primitive as was the nuclear state's routine aspersion of them.[8] In this chapter, we explore the territory of knowledge acquisition by beginning with a discussion on the tenacious hold of statistics, before accounting for grassroots plans for a survey of radiation levels and health problems in villages in the two districts of Kanyakumari and Tirunelveli.

To Know or Not to Know

Overwhelmingly, statistics have become an instrumental part of first apprehending and then controlling the world. They play an integral part in governmentality—described by Michel Foucault as capillary connections between 'the governance of the self' and 'government of the state' to describe political sovereignty over a territory and its population.[9] Describing this phenomenon as correlated with a 'statistical gaze', the historian Gyan Prakash notes how in the colonial Indian context, officials tried to spread the 'net of statistics' as widely as possible in order to collect as much information as possible about the populace so as to regulate and control.[10] The statistical gaze continued into the modern nation-state, where science was authorized with a special status to lift postcolonial India out of its economically depressed condition into one where high levels of economic growth, standards of living, employment, and the manufacture of goods and services could be realized. Its administration had to continue with statistics through census, surveys, and their influence on developmental programmes.[11]

Despite state ideals for a totalizing mesh of governance, the survey protagonists that I met in south India felt that statistics provided by government and related authorities were unreliable if at all available. Even with their authoritative address they were seen to be manipulated, and at other times, merely incomplete or incompetent. Equally difficult were diachronic comparisons of radiation-related illnesses such as data on cancer.[12] By virtue of the available records of deaths in the district, certain diseases were, as one health expert put it, 'not notifiable because they are not communicable. Cancer is included here. So records are poor'. He informed that pre-1950s records on cancer statistics as a base line to compare later data after the sand mining companies had been established in three villages on the Malabar coastline straddling the states of Tamil Nadu and Kerala—Aluva, Chavara, and Manavalakuruchi. These records, however, lay only with the DAE, to which the public had little access.[13]

Statistics then were less a part of a *discursive regime* that regulated them, and more a *discursive web* replete with its gaping holes and flimsy, gossamer tendrils that my interlocutors sought to straighten out themselves. Paralleling anthropological observations on the constitutive split of the sublime and profane aspects of the state by Thomas Blom Hansen, the sublime imaginaries of statistics wrestled with the actuality of profane practice.[14] As the public received information sporadically, not coherently or systematically, it meant that any effort to campaign for better health and safety conditions and any related compensation could not proceed far as it would be based on speculation, rumour, and insubstantial data. Concerned residents felt that they had to operationalize the logic and authority of science and statistics in order to prove that there were compelling links between radiation and health problems in the region, a relation that was consistently denied by the authorities.[15]

Even when it was understood that there were several problems with surveys on public health, why would statistics continue to have such sublime resonance? Recalling Weberian analysis of the rational bureaucrat functionary, statistics as part of 'scientific rationalities' are both abstracted and yet contingent on politicized contexts.[16] They are both universal and particular, rarefied as above politics but in effect exercising 'politics by other means', to draw on ethnographic insights by Simone Abram.[17] Here, we have an example of members of local communities adopting scientific rationalities otherwise associated with state or institutional structures; first, where they subscribe to an objective discourse that seemingly transcends self-interest; and second, in a bid for self-empowerment and therefore a consequence of subjective rationales. The anthropologist Eeva Berglund observes in another case study: 'the less obviously political they are and the more "commonsensical" they appear, "facts" ... seem to speak for themselves, providing the grounds for objectivity'.[18] 'Scientific' was seen as an authoritative word that implied hard data, neutrality, rigour, and fact, in contradistinction to the soft underbelly of rumour or impression.

For these dual purposes—of interest and interest-independence—statistics were part of a powerful discourse such that few were immune to its seductive grip in understanding and acting in the world.[19] They enabled a means of managing the uncertainties and risks of high levels of radiation in the locality. Moreover, they enabled people to imagine themselves negotiating with the nuclear authorities. Such an initiative could be argued as another instance of 'returning the gaze' or as the anthropologist Martin Webb puts it, 'disciplining the state' enabled by recent legislation, the Right to Information Act (RTI Act, 2005).[20] With this law, applications to state agencies can be made for information in the public interest of greater transparency and accountability. However, the nuclear authorities have placed themselves outside the act's remit due to their national security prerogative.[21]

The overriding consensus amongst local residents was that surveys had not to be done *on them, but for or by them*—that is, the tactical collation of statistics and 'facts' that they could apply to their own agendas. Statistics collated by local researchers rather than through the parachuting of state officials into the region was paramount. This way they could at least ensure that the data acquired was not manipulated for the sake of public presentments, and thus could aspire to raise the 'profanity' of the available statistical data to the sublime levels of transparent representation. The endeavours constitute an effort to equip oneself for 'information warfare' according to scholar and activist Sabu Kohso, where the collation and distribution of data becomes crucial in the aim to understand dangerous radiation and to inform opposition against the expansion plans of nuclear authorities.[22]

Information Warfare

To convince the authorities of the hazards of living with high levels of background radiation, Dr Samuel Lal Mohan, a retired marine biologist, stressed the need for an independent survey. Born and resident in Nagercoil, he was concerned about the health consequences of sand mining and the nuclear plant. He believed that surveys and statistics could be one way of 'rationally' presenting their case against nuclear expansion. He became a consultant to those who conducted the survey.[23]

Through literature available in his compact library, Lal Mohan was aware that studies conducted by government-funded scholars and agencies in India conclude that no significant correlation exists between thorium extraction and cancer or intellectual disabilities.[24] This then precludes IREL from accepting any responsibility for health problems due to sand mining for radioactive minerals. A prime site for these studies in the region funded by the DAE is the Regional Cancer Centre, which was established in Thiruvananthapuram (Trivandrum) in the neighbouring state of Kerala, and BARC in Greater Mumbai, Maharashtra.

Conversely, concerned experts argue that there is a strong correlation between background radiation in the environment and human health. A national conference was organized in 2002 in Nagercoil on the health hazards of radiation by the Nuclear Power Awareness Committee. According to seminar proceedings, Dr Indira Surendren from the Gopala Pillai Hospital talked about new unnameable diseases, especially of the skin, among people living around coastal sand mining. To similar ends, Dr Jeyakumari Jacob from the Institute of Indian Technology in Chennai explained the radiation hazards of the alpha-emitter thorium along the coast. She highlighted health hazards on sand mining labourers and residents and informed that cases of cancer and congenital diseases were more frequent in these areas. Environmental scientists have also demonstrated interest in the Malabar coast due to its high natural radiation: their results seem to favour that the environment does play a significant part in public health.[25] Invariably government officers dismiss such studies for being biased, methodologically flawed and so forth, even while their own studies have been held to be 'incomplete and contradictory' according to an assessment by Ramana and Gadekar, if they are not kept away from public scrutiny altogether.[26]

It is true that state-endorsed health reports have not been published in publicly available outlets in their entirety where independent experts could analyse their methodology, findings, and conclusions. Information may well be available somewhere in the public sphere, but it could be spread across various sites or annual or monthly reports.[27] Studies were known to be done on radiation exposure with a few focusing on health impact along the coast in Kerala but not the adjoining Kanyakumari District in Tamil Nadu.[28] One in particular was believed to have been conducted by the DAE on the high rates of children with MR in Kerala that the survey protagonists wanted to see, MR being the locally used term for 'mental retardation'. The report was rumoured to be available with BARC in Mumbai but nobody in the district could gain access to it. 'They will say that we can't have it because it is a national security issue', explained one disability activist.

When it came to any queries regarding data on health to do with the nuclear industries in India, national security was bandied around by officials in an almost frenzied manner. Anything in the public interest was not released for fear that it could unleash a backlash against nuclear plants or sand mining and mineral processing of monazite sands—the cause of increased radioactivity in the locality. Some of my interlocutors entertained ideas of using the RTI Act to gain access to health reports, arguing, 'How could they deny us this report. This is nothing about defence. It is not a threat to national security. It is about people's security.'[29] 'National security' acts like a black box into which anything could be thrown; or perhaps, as we shall see in Chapter 10, a black hole that consumes everything and crushes it beyond recognition and existence. Lal Mohan quipped that 'they won't release reports because it is about *their* security'—that is, the security of nuclear officials.

Silence spoke of suppression as information that could be even remotely connected to nuclear industries was severely curtailed. Based on his studies on the monazite beaches of coastal Kerala, Itty Abraham notes that even though public health of the community was taken seriously by the state,

> Such care ... has an inherent limit. It must stop short of acknowledging the dangers inherent in exposure to radiation, while maintaining the

well being of residents. Making 'life live', in this context, means stopping just short of (radiation-induced) death.[30]

In countries of the global south, public health has been irrevocably influenced by, as Gabrielle Hecht states, 'colonialism, missionary work, mineral extraction, and other external interests'.[31] Such countries then become caricatured as countries of disease, poverty, and overpopulation to the 'near-total absence of national cancer and tumor registries'.[32] In the case of India, the colonial mentality extended to ensure that registries or survey results to do with radiation-induced ailments do not enter the public domain, and if they do, that they be doctored to appear inconsequential. If public health records were, on the one hand, non-existent or inaccessible, and on the other, incomplete and/or partial, the order of the day was to try and reassess and conduct a reliable databank for oneself.

Planning the Survey

Apart from the consultant, Lal Mohan, others who worked on the survey included a manager of a home for people with intellectual disabilities who I have called Martin. While Martin became the project co-ordinator, eight women from neighbouring villages were recruited as fieldworkers as they were to go out into the villages to conduct the surveys. Women from the locality were favoured as it was felt people would be more willing to discuss health issues with them than had they been men. Their ages ranged from the mid-20s to the early 40s. All were educated to the twelfth standard at the age of 18, with one of them having completed a degree in the arts. All were known to Martin due to their prior work with children with intellectual disabilities and from other field studies that he had conducted with people with intellectual disabilities in the region.

It was agreed by the survey organizers that the three most significant indices for health problems due to radiation were (a) cancer; (b) infertility and spontaneous abortions after the first trimester; and (c) MR and variations of intellectual and physical disabilities thereof. First, the survey would be conducted to assess the health situation in the coastal mining village of Manavalakuruchi. Levels of natural radiation were expected to be high here due to the presence of monazite and IREL sand mining that would entail more contact with the alpha radiation emitter, thorium. This data was to be compared with that from an inland village about 28 km away set in agricultural lands, Thovalai, 'the control site', where negligible levels of radioactivity were assumed to be present.

Simultaneously, sand was to be collected by the project co-ordinator from several designated places in the villages and sent to a nearby laboratory scientist who had offered to analyse it for levels of radiation in his free time. Only alpha and beta radiation could be tested due to a technique that follows a scintillation mode for alpha detection; and where Argon gas was passed over the sand samples, which were then weighed for changes for beta detection.[33] Gamma radiation could not be analysed with this method, and as nobody in the town had a Geiger counter and nor could they easily get access to one, this enquiry had to wait for another occasion. Radiation results were to be tabulated and then cross-referenced with the health data from the areas in the respective villages by Martin. While ascertaining causation was out of the question, the aim was to identify clusters so as compelling correlations might be made between readings on radioactivity and health problems.

Each of the fieldworkers received a week's training as to survey objectives, radiation-related health problems, methodology, and details about the villages. According to panchayat (village council) figures in 2006, Manavalakuruchi is a village of 4.2 square miles with a total population of 10,412 (including the adjoining coastal hamlets of Periavilai and Chinavilai). This approximates to 2,866 houses. Thovalai is a smaller village with a population of 4,605, and the population of Kuttankuzhi—a village that was selected after the public hearing described in the previous chapter—is 4,537, both villages consisting of roughly 1,000 households each. Outwardly, the survey was focused on MR with intentions to find out about primary disabilities—hearing and/or visually impaired, orthopedically disabled, Down's Syndrome, and Cerebral Palsy (locally referred to as CP). Unofficially, especially in Manavalakuruchi, it was also about incidents of cancer and other signs of genetic mutation. Wherever possible, data was also to be sought about spontaneous abortions, infertility, and pregnancy-related health problems, but it was appreciated that this was a sensitive area, and many people would not readily divulge such information to strangers even if the surveyors were female.

The four fieldworkers were assigned to conduct a survey in each of the two main villages over a one-month period. The objective was to conduct a door-to-door enquiry, identify households with people with MR, cancer, or any problems to do with infertility or pregnancies, followed with the completion of a four-page questionnaire, a pro forma, where a health problem was reported. The pro forma to be used by the fieldworkers was adapted from a previous 'health audit survey' conducted by BARC's Low-Level Radiation Research Lab in Kerala that Lal Mohan had come across, further underlining the modified reverse-mirrored mimicry of their tactical survey.

Daily reports were to be given by each of the fieldworkers to their team leader who then briefed Martin. The results were to be tabulated by Martin on a weekly basis. At the end of each week, a small group including a doctor were to visit some of the main cases of people with health problems to verify the data collected in a medical camp. These medical camps were set up by local contacts—usually a room with a desk where individual families were seen by the doctor who later accompanied the survey protagonists. Several weekly meetings were also to be held including Lal Mohan where fieldworkers provided a report on their progress and any problems that they encountered. With this data, they imagined that officials would sit up and take notice. A 'seminar' (the local appellation for an afternoon symposium) to publicize the results of the survey was planned with this in mind along with ambitions to invite the District Collector, the District Health Officer, IREL personnel, and the local media. I became involved in the two-month long survey, attended meetings, and accompanied the project co-ordinator and two doctors to the fortnightly medical camps. My views on the survey work were also solicited on occasion, although this was more for opinion than determining the direction of the survey.

Ethnography of a Survey

Gandhi Day on 2 October was chosen as the day to start the survey: 'a good day to start' explained Lal Mohan in view of his thoughts on Mohandas Karamchand Gandhi as a 'person of the people'. The sari-clad fieldworkers caught their respective buses to Manavalakuruchi and Thovalai early in the morning from the central bus station in the town. With the help of local maps, the two team leaders gave instructions for the day and organized the paths of field-workers who worked in pairs throughout the villages.

On average, a total of about 70–150 houses were visited daily by the four fieldworkers in the two villages. At a project meeting at the end of the first week, team leaders aired any problems that they had encountered. In Manavalakuruchi in particular, householders were willing to give information about children with MR as they had not then made a correlation between this affliction and radiation. However, when it came to other questions such as cancer and problems to do with infertility and miscarriages, they were less obliging—the former largely due to the imposing presence of the IREL and the latter mainly due to social stigmas, an anxiety that also looped round to suppress cancer talk.

The IREL had such an imposing presence in the village that to openly talk about radiation-related illnesses was to invite castigation, particularly if they worked for the company. Those who had developed cancer in Manavalakuruchi did not wish to draw attention to it for another reason. As the philosopher and political activist Susan Sontag has argued, a cancer diagnosis meant not just the degradation of the physical but also the social body.[34] Similarly, in peninsular India, the cancer stigma extended to members in their families who were seen as likely to share or inherit a damaged gene pool. Such declarations jeopardized the marriage of young people in a region where practically all able individuals with means were wed by the age of thirty. As a disease whose causes are rarely accurately known, it was subject to deep-rooted fears of genetic as well as social contagion. To contract cancer was tantamount to being served a 'death sentence', the imminent death of not just the physical but also the social being of the afflicted and their next-of-kin.

In light of public reticence to discuss various afflictions in Manavalakuruchi, two tactics were adopted. One was to change the field team leader and replace her with the more experienced one that had been working in Thovalai. She had been working in a home for children with special needs for 21 years and was highly attuned to ascertaining sensitive information due to previous fieldwork experience with intellectually challenged children. Emphasizing the anonymity of participants, focusing on related concerns to do with children, and comparing the total number of pregnancies and the total number of children for any one mother were a few ways information could be gathered. It was also considered that Martin could go out to the village himself and familiarize himself with village residents and the local ramifications of IREL politics, thereby developing a relationship of trust with the residents and attaining other kinds of information that could supplement the figures collated by the main fieldworkers.

During the first week of the survey on 6 October a public hearing was held by the Nuclear Power Corporation of India Limited in order to pass the construction of four more nuclear reactors at Kudankulam as we saw in the previous chapter. Coastal residents shouted vehemently at the nuclear officials at the public hearing as they blamed sand mining for health problems in their families. Here the sand mining was not at the behest of the IREL, a government undertaking, but the private venture, V.V. Minerals, known also as the 'sand mafia' due to their goondha (gangster) activities.

This event that Lal Mohan too attended was to have a marked influence on the survey process. Learning about the high number of disabilities in the coastal village of Kuttankuzhi, Lal Mohan advised that fieldworkers should collect data from this site as well in the ensuing week in order to gauge as to what extent their health was affected by sand mining in the vicinity. The survey fieldworkers were warned about the company and their henchmen, their safety ensured through the help of the village priest. It was decided that for two days, all eight fieldworkers would go out and gather data: it was to be a 'quick in, quick out' approach. No medical camp was organized here for that would be to invite too many queries and possibly even trouble.

Located in Tirunelveli District, Kuttankuzhi consisted of poorer households than those in the villages of Kanyakumari District. There was no hospital or doctor in the village, only a clinic and often the health worker would not be present. At the project meeting after the second week of fieldwork, team leaders reported that in Kuttankuzhi, many families wanted to *exaggerate* afflictions to do with physical disability. The residents assumed that the fieldworkers were officials who could exact funding for such ailments.[35] They were aware that the government provided social welfare funds for the physically disabled. Thus a lot of information was attained on CP and less on MR for, at the time, welfare for people with MR was comparatively lacking. However, CP too could be due to environmental factors, speculated the survey protagonists. Again, residents were reluctant to discuss cancer and infertility or pregnancy-related problems. In comparison to what Lal Mohan had encountered at the public hearing, residents would not only identify the sand mining as a possible cause for children with MR, but also sometimes in the same breath, dismiss them as due to fate or 'the woman's fault'. Blaming the woman for defects of fertility or the health of a child was indeed common across India.

Drops of Experience

Due to reasons of length, I will briefly summarize the main 'drops of experience' or 'actual occasions' as the philosopher and mathematician Alfred North Whitehead describes them: those moments deemed as significant by the survey practitioners in the process of establishing the 'concrete'.[36]

On alternate weeks, medical camps were held with a doctor to confirm the survey findings in the two main villages. At the first medical camp held in a village school office in Thovalai, we saw some of the people with MR identified by the fieldworkers one by one, after which Martin distributed a packet of biscuits to the children. There were in total seven children who came to this first camp, three girls and four boys. Some had mild MR. One was extremely severe—a combination of CP, MR, and microcephalus (indicated by an undersized head). The mother of the girls who were all sisters also had mild MR so the doctor established the cause as 'hereditary'. In another case, a woman from an agricultural Vellalar caste admitted that she had married her cousin, which the doctor attributed as conclusive proof for MR in the child. With other cases, causes were less certain. The doctor would ask questions, be none the wiser and put it down to other reasons such as late delivery. 'If a baby is born an hour or so late, it can develop MR' was her explanation.

In Manavalakuruchi, the onus was to visit houses spread across its widely dispersed geography rather than organize a medical camp in a central location. When we moved down 'IREL Road'—the road that leads to the sand mining company on the beach—we could hear the hum and drone of machinery dredging the sands. Outside, the road was covered with blackish sands from the trucks that regularly came with their loads of mined sand. We entered a house from which we could directly see the frequent passing of sand-laden trucks to and from the IREL company. The child here was aged one and had problems with his bowels. His father was a 'coolie'—that is, he worked as a casual labourer including for the IREL. The mother was aged 40 when she had the child, after eight years of marriage—late conception being the likely cause of this health problem reasoned the doctor.

Further into the copse on the beach, we met the family of a nine-year-old girl. Her two other sisters were of good health. She was the first child, and her mother claimed it was because her husband would get drunk and beat her that she gave birth to child with MR. Another slightly more upmarket house was located on the other side of the dusty IREL Road. Here, a thirteen-year-old girl had CP from birth. The mother reckoned that during birth, the baby suffered from 'some suffocation'. The survey protagonists speculated otherwise.

In a fenced-off house with goats, chickens, and ducks, we entered a dark house where there lived a family with a nine-year-old girl with Down's Syndrome. She hardly went out. 'If she goes out people gossip in a bad way', explained the mother: 'They think she should stay indoors and not scare the children'. The mother elaborated that Down's Syndrome was due to a curse or a spell—an understanding that the survey organizers were keen to dispel while the doctor saw it as further evidence of their ignorance.

At another medical camp held in Manavalakuruchi, there were five known cases of consanguinity that could be associated with the birth of children with disabilities. However, the afflictions of the rest of the children with intellectual and physical disabilities could not be gauged. The major revelation here was encountering an IREL worker who estimated that there were about 40 people with cancer in a total of about 300 people who lived in Periavilai, the hamlet directly next to IREL sand mining factories, and who provided daily labour for the company. Martin asked one local fisherman to get names and details of those with cancer, 17 of which he provided a week later.

It became clear that IREL did not systematically inform casual/daily labourers (hired by a sub-contractor) of the hazards and protective measures, nor were they sufficiently resourced with training and protection against the hazards of radiation, as we saw in Chapter 4. They had also signed away their houses to the IREL in order to move to replacement accommodation inland should the company want to expand in the future. Discontent was brewing amongst the village residents against the IREL but, at the same time, people were in a double bind in that they had become dependent on the company in various ways. Such discontent led some casual labourers in Periavilai to support the survey initiative, believing that it could improve their work and residential conditions. The epidemiological approach adopted by the survey protagonists was to identify clusters of intensities to do with high levels of health problems and radiation, a strategy to which local village residents also added any relevant data on health problems.[37]

The *a priori* assumptions of the two doctors that were engaged for the medical camps presented somewhat of a stumbling block. While the doctors' knowledge was impressive, it mainly consisted of 'text book' diagnoses, relying a lot on 'Davidson's book on medicine', a primary text for students of medicine in India.[38] They understood intellectual disability as the result of a competition between recessive and dominant genes, consanguinity, complications during birth, or due to the mother's late-in-life pregnancy. In their eyes, even the age of 31 was seen as late to have a child. Be that as it may, the crippling pressures of dowry payments led many poorer families to delay their daughter's marriage so as a dowry could be collected. Consanguinity occurred frequently in the region so as to avoid dividing property when marrying off a sister's daughter amongst Muslim, Hindu, and Christian families alike.[39] Yet these aspirations were antithetical to general medically informed thinking that cross-cousin marriage was to risk genetic malformations amongst offspring.

In addition to biomedical causes, doctors cited reasons to do with lifestyle such as any signs of drinking and smoking that could seriously undermine reproductive health. They attributed ignorance to laypersons who stated that disabilities and other afflictions were due to a curse or a spell. While they noted limitations in other people's understandings, they could not see their own: their biomedical textbook weighted interpretations of physical afflictions made it difficult to assess the viability of other possible causes such as the environment.

The doctors' initial attitude to the survey represented a strand that was prevalent amongst other state-backed officials: when it comes to health, they emphasize the predominance of science and lifestyle over environmental considerations. Lal Mohan had encountered this limitation before as he tried to attain information from medical experts from the state-funded Trivandrum Regional Cancer Research Centre where they attributed cancer to factors to do with genetics and lifestyle, but not the environment. As Lal Mohan commented at a project meeting: 'Whatever their credentials, the majority of doctors are not well-qualified to think outside the box'.[40] Sometimes the survey doctors had to write 'no information' on the pro forma that they were likely to dismiss as 'not relevant'. But, of course, these mystery cases were of extreme relevance to the project organizers. If a lifestyle, biomedical, or genetic predisposition could not be ascertained, could the reason be something to do with the environment?

Nevertheless, despite differences in perspectives, the survey organizers deftly utilized the role of the doctors for their own agendas. Anything that the doctors dismissed with a statement such as 'no information' was taken seriously by the fieldworkers. In addition, throughout the course of conversations with Lal Mohan and Martin, and on realization of the scale of the health problems and radiation readings, the doctors began to moderate their opinions and appreciate more and more the contributory factor of the ecology. On one occasion, for instance, on the way back from the first medical camp in Thovalai, Martin had to make an impromptu stop at the home for people with special needs that he managed. While we waited for him in the garden, the doctor was approached by about 20 people with MR who all gathered around her out of curiosity. I had already familiarized myself with the occupants, but she was visibly unnerved by the concentrated number of people with intellectual disabilities in one place, an experience that she had not had before. Later in the car, as she learnt more about their lives, she was enraged by the social prejudice and neglect that the residents of the home had to undergo. Her anger led her to reassess her earlier views on causation and to move away from ignorance and lifestyle factors alone. Seeing the children together triggered off a chain of emotions—sympathy, incomprehension, fear, anger—that then got directed at officials who may contaminate the air, land, and sea with more radioactivity that could lead to increased genetic mutation.

This turn of opinion was an important development in view of the fact that scientific and medical experts in general held a high status in the region and many people looked to them for clarification and guidance. They were trusted communicators who also provided a stamp of verification that the survey was conducted on credible grounds and the health afflictions were as identified. For these reasons, their role was seen as crucial to the project's credentials and its broader impact.

'Numbers Are Required, Not Reasons'

Sieving out the 'social' from the numerical was both imperative and overly ambitious. After a month of fieldwork, a final count was made: 109 cases of physical and intellectual disability were identified in Manavalakuruchi, 66 in Thovalai, and 154 in Kuttankuzhi. Manavalakuruchi was the larger village, and the two others were roughly comparable in size, Kuttankuzhi registering more than double the number of disabilities when compared to Thovalai. Even though information on spontaneous abortions was least forthcoming, from the available figures, Lal Mohan reasoned that in Manavalakuruchi, the total number of pregnancies in 109 households was noted as 483, yet the total number of children in these households was 403. Therefore there were 80 losses, 17 per cent. However, despite this striking figure, it was inconsistent with the findings in Thovalai, a village away from the coast and therefore away from high radiation zones: out of 66 households, the total number of pregnancies was 231, and the total number of children was 186. Here there were 45 losses, an even more dramatic percentage of 19 per cent. Lal Mohan concluded that this could be due to their impoverished plight and the fact that there were no nearby hospital facilities in Thovalai. Another more likely reason suggested by Martin was that the residents in the village were more helpful as they were familiar with the fieldworkers and therefore were more willing to disclose information about health problems. With this reasoning, Martin implied that there would be more 'hidden cases' in Manavalakuruchi. By comparison, in Kuttankuzhi, the total number of pregnancies was 501, the total number of children 492 out of a total of 105 households. Nine losses led to a percentage of 2 per cent. The comparatively low figure next to results from Thovalai surprised the project organizers in view of their recollection of the residents' complaints at the public hearing about the health impact of toxic industries. Similarly, surveyors concluded that perhaps not 'all relevant information was disclosed'—that is, they did not question the mode of surveying, only that some data could not be sought due to various 'social factors'.

Results on cancer were sporadic, as several people chose not to talk about it mistrusting the destination of the information and fearing stigmatization. The only significant findings on incidents of cancer were concentrated in the hamlet of Periavilai next to the sand mining company in Manavalakuruchi. Here one of the acquaintances from the fishing community provided an indication of the numbers—40 (from which details of 17 people were provided) in a total of about 300 people where men provided daily labour for the company, a percentage of 13 per cent. The incidents of cancer were mainly of the mouth amongst men and breast cancer amongst women. Mouth cancer was explained as due to the *paan* that they eat: the betel leaf and lime used in this digestive preparation contains radioactive minerals passed through the soils. The ages for breast cancer were comparatively young, as it was for thyroid cancer present in a few individuals who were in their thirties. There was also an incident of prostate cancer. It was acknowledged that there could be more maladies as many men with prostate cancer might not do regular checks or not mention it, only seeking advice and help if the condition manifested itself in other problems such as in the act of urination when it would invariably be too late to stem the spread of the disease. Based on such findings, Lal Mohan suggested that in future it may be useful to choose sympathetic residents to gather information in their neighbourhood that could then be compared with that of the fieldworkers.

When considering radiation readings in the area, levels in Kuttankuzhi and Manavalakuruchi were extremely high. They even indicated the presence of beta radiation, the source of which the project organizers could not then get any further information and concluded that it was from other elements in the sand such as the radioactive potassium-40 isotope and/or part of the decay material of thorium as with polonium. Whereas the highest readings for the inland village of Thovalai were 260 Bq/kg for alpha radiation and

3,920 Bq/kg for beta radiation, those in Manavalakuruchi were dramatically high. Periavilai hamlet was reported to contain as high as 9,780 Bq/kg dry weight of gross alpha radiation and 69,260 Bq/kg dry weight of beta radiation. In Kuttankuzhi, the readings were even higher—15,000 Bq/kg for alpha radiation and 99,040 Bq/kg for beta radiation.

Radioactivity was almost given tangible form through the numerical readings, the higher readings transmitting ripples of anxiety amongst the survey organisers for having exceeded their expectations. They indicated extremely high levels of radiation in both the coastal villages of Manavalakuruchi and Kuttankuzhi, some of which were deemed dangerous for human habitation if the alpha and beta radionuclides were to enter the body and food chain.

While statistical collation seemed relatively easy through tests in laboratory conditions, a context where obstructive social factors were deemed to be minimal, when it came to the village surveys, numerous factors and limitations became apparent.[41] The survey data looked impressive on paper for its comprehensive coverage, but in reality the force of social hierarchies, conventions, stigmas, and the politics of place came into play in the concrescence of the survey process.[42] Lal Mohan and Martin reflected on these concerns as a problem in practice, 'social factors', but it did not dent their faith in the principle of comprehensive survey work to ascertain statistical probabilities. They reasoned that they had dug up a veritable minefield of useful information, a database that could be further developed in an additive, rather than revised, frame of logic.[43] What may be called scientistic (science-statistic) methods and the facts that they built continued to hold a sublime grip on them with which to make validity claims.[44] The methodological problems in dealing with different contexts were quantified; and the project organisers estimated that there would be about 'a 5–10% margin of deviation' in the final results.

Public Presentations

Information had to be mobilized to have any impact. To these ends, a report was written and district dignitaries were invited to speak on the subject of radiation and health, after which the results of the survey were disseminated. In a matter of days, Lal Mohan wrote up the report in Tamil and English. With an introduction, and sections on methodology, findings, a discussion, and recommendations, the report also incorporated appendices with maps, photographs, and tabulated results from all three villages alongside radiation readings from specific sites in those villages.

The report highlighted the rigour (quantitative methodology and statistics) and minimized the 'fluff' (process and context): it presented survey findings as 'facts' in that any 'background noise' that obscured any of the results was omitted by concentrating more on incidents of MR and cancer from the small hamlet next to IREL. In so doing, it exemplified what the anthropologist Alan Rew has identified as the 'iceberg axiom' that characterizes report writing in general.[45] The survey process and report illustrated an attempt to enlist the discourses of modernity, rationality, and objectivity to their cause of highlighting local health concerns that were being dismissed by the authorities. Lal Mohan concluded the report as follows:

> The survey has shown that incidents of MR persons and cancer are significantly higher in coastal villages in Kanyakumari and Tirunelveli districts due to the high background radiation. This is increased in places where there is sandmining.

Alongside the report, a seminar was organized and publicized to coincide with Children's Day on 14 November 2006. They included speeches by the chairman of the municipal council, a member of the legislative assembly, and another local doctor who spoke about congenital anomalies and birth problems in the region, comparing the cumulative effect of low radiation dosage over many years as comparable to high doses received amongst the survivors of the Hiroshima and Nagasaki atom bomb attacks.[46] The survey findings were then read out by Martin. They were presented with little contextualization, only to mention that each house was visited in the three villages, followed by a reeling off of numbers to do with radiation readings and health problems in these locations including the separate data on cancer in Periavilai.

In all, about 80 people attended the seminar. These included some of the householders who were included in the survey as well as the young people who we had met at the home for MR. Whereas many of the attendants understood radiation, most of those that I talked to after the seminar did not fully understand the effects of radioactivity or units of measurements. However, conjoined with high numerical figures, they saw it as testament to the convincing case for the hazards of living with high background radiation in the region. On the basis of the 'comprehensive survey' and the statistical information gathered, and their conjunction with experts or figures of authority standing on the stage delivering presentations, the information sounded impressive and many of the audience members became convinced of the links between not just radioactivity and cancer, which was well-known albeit through hearsay and speculation, but also radiation and genetic mutation as expressed through MR. Several of them began to consider people with MR in a different light, adopting a more Euclidean gaze and allocating possible causes to the environment rather than to family history, fate, a curse or a spell. The information from the survey left a deep impression and became a filter with which to view phenomena around them, transposing socio-cultural explanations to that of scientistic discourse about health risks in the locality.

One attendant was particularly impressed by the doctor's presentation on pregnancy and congenital abnormalities and wove this information with a bricolage of thoughts influenced by the survey results. She recalled how her friend had a couple of spontaneous abortions in the spate of a few years. Even though her friend did not live on the coast and background radiation readings were not known for her village, the attendant began to suspect that environmental radiation could well be the cause of the miscarriages, not what she had earlier thought—stress and anxiety due to personal problems with her husband. The feedback loop was triggered by articulate observations and charismatic commentaries coupled with the spectral presence of assembled numbers that resulted in a certain reflexivity on the parts of the attendants. While numbers appeared to provide concretising information, they could also take on a magnetic and libidinal force as they drew other kinds of information around them, and shone light onto otherwise obscure associations. They could become alchemical agents as individual interpretations and extrapolations of statistics were woven in with social experience, understanding and intention. In the process, the representative and containing function of the statistical information mattered little when they shape-shifted and settled into other forms to do with this person's experiences and thoughts.[47] Numerical data may demonstrate an attempt at mastering the lifeworld, but in the end, the lifeworld got the better of numbers.

Speaking to Power

According to the official script, nuclear plans empower, defend as well as strengthen the country with its civilian and military offshoots. Kudankulam in particular is deemed 'India's nuclear pride', a pride that comes with much prejudice against anyone who cares to differ.[48] What is not officially acknowledged is how it may affect people's rights and routes to social justice, desecrate the environment, and have deleterious effects on people's lives over the generations—the 'slow violence' of attritional ecology and health as the environmental humanities scholar Rob Nixon describes such phenomena.[49] Any reports or surveys that officials have conducted are either kept out of the public eye or presented in a censored or distorted fashion to obscure anything that could verify radiation-induced health and environmental hazards.

The survey conducted in peninsular India aimed to redress this imbalance to seek more clarity as well as certainty. Whereas there is little in terms of an ethnography of a statistical survey in the literature, a parallel may be noted in the collaborative year-long fieldwork by the anthropologist Bruno Latour and sociologist Steve Woolgar on laboratory practice.[50] They demonstrated that the actualities of laboratory practise are far from the truth regimes enshrined in the discipline of science. While scientific method was premised on experiments as proving grounds for theories, laboratory practice was not as systematic and objective as scientists would like to claim. Inconclusive data was attributed to the failure of the experimental method or apparatus, and subjective decisions as to which data to retain were more than frequent. I too observed the truth regimes statistics informed, a sublime goal for there were a number of difficulties, hindrances, and inconsistencies in collating statistical data for survey protagonists.

Diverging from Latour's and Woolgar's deconstruction of dominant truth regimes, however, science as a close kin to statistical discourse could also be powerfully counter-hegemonic. On a related point, Ulrich Beck observes that while scientists are obsessive about making causal connection, such correlations can also be useful for environmentalists to fight their cause.[51] Beck gives the example of a Japanese court case where statistical data was held as demonstrating proof for environmental pollution.[52] It was a similar impulse that drove my interlocutors to organize, conduct, and disseminate their own survey.

Statistical data was injected with a vitality such that it became 'facts that speak' or to cite the anthropologist Talal Asad, 'a strong language', around which to weave impressions of dangerous radiation.[53] A strong language imbued confidence in that everyone knew that in political corridors, it is statistics and science that matters in the creation of an evidence base about the incidence of health problems and radiation levels. With such data, campaigners could enter more forcefully into debate with state representatives. They could attempt to speak truth to power and domination.[54] Even if there were certain disappointments and inconsistencies in the results, these paled next to the consistent findings that were deemed enough to make the survey worthwhile and 'evidential', lending authority and agency to people's negotiation with state agencies. Such an attempt at creating hard facts was an integral part of intensifying spaces of criticality, one that five years later had hit a live wire in the wake of the triple unit disaster in Fukushima involving an earthquake, tsunami and nuclear meltdown.

Notes

1. Jain, Lochlann S. *Malignant: How Cancer Makes us.* Berkeley: University of California Press, 2013, p. 2.
2. Ramana, M.V. and Surendra Gadekar. 'The Price we Pay: From Uranium to Weapons'. In *Prisoners of the Nuclear Dreams*, edited by M.V. Ramana and C. Rammanohar Reddy. New Delhi: Orient Longman, 2003, p. 419.
3. Joseph, Manu. 'Spot the Indian Nuke Scientist'. *Sunday Times of India*, 12 March 2006, p. 13.
4. E. Cardis, M. Vrijheid, M. Blettner, E. Gilbert, M. Hakama, et al. 'Risk of Cancer after Low Doses of Ionising Radiation: Retrospective Cohort Study in 15 Countries'. *BMJ* 331, no. 7508 (2005): 77–83, p. 83. Available on http://www.bmj.com/content/331/7508/77, accessed 24 January 2019. See Banerjee, Esha. 'The Radioactive Silence around Kalpakkam'.

Pensieve, 15 April 2015. Available on https://banerjeeesha.wordpress. com/2015/04/15/the-radioactive-silence-surrounding-kalpakkam/

5. Jain, *Malignant*, p. 190.

6. The report is in three parts. The first concentrates on (*a*) a review of geological and oceanographic studies; (*b*) the provision of fresh water; (*c*) limestone mining that was banned by the Atomic Energy Regulatory Board; (*d*) the construction of a township for tsunami survivors near to the plant; (*e*) an analysis of expert statements; (*f*) the possible effects of space weather anomalies. The second part focuses on low-level radiation and epidemiological studies in (*a*) Hiroshima and Nagasaki; (*b*) among the down-winders of Madras Atomic Power Station in Kalpakkam; (*c*) in nuclear facilities; and (*d*) among those living in high natural background radiation areas in Kerala. The third part includes studies of 'the effects of releases of radioactivity and 700 billion litres of hot water from each reactor every day on the marine eco-system'. Available on http://www. sacw.net/article2621.html, accessed 21 November 2017.

7. On the disjunctions between interpretations and practices as they apply to surveys on people, see Ratcliffe, J.W. 'Analyst Biases in KAP Surveys: A Cross-Cultural Comparison'. *Studies in Family Planning* 7, no. 11 (1976): 322–30; Stone, Linda and J. Gabriel Campbell. 'The Use and Misuse of Surveys in International Development: An Experiment from Nepal'. *Human Organization* 43, no. 1 (1984): 27–37. On the implications of surveys for ethnography, see Raminder Kaur. 'The Power and Limits of Numbers: An Ethnography of a Survey on Background Radiation and Health'. *ASAOnline* 7, https://www.theasa.org/publications/asaonline/ articles/asaonline_0107.shtml

8. On the attempt to create earlier databases relating to radiation and health, see Anumukti. Available on http://members.tripod.com/~no_nukes_sa/ anumukti.html, accessed 18 January 2019.

9. Foucault, Michel. 'Governmentality'. In *The Foucault Effect: Studies in Governmentality*, edited by Graham Burchell, Colin Gordon, and Peter Miller. Chicago: University of Chicago Press, 1991. See also Hacking, Ian. *The Taming of Chance*. Cambridge: Cambridge University Press, 1990; and Miller, Peter and Nikolas Rose. *Governing the Present: Administering Social and Personal Life*. Cambridge: Polity, 2008, pp. 87–104.

10. Prakash, Gyan. *Another Reason: Science and the Imagination of Modern India*. Princeton: Princeton University Press, 1999, p. 135.

11. See Murdoch, Jonathan and Simone Abram. *Rationalities of Planning: Development versus Environment in Planning for Housing*. Aldershot: Ashgate, 2002.

12. On a related point, there are no baseline studies at Jadugoda where the environment is carelessly ravaged by uranium mining and tailings; and there exist no monitoring of radiation of temporary workers even in nuclear research centres such as in Trombay. Ramana and Gadekar, 'The Price We Pay', pp. 425, 437.

13. A 1991 study of public health on five villages has been known to be conducted around the Rajasthan Atomic Power Station in Rawatbhata. Details about training of employees are also kept off-limits, as are details about reactor engineering and related nuclear issues. Ramana and Gadekar, 'The Price We Pay', pp. 415, 430.

14. Hansen, Thomas Blom. 'Governance and State Mythologies in Mumbaï'. In *States of Imagination: Ethnographic Explorations of the Postcolonial State*, edited by Thomas Blom Hansen and Finn Stepputat. Durham: Duke University Press, 2001, pp. 221–54.

15. This study then is less about 'biological citizenship' or biopower as the basis for making scientific and health claims in Adriana Petryna's study on victim-survivors of the Chernobyl disaster in Ukraine as the country transitioned from socialism to global capitalism. It was more about countering what might be described as 'agnatological necropower' that enables a privileged few to decide the fate of others while keeping the latter in the dark as we saw in Chapter 4. Petryna, Adriana. *Life Exposed: Biological Citizens after Chernobyl*. Princeton: Princeton University Press, 2002.

16. Weber, Max. *Economy and Society, Volume 3*, eds Guenther Roth and Claus Wittich. New York: Bedminister Press.

17. Abram, Simone. 'Introduction: Science/Technology as Politics by Other Means'. *Focaal: European Journal of Anthropology*, no. 46 (2005): 3–20.

18. Berglund, Eeva. 'Facts, Beliefs and Biases: Perspectives on Forest Conservation in Finland'. *Journal of Environmental Planning and Management* 44, no. 6 (2001): 833–49, p. 847.

19. See Strathern, Marilyn. 'Introduction: New Accountabilities'. In *Audit Cultures: Anthropological: Studies in Accountability, Ethics and the Academy*, edited by Marilyn Strathern. Abingdon: Routledge, 2000, p. 8.

20. See Webb, Martin. *Boundary Paradoxes: The Social Life of Transparency and Accountability Activism in Delhi*. PhD manuscript, University of Sussex, 2010.

21. In an unprecedented move in 2012, the RTI Act was successfully used by S.P. Udayakumar as part of a second appeal to obtain information from nuclear authorities. The eventual release of some information by NPCIL was in direct response to the Central Information Commission's order in April 2012 to provide documents that are relevant to public safety

concerns regarding the two reactors at the Kudankulam Nuclear Power Plant. Even though Sections 8(1)(a) and (d) of the RTI Act exempted any information that might compromise the 'security, strategic and scientific interests of the State' or 'commercial confidence' of the parties involved, no information was provided to support the NPCIL's claims at the hearing. Subsequently, the NPCIL had to release one of the two requested reports, the Site Evaluation Report (SER), on 18 May 2012. But it was a heavily truncated, barely legible report that appeared to rely on outdated information as it refers to the 'Soviet Union' at several points. The NPCIL denied the public release of the Safety Analysis Report stating that 'is a "third party document" to do with Atomstroyexport and Atomenergoproekt (AEP), under the state corporation, Rosatom'. See Central Information Commission Decision No. CIC/SG/A/2012/000544/18674 Appeal No. CIC/SG/A/2012/000544. Available on http://humanrightsinitiative.org/programs/ai/rti/india/national/2012/email_alearts/May/02/CIC-KKNP-Site&SafetyReportscase-order-Apr12.pdf; and PMANE. 'Koodankulam Site Evaluation Report—An Analysis by PMANE'. *DiaNuke*, 24 May 2012. Available on https://www.dianuke.org/koodankulam-site-evaluation-report-an-analysis-by-pmane/, accessed 31 January 2018.

22. Kohso. 'The Age of Meta/Physical Struggle'. *Cultural Anthropology, HOT SPOTS, 3.11 Politics in Disaster Japan*. Available on http://www.culanth.org/?q=node/422

23. On other studies on the effects of radiation on health based on the Chernobyl nuclear disaster, see Bennett, Burton and Michael Repacholi. 'Health Effects of the Chernobyl Accident and Special Health Care Programmes: Report of the UN Chernobyl Programme Expert Group "Health"'. Geneva: World Health Organization, 2006. Available on http://www.who.int/ionizing_radiation/chernobyl/who_chernobyl_report_2006.pdf, accessed 11 March 2012. This is a conservative analysis which is essentially compatible with the Indian government view. For alternative views on the irrevocable damage of exposure to even low-dose radiation, see Busby, Chris and Alexey Yablokov (eds). *ECRR [European Committee on Radiation Risk] Chernobyl: 20 Years On*. Available on http://www.nonuclear.se/ecrr2006chernobyl2ed. Fairlie, Ian and David Sumner. *The Other Report on Chernobyl*. (Government of India: Department of Atomic Energy. *Torch*), 2006. Available on http://www.chernobylreport.org/?p=summary, accessed 25 May 2012.

24. See Abraham, Itty. 'Geopolitics and Biopolitics in India's High Natural Background Radiation Zone'. *Science, Technology and Society* 17, no. 1 (2012): 105–22, pp. 111–13.

25. Forster, Lucy, Peter Forster, Sabine Lutz-Bonengal, Horst Willkomm, and Bernd Brinkmann. 'Natural Radioactivity and Human Mitochondrial DNA Mutations'. *Proceedings of the National Academy of Sciences* 99, no. 21 (2002): 1–5, p. 5.

26. Ramana and Gadekar, 'The Price we Pay', p. 418.

27. See Ramana, M.V. 'India's Nuclear Enclave and the Practice of Secrecy'. In *South Asian Cultures of the Bomb: Atomic Publics and the State in India and Pakistan*, edited by Itty Abraham. Bloomington: Indiana University Press, 2009.

28. See Abraham, 'Geopolitics and', p. 111.

29. See Moolakkattu, John S. 'Nonviolent Resistance to Nuclear Power Plants in South India'. *Peace Review* 26, no. 3 (2014): 420–6, p. 424.

30. Abraham, 'Geopolitics and Biopolitics', p. 108.

31. Hecht, Gabrielle. *Being Nuclear: Africans and the Global Nuclear Trade.* Massachusetts: MIT Press, 2012, p. 42.

32. Hecht, Gabrielle. *Being Nuclear.*

33. Karthiga, R. *Studies on Background Radiation in the Southern Coast of Kanyakumari District with Special Reference to Cancer.* Postgraduate Department of Physics, Women's Christian College, Nagercoil, 2006, pp. 23–39.

34. Sontag, Susan. *Illness as Metaphor.* London: Allen Lane, 1979.

35. On an account of village inhabitants' engagement with a government survey, see Sharma, Aradhana and Akhil Gupta. 'Introduction: Rethinking Theories of the State in an Age of Globalization'. In *Anthropology of the State: A Reader,* edited by Aradhana Sharma and Akhil Gupta. Oxford: Blackwell, 2006. pp. 14–16.

36. Whitehead, Alfred North. *Process and Reality: An Essay in Cosmology.* Gifford Lectures Delivered in the University of Edinburgh during the Session 1927–28, Glencoe, IL: Free Press, 1978.

37. On an investigation of such variability, see Petryna, *Life Exposed*, p. 100, pp. 149–90; and Abraham, 'Geopolitics and Biopolitics', pp. 112–13.

38. Boon, Nicholas A., Nicki R. Colledge, Brian R Walker, and John A.A. Hunter. *Davidson's Principles and Practice of Medicine.* London: Churchill Livingstone, 2006.

39. See Cecilia Busby's observations on the Mukkuvar fishing community in Kerala. Busby, Cecilia. *The Performance of Gender: An Anthropology of Everyday Life in a South Indian Fishing Community.* Oxford: Berg, 2000, p. 83.

40. See Good, Byron. *Medicine, Rationality and Experience: An Anthropological Perspective.* Cambridge: Cambridge University Press, 1994.

41. See Latour, Bruno and Steve Woolgar. *Laboratory Life: The Construction of Scientific Facts*. Princeton: Princeton University Press, 1988.

42. See Lefebvre, Henri. *The Production of Space*. Oxford: Blackwell, 1991.

43. Even though sympathetic NGOs professed to do more work on the area, to date, the only subsequent survey has been at the behest of a local Nagercoil-based private hospital in 2012. The project was funded by the DAE and not made publicly accessible at the time of writing. Through his contacts, Lal Mohan had heard that the surveyors, not surprisingly, had concluded that there were no substantial health risks from high background radiation in the coastal areas of Kanyakumari District. He surmized that the survey was just for 'namesake'—that is, to log and register a 'counter-survey' to their own.

44. 'Scientistic' is a term I have used to indicate the compatibility of discourses about science and statistics. It is not intended as a means to describe the survey work as somehow inferior or a simulacrum of 'real' science.

45. Rew, Alan. 'The Organizational Connection: Multi-disciplinary Practice and Anthropological Theory'. In *Social Anthropology and Development Policy*, edited by Ralph Grillo and Alan Rew, pp. 185–197. ASA Monographs 23, London: Routledge, 1985. See also Good, Anthony. 'Writing as a Kind of Anthropology: Alternative Professional Genres'. In *Critical Journeys: The Making of Anthropologists*, edited by Geert de Neve and Maya Unnithan-Kumar, pp. 91–116. Aldershot: Ashgate, 2006.

46. Thomas George notes that in the case of Hiroshima it was blood cancer that first affected the victim-survivors. In the long-term, breast cancer was more frequent. Survivors showed decreased fertility and stillbirths were born with several defects including cataracts, and several physical and intellectual disabilities. George, Thomas. 'The Last Deadly Sin: Effects of Nuclear Weapons on Humans'. In *Prisoners of the Nuclear Dreams*, edited by M.V. Ramana and C. Rammanohar Reddy. New Delhi Orient Longman, 2003, pp. 448–9.

47. Peter Miller and Nicholas Rose discuss a similar process describing it as 'thixotropic associations' where, for instance, scientific/statistical configurations shape-shift and settle into other forms. Miller, Peter and Nicholas Rose. *Governing the Present: Administering Social and Personal Life*. Cambridge: Polity, 2008.

48. 'Kudankulam Nuclear Power Plant—India's Nuclear Pride'. Available on https://www.youtube.com/watch?v=132JL_C4K5Q, accessed 21 January 2019.

49. Nixon, Rob. *Slow Violence and the Environmentalism of the Poor*. London: Harvard University Press, 2011.

50. Latour and Woolgar, *Laboratory Life*. See also Traweek, Sharon. *Beamtimes and Lifetimes: The World of High Energy Physicists*. Cambridge: Harvard University Press, 1992; Gusterson, Hugh. *People of the Bomb: Portraits of America's Nuclear Complex*. Minneapolis: University of Minnesota Press, 2004; Cetina, Katrina Knorr. 'Culture in Global Knowledge Societies: Knowledge Cultures and Epistemic Cultures'. *Interdisciplinary Science Reviews* 32, no. 4 (2007): 361–75.

51. Beck, Ulrich. *Risk Society: Towards a New Modernity*, trans. Mark Ritter. London: Sage, 1992, pp. 63–4.

52. On the disjunction between biomedical and legal contexts too, Jain states 'whereas the law requires proof of causality, biomedicine can only generate statistical probabilities'. Jain, Lochlann S. 'Living in Prognosis: Toward an Elegiac Politics'. *Representations* 98, no. 1 (2007): 77–92, p. 87.

53. The former quote is derived from Berglund, 'Facts, Beliefs and Biases'. Asad, Talal. 'The Concept of Translation in British Social Anthropology'. In *Writing Culture: The Poetics and Politics of Ethnography*, edited by James Clifford and George Marcus, pp. 141–164. Berkeley: University of California Press, 1986.

54. See Scheper-Hughes, Nancy. 'Hungry Bodies, Medicine, and the State: Toward a Critical Psychological Anthropology'. In *New Directions in Psychological Anthropology*, edited by Theodore Schwartz, Geoffrey Miles, and Catherine Lutz, pp. 221–250. London: Cambridge University Press.

8

An Unlikely Powerhouse

I wonder about my future—what would I like to be when I grow up. Surely a fisherman. I love to see my brothers Michael and Androo get up early morning and go to the sea. Sometimes I accompany them to the sea shore and watch as they take the boats over the waves. When I return in the evening I see the fruits of their labour in the nice sardine curry and the occasional luxury of the prawn fry. I see my mother and sisters smile on the day there is a good catch and the sales at the local fish market is stable. I know it also in the laughter from neighbouring houses where my mother would have reached a share of the fish. But now, I hear my brothers and their friends talk about the drop in fish catch. They often speak about how the police from the KKNPP [Koodankulam Nuclear Power Plant] shout at them to move away from the fishing zones close to the shore where an abundant catch is assured. They speak about the desalination plants that will spew out hot water into the sea killing all life. I do not want a Plant to come that will take away not only fishes and shells but also a way of living that has been ours for so long [sic]. (Ignatius, 2012)[1]

Entering the village of Idinthakarai is like entering any other along the south Indian coastline. *Pukka* or brick houses flank the main road combined with kiosks selling snacks, drinks, tobacco, and other sundries down the main road from the Kudankulam intersection on the national highway. Venturing further towards the coast, a school stands to the left just before the main bus terminal that is straddled by small shops. They surround a large square with, on the right, a striking grey building with two Indo-Gothic spires, Our Lady of Lourdes Church, otherwise known as Lourdes Matha Church (Figure 8.1). The century-old church is fronted by a large pavilion (*shamiana* or *pandal*) made of bamboo and palm mats to provide shade from the unrelenting sun. There is a nunnery to the other side of the quadrangle, and a Hindu temple dedicated to the elephant-headed god, Vinayaka or Ganesh. In the near distance lie more *pukka* houses, fronted with palm-leaf thatched verandas. Through narrow alleys between the houses are glimpses of the sea, the Gulf of Mannar, lying between the southern coast and India and the island of Sri Lanka, a channel that is home to an ecologically diverse biosphere. However, its exquisite biodiversity is now threatened. From virtually every corner of the village, the towering domes of two nuclear reactors are visible about a kilometre or two away. They stand like tall sentinels as a striking reminder of the might of the nuclear state (Figure 8.2).

Figure 8.1 Our Lady of Lourdes Church, Otherwise Known as Lourdes Matha Church, Idinthakarai, 2012.
Source: Raminder Kaur.

Figure 8.2   Living in CASA, A Tsunami Rehabilitation Compound Built on the Outskirts of Idinthakarai Near the Nuclear Power Plant, 2012.
Source: Raminder Kaur.

A breakwater of boulders on the beach has led to the name of the village, Idinthakarai, literally meaning 'broken shore'. It hosts about 12,000 people with a constant stream of people migrating in and out to other regions of India and across the seas. A large proportion of the households are engaged in fishing or in fishing-related industries with a fair amount of women who earn a slender income from small-time trade such as marketing fish and making *beedi* (tobacco 'rollies').[2] The majority are Roman Catholics. Most of the rest had converted to Hinduism around 1967 when the fishing community revolted, objecting to rising church taxes on their earnings. Without turning their backs on Christianity altogether, they built their Vinayaka temple *mandir* directly across the large square to the main church.

Despite its fairly ordinary features, Idinthakarai became part of an extraordinary phenomenon from 2011. A few months after the Fukushima Daiichi nuclear disaster, it grew to become the centre of India's largest anti-nuclear power movement. The parish house with its series of partitioned rooms became the home for the People's

Movement Against Nuclear Energy (PMANE), and its canopied back-yard beyond the kitchen, the forum for coordination meetings. The dais on top of the steps in front of the Lourdes Matha Church would double up as the main stage for major events including speeches, con-ferences, cultural performances, press meetings, candlelit vigils, and public acts of civil disobedience for thousands of people (Figure 8.3). At other times, the porch was a central point in daily lives whether it be to sit and talk, knit, or play games, attend Mass in the morning, and return again after daily chores for the evening to pray and sleep.

As campaigners upped the anti-nuclear ante, Idinthakarai swiftly became known as the 'epicentre' of agitation: even as one news report by Tariq Abdul Muhaimin puts it, 'the nucleus of anti-nuke struggle against Koodankulam Nuclear Power Plant'.[3] Deriving from Greco-Latin, epicentre confers a cardinal point to the earth's surface above where an earthquake fault begins to rupture. Idinthakarai began to be seen as an epicentre of anti-nuclear energies. Some referred to the village as the 'ground zero' for protest, a term usually used to describe the point closest to a nuclear detonation, earthquake, or any other disaster as with the area where two aeroplanes struck the World Trade Centre Towers in New York in 2001. With good or bad intention, all

Figure 8.3   Candlenight Vigil on Hiroshima Day in Idinthakarai, 2012.
Source: Amirtharaj Stephen/Pep Collective.

conferred bold centrality to a modest village in the goliath shadows of the Kudankulam Nuclear Power Plant.

By mid-2011, the conditions of ready-and-waiting, permanent provocation and direct confrontation observed in Chapter 6 had flourished into a remarkable agitation. Cinematic in its collective show of strength, the movement drew support from across the nation that resonated abroad. It accommodated both exceptional and everyday forms of resistance in an effort to call nuclear state officials to task, while insisting on people's right to a healthy life and livelihood. Given the circumstances, the more irregular and capricious *space of criticality* came to a head to consolidate a *state of criticality* in this nucleus, epicentre, or ground zero for a few years at least. This provisional and parallel state sought sovereignty and justice for its people—a contingent collective that although numerically dominated by Christian Paravar fishing communities, embraced people from all walks of life and of varied religious–caste–class–ethnic identities. Fishing and farming communities, environmentalists and activists, scholars and lawyers among many others came together to propose alternatives for the country's energy needs and development plans, while all along heralding a campaign for deep democracy that incorporated the voices and rights of the marginalized multitude.[4] To perhaps overstretch the nuclear metaphor, we could call it a combined state of fusion and fission, allying and disintegrating, empowering while always teetering on the edges of rupture, division, and even explosion.[5] Altogether, it demonstrated an agonistic equilibrium against an antagonistic nuclear state.

In order to consider the trajectory of the rise and demise of this hub, we may tend to this third phase of our genealogy of the anti-Kudankulam Nuclear Power Plant movement in terms of constitutive yet overlapping periods. The first can be identified from around May 2011, which saw the proliferation of anti-nuclear civil disobedience in the village and across the region. The second stems from March 2012 after a police and paramilitary clampdown when protest activities, clashes, allegations, arrests, and other oppressive measures escalated. The third marked the demise of the movement that will be resumed in Chapter 11. It may be pinpointed from March 2014 after PMANE coordinators decided to canvas as single-issue candidates and venture out of the village for the national elections for the Tamil Nadu unit of the Aam Aadmi Party under the national leadership of Arvind Kejriwal who campaigned for an ethical politics to eradicate corruption in India.[6]

A Siren Call

Looking back at the recent past is to recount narratives riven with deception and non-disclosure on behalf of the nuclear state. From the late 1980s, Idinthakarai residents had been told that a 'plant' or a 'factory' would be built in the vicinity. The true nature of the construction or its potential environmental hazards were not revealed, and word only got out that it was to be the site of nuclear reactors through horizontal networks with people from other villages and towns. Pacifying measures by NPCIL such as arranging trips for those along the coast to visit fishing communities around the nuclear plant in Kalpakkam on the western coast did not lessen their distrust, nor indeed their nuclear anxieties.[7]

After the 2004 tsunami, families were moved from the centre of the village to its outskirts in a neighbouring hamlet Church's Auxiliary for Social Action (CASA). As its alternative name goes, the Tsunami Rehabilitation Village had been developed for those whose houses were destroyed in Idinthakarai by the high waves. However, this was at a time when the nuclear plant was under construction and regulations stipulated that no house should be built within 1.5 km of the plant in this 'sterilisation zone'. Nevertheless, such protocols were not heeded as new houses were built less than a kilometre away from the nuclear plant. Residents were only given 'conditional ownership' of their homes as their title deeds state that if necessitated, they must evacuate the building with no further details about their right to compensation or return.[8]

A few of the older village residents remembered joining agitations alongside kindred fishing communities, and the injuries and arrests that they endured, as was recorded in Chapter 3.[9] This created an uneasy legacy peppered with anger and bitterness to add to the countless other ways that coastal communities were socially and politically marginalized. Some of them had also participated in the public hearings organized by the NPCIL that we learnt about in Chapter 6. They themselves organized a protest fast and cultural event for thousands in the village to raise more public awareness.[10] But it was not until 2011, that their village became designated as the prime place for a 'protest camp' due to its symbolic significance next to a nuclear plant and its strategic position on the coast.[11]

Idinthakarai residents recalled how in March 2011 they had heard about the triple disaster in Fukushima—earthquake, tsunami, and a meltdown in three nuclear reactors that led to the release of radioactive material into the environment. Ignatius, a 12-year-old boy, recounted how he felt when he watched television footage:

> More than how and why [the Fukushima Daiichi disaster] happened, we understand how weak the nuclear power plants are. Any small delay in getting the right amount of water, any human error in turning on or off a switch, a short tremor or shift in the earth, a wave from the nearby ocean that rushes in astray, a valve with a rusted nail in it can all start a disaster. And we realize that we are just 1.5 km away from a disaster.[12]

The simultaneity of their lives with those in Fukushima—next to a nuclear plant, a region on a low seismic fault line, on the coast, and to boot, having already been subjected to a tsunami in 2004—fuelled much concern. The incident became the tipping point from which to mount further agitation.

For all its woes in human and environmental costs, the disaster in Fukushima presented a millennial crossroads for nuclear industries across the world. M.V. Ramana outlines three main responses: those countries that phased out or abandoned plans for nuclear plants; those governments who were forced by public pressure and protests to change their policies; and state representatives who dismissed the accident as inapplicable to their own countries while going into overdrive with respect to propaganda campaigns about their superior reactor designs and the urgency of their country's energy needs among other messages.[13]

India falls into the latter response. Even though the then prime minister Manmohan Singh assured everyone in parliament that the DAE would conduct a safety review, he also stated that Indian plants were safe and sturdy.[14] For the most part, it was nuclear business-as-usual. Officials even maintained on air that what had happened in Fukushima was *not* a nuclear accident.[15] As they are wont to say, nothing happened, or at least nothing of consequence. Three days after the disaster, the NPCIL chairman and managing director S.K. Jain changed his tune somewhat. He declared:

It is a well planned emergency preparedness programme which the nuclear operators of the Tokyo Electric Power company are carrying out to contain the residual heat after the plants had an automatic shutdown following a major earthquake.[16]

Remarkably, Indian officials were defensive about reactors over which they had no jurisdiction. At no cost could they point to any chinks in their enchanted atomic armour. As we have already seen in Chapter 4, from 'nothing happened' came the discourse of nuclear manageability—this time by way of association with an imagined transnational nuclear brotherhood.[17]

Not many could give credence to what appeared as overblown ambitions of reactor control, especially not in Idinthakarai. Village residents had learnt about the Japanese disaster through other media outlets and spokespeople, and were incensed with what they saw as yet another display of duplicity by Indian nuclear officials.

The repercussions of Fukushima Daiichi led to a multiplier butterfly effect.[18] It did not directly stimulate anti-nuclear activism in the subcontinent for it was already simmering, but the disaster fed into a host of domestic factors that boiled over into a phenomenal mass movement. Residents called Michael Pushparayan to the village, a former parish priest and convener of the Coastal People's Federation, a network of activists and fisher leaders (Figure 8.4). A 'Struggle Committee' was formed comprising various members from farming, fishing, and ecclesiastical backgrounds across the peninsular districts. Aside from Pushparayan, this committee included Sivasubramanian, Father Michael Pandian Jesuraj, R.S. Mugilan, V. Rajalingam, Peter Milton, and the Idinthakarai parish priest, Father F. Jayakumar. They were joined by Dr S.P. Udayakumar who came from the town of Nagercoil. Udayakumar had already lectured in the region to deliver information about nuclear hazards as the convener of PMANE. His reputation as a well-informed, charismatic, anti-nuclear activist went before him.

Set up in 2001 by Udayakumar, the umbrella organization, PMANE was less an organization and more orientated around one man diligently working from his home when he was not teaching,

Figure 8.4   Michael Pushparayan on a Boat with Supporters, 2012.
Source: Amirtharaj Stephen/Pep Collective.

lecturing, writing, networking, and organizing activities with other individuals and groups. With a postdoctoral scholarly standing, the movement's stated aims and objectives were informed by years of analytical thought and its dissemination across numerous towns and villages disconcerted nuclear authorities. To add to the principles outlined in Chapter 1, PMANE decried that it 'does not recognise the two VVER-1000 NPPs [nuclear power plants] at Kudankulam'. As stated in a leaflet, this requires a strident politics of 'anti-nuclearism':

> PMANE holds that nuclear power and nuclear bombs are intricately linked and therefore it opposes all nuclear energy programs and projects. *Anuvilakku*, Atomic Prohibition, is our principle and no compromise whatsoever on that.

With the appeal that 'development should be democratic', PMANE underscored that it was not against India's development per se, but against undemocratic development where project-affected people were not attributed their rights, consulted, and/or duly and adequately compensated.

Udayakumar's eager reception in Idinthakarai at a critical moment after the Fukushima Daiichi nuclear disaster provided a vital boost to PMANE's mission. Situating PMANE in the coastal village enabled a nexus of hardy supporters that could spread the message near and far. 'Nuclearism' was all too evident around them. All it needed now was people to operationalize principles of anti-nuclearism, or in other words, *anuvilakku*.

But against powerful institutions backed by the state, the question was how? The answer lay in the power of numbers and non-violence. This approach was seen as not only tactical against a strong and unyielding state thereby aligning the struggle with the pedigree of national liberation, but also ethical and emotive that could appeal to wider sensibilities.[19] Based on past precedent, campaigners knew full well how fishing communities could be repressed, harmed, and/or imprisoned by the authorities. Their priority was to retain the higher moral ground through the power of people and the force of non-violence even while the state prioritized muscle and mammon. At no point, could they give viable cause to the authorities for punitive and aggressive responses. Simultaneously, they wanted to ensure as many people as possible knew about their appeal through internet campaigns and media-orientated stunts, tactics that were highly effective for a considerable period of time.

Several meetings were held within the PMANE Struggle Committee and with the larger village. This led to the implementation of a robust and transformational PMANE agenda, and formidable acts of non-violent civil disobedience were unleashed across the region. Life, livelihood, and social justice were at the crux of the matter. Parallels with the circumstances in Chernobyl were all too striking. One of the village residents had been in the Indian Merchant Navy and, among his world tours, had the chance to visit Chernobyl in the years after the 1986 nuclear disaster. He recalled its eerie character with its deserted buildings, streets, and fairground. He feared the worst, should there be a disaster in Kudankulam. The fact that the reactors were of a similar design was troubling. In view of Soviet neglect and non-disclosure, he suspected that Indian authorities too would be unprepared for dealing with a nuclear disaster, nor could they adequately and effectively instruct nearby residents about safety and evacuation procedures.[20]

By May 2011, people gathered round to take a firmer stand against the nuclear plant. They organized a demonstration demanding the shutdown of not just the Kudankulam but also the nuclear reactors located in Kalpakkam near the state capital, Chennai. Instead, the NPCIL pressed on. In July, nearby residents were subjected to never-ending loud sirens coming from the Kudankulam plant as there was a 'jet flight flying over our heads' in the words of one woman. There was smoke amidst the loud sounds as steam relief valves were tested and vented. Seeing this smoke added to the anxiety and anger fuelling in the village. The NPCIL was conducting a 'hot-run'. The test involved the use of dummy fuel that imitated enriched uranium in primary coolant water, which is heated to the operating temperature of 280 degrees and above.[21]

This was followed by a NPCIL notice on 13 August about a 'mock drill' and what to do in case of an emergency. As another sign of their irregular approach to public protocol, the notice appeared in the Kanyakumari District edition of a local daily but not the Tirunelveli District where the plant is located. The reporter Sumana Narayana states: 'The public was instructed to find shelter at the first bell, and consume iodine tablets among other things at the second bell.'[22] This call for emergency preparedness for workers and the public was intended to create more confidence in the plant's management in the wake of the Fukushima Daiichi disaster. With instructions such as 'run in a direction away from the plant with eyes and face covered in case of any symptoms of discomfort and reaction', the notice had in fact the opposite effect.[23] For many, the mock drill was yet another match to the powder barrel. It was time to make a public mockery of their mock conduct in what many held was only a 'paper demockery'.

On 15 August, three clusters of village councils (*gram sabha*) passed resolutions for the plant's closure. This was followed by an 'indefinite fast until death' unless their demands were met. The fast began ritualistically in the church pavilion, an occasion at which about 10,000 people gathered. On a day when the rest of the nation celebrated Independence Day, they highlighted the continuity of colonialism in the contemporary era.[24] After an 11-day fast followed by failed talks with officials, another one was initiated in September with more than a hundred volunteers. on receiving news that the first reactor would be commissioned that month.[25] Alongside the protest, PMANE issued 13 demands that centred on the need for construction protocol, mandatory reports, public consultation, alternative energy options, and coherent evacuation plans for the million or so people who lived near the plant.[26]

At a time when Fukushima was in the minds of many, the hunger strikes received much media coverage and compelled central and state officials to request talks with the activists. It led to a 'stay' on all work at the nuclear plant until the fears and concerns of the local people were allayed as requested by the Tamil Nadu state cabinet under Jayaraman Jayalalithaa from the AIADMK. The cabinet resolution meant that central government under the Congress leadership of Manmohan Singh (as part of the United Progressive Alliance) had to issue a command to stall ongoing work at the plant.[27] This 'stay' was received with wary delight by those in Idinthakarai, and their supporters. They were fully aware that it was under Jayalalithaa's rule that the reactors were first inaugurated in 2002, so stalling construction was at best a very tenuous triumph; at worst, a cynical ruse in view of the AIADMK's desire to gain more support among the people of Tamil Nadu.

Tamil grievances played into state cabinet machinations with the central government. A parallel series of events across the seas bolstered the Tamil fist against the heavy hand of a centralized state. News from across the Gulf of Mannar in Sri Lanka had been making itself fiercely heard. Tamil communities in the north east of the island were subjected to a brutal campaign of shelling, torture, and extra-judicial executions in the concluding months of the Sri Lankan civil war from 2008–2009.[28] The atrocities led to a resurgence of sympathies for the plight of Sri Lankan Tamils. Coastal Indians helped refugees from the island while they continued to fight on behalf of those Indian fishermen who had been detained by the Sri Lankan navy for allegedly fishing in their territorial waters. Many were mortified on seeing documentaries about the bloodcurdling ordeal of Tamil people in the hands of the Sri Lankan army as is evident in Channel Four's *Sri Lanka's Killing Fields* (2011, dir. Callum Macrae) and Tamil versions thereof that were aired on television monitors outside Lourdes Matha Church. Young men with T-shirts of the roaring visage of a Tamil Tiger and the Liberation Tigers of Tamil Eelam (LTTE) leader, Thiruvenkadam Velupillai Prabhakaran, openly declared their affiliation along the southern coast as they participated in anti-nuclear activities. They represented the Save Tamil Movement that later became Ilam Tamilagam, and the May 17 Movement marking the last day of the bloody defeat of the Sri Lankan civil war.

These events hardened Tamilian resolve to not be victims of what they felt to be central state shenanigans long before the disaster in Fukushima. They felt a keen sense of oppression from the centre (based in New Delhi), transnationally (through the government's nuclear deals with Russia among trade with other foreign countries for a nuclear India), and through fighting regional claims in the sharing of river water with adjoining states, Karnataka, Kerala, and Andhra Pradesh as well as fishing zones with Sri Lanka to add to the continuing persecution and discrimination of Tamil people on the island itself. Even though PMANE stressed the pan-Indian and global importance of rallying against nuclear expansionism, a reignited Tamil nationalism was irrepressible such that members of the Struggle Committee too capitalized on it from time to time depending upon their audience. There was therefore a striation of regional, national, and transnational orientations, some coming to the fore at any one moment, while others lay in the background as part of their repertoire of resistance.

Open-Air Jail

The relatively peaceful and prolific months of the struggle from mid-2011 were nevertheless riven with trepidation. Exhilaration and expectations were running high but with an anxious sense of foreboding for PMANE supporters had already been subject to incriminating measures through allegations of 'foreign-funding' and the indiscriminate filing of police charges. People knew full well that the Tamil Nadu chief minister who held the reins of decision-making power, Jayalalithaa, could change her mind to reinitiate reactor construction at any time. Struggle Committee meetings late into the night would be wracked with such discussions, attendants speculating on what to do next. They talked confidently yet cautiously as if they were in the calm eye of a brewing storm.

They had organized a two-day national cultural programme on 17–18 March 2012 that involved many artists, presenters, and supporters travelling to Idinthakarai to express their solidarity. Invitations were taken up across the country, some in journeys from north India that lasted as much as three or four days by train. Visiting this epicentre was empowering even in the knowledge that it was a risk to do so for fear of intimidation by coming under the stepped-up radar of state surveillance. Nevertheless, with a steadfast belief in the right for democratic dissent, many came and went away feeling inspired that non-violent resistance could be effective against powerful paragons of the state. As a testimony to people's power, it was an electrifying sensation.

When most of the crowds had dissipated after the programme, the police racked up their already stringent surveillance of the village. On the evening after the national event was over, a police car drove into the village to check at the parish whether there were any 'outsiders' still in the village. A few visitors had stayed over after the cultural programme but, on learning about the police call, they decided to leave that night.

Earlier in the day, the personal assistant to the Tirunelveli District Collector requested that Udayakumar and Jayakumar meet him in his office in the morning. This was followed by two calls from the District Collector himself, R. Selvaraj. Something seemed to be fermenting. Perhaps, this series of requests was to lead to the restart of reactor construction, speculated members of the Struggle Committee. How events transpired soon after caught everyone with a corrosive combination of surprise and shock.

The following day, roughly 6,000 armed police and paramilitary including the Central Industrial Security Forces from Thiruvananthapuram were swiftly relocated to Kudankulam shortly after a local by-election.[29] With sturdy barriers and armoured vehicles, they blocked off the main road into Idinthakarai. They stopped all bus services and supplies of water, food, milk, and fuel that coastal residents relied upon for their daily needs. They imposed Section 144 of the Code of Criminal Procedure (1973) on the Radhapuram taluk in which Kudankulam and Idinthakarai are sited. This hangover of a colonial order prohibited any meeting exceeding four people, and entry or exit out of Idinthakarai. They also detained over 180 people across Tamil Nadu, some of whom were held for sedition and inciting communal violence only to be later released.[30]

Criticism of the crackdown spread like wildfire. Solidarity statements were released across the board ranging from national organizations such as the National Alliance of Peoples Movement and the South Asians Citizens Web to petitions from international observers (Figure 8.5). This was accompanied by further protests where hundreds were arrested for marching towards Idinthakarai and a man was shot dead by the police in a rally in Thoothukudi.

Public space was visibly militarized (refer to Figure 10.2 of this book). Police officers stated that if the blockade was to be lifted at Idinthakarai, PMANE leaders, Udayakumar and Pushparayan, had to be first handed over to them. Marked by a deep trust deficit, village residents responded with a chorus of rebuttals, saying that they should arrest each and every one of them, if sedition is to be the charge against them.

In an open letter, Udayakumar vividly recalled the series of events after the phone calls from the Collector:

A warning bell rang in my mind and I told my friends that we were all going to be arrested. My intuition proved to be right; some 200 of our friends from Kudankulam, Koottapuli, Chettikulam and Erode were arrested. Rayan and I and 13 others embarked on an indefinite hunger strike demanding our friends' immediate and unconditional release.

The Superintendent of Police, Mr. Vijayendra Bidari, called me on my mobile on March 19th evening and asked me to surrender. With him still on the phone, I asked the thousands of people who had gathered there

Figure 8.5 One of many National and International Solidarity Posters, Idinthakarai, 2012.
Source: Raminder Kaur.

for their permission to surrender and they all shouted down the idea. I asked the SP [Superintendent of Police] to send enough vehicles and two officers with the arrest warrants so that we all would get arrested en masse. He did not like that idea and hung up by saying, 'This is the last time I speak to you.' We used to speak to each other quite often as I got his oral permission for all our rallies, campaigns and public meetings.[31]

Udayakumar's appeal to the people that he should go outside the village was met with emotional scenes of hugs and tears. People held tightly on to him then as they did on a later occasion when he considered surrendering after thousands camped outside the nuclear plant compound walls defying the prohibition of assembly order in September 2012. Women shook orthodoxy by the head and hugged their adopted brother in public, not wishing to let him go (Figure 8.6). It did not matter that he was unrelated or from a different caste or religious background. One of them, Tamil, reflected on her lament as she talked to the environmental educator and activist Anitha S.

Figure 8.6 Dr S.P. Udayakumar Considers Surrendering to the Authorities After an Attack on their *Satyagraha* or Peaceful Act of Civil Disobedience against the Kudankulam Nuclear Power Plant, 2012.
Source: Amirtharaj Stephen/Pep Collective.

His words seemed to me like a farewell. I just could not take it. I felt that someone has departed forever. I cried like I have not in years. And without me knowing I ran up the steps to the stage and embraced him, my dear brother.[32]

A solidarity visit by Kejriwal sealed their collective commitment, and Udayakumar and Pushparayan decided to not surrender to the authorities.[33] Instead, with renewed vigour, they intensified their agitation convinced by their right to peacefully dissent.[34] While they held the fort surrounded by a phalanx of supporters, Idinthakarai became an 'open-air jail' for them: they could organize and walk around its central environs, but could not venture out without risking arrest.[35] From this base, they continued with their peaceful acts of civil disobedience.

As protests swelled against the crackdown on a non-violent social movement, so did the framing of charges against anyone that came into the fray. Acts of *civil* disobedience were interpreted and reframed as acts of *legal* disobedience in order to criminalize dissent. Allegations in First Information Reports (FIRs) piled on top of each other with named individuals—invariably including Udayakumar, Pushparayan and others from the Struggle Committee—along with a couple of thousand or more unspecified others. Even though riddled with farce, the tailspin of allegations were severe:[36] 'sedition' and 'waging, or attempting to wage war, or abetting waging of war against the government of India' with recourse to laws such as Sections 124A and 121 of the Indian Penal Code; 121(A) referring to 'a conspiracy to overawe, by means of criminal force or the show of criminal force, the central government or any state government', 147 with respect to rioting; 148 in relation to rioting with deadly weapon; 149 on unlaw-ful assembly, 307 in relation to the attempt to murder, 353 implying 'assault or criminal force to deter public servant from discharge of his duty'; 395 with respect to dacoity (armed banditry); and Section 3 of the Tamil Nadu Public Property Damage Act.[37] The list went on.

Children too were not left without charges: they were accused of attempting to wage war against the state, sedition, rioting with deadly weapons, assaulting public servants, and registered under Sections 121, 124, 147, 148, 353, with another one to book - 294(b) with respect to 'sings, recites or utters any obscene song, ballad or words, in or near any public place'. Most curiously was a charge under Section 4 of the Endangered Species Act for the protection of rare and endangered life forms.[38] Without rhyme or reason, it was as if the entire book of commandments was thrown at campaigners, not unlike an exasperated parent against an errant child.

It was time for state payback. On the specific charge of sedition, Itty Abraham notes that it 'can be read both as [the state's] re-appropriation of lost sovereignty as well as a performative response to the wound produced by the discursive outcomes of popular speech'.[39] Its punitive comeback owes to its need to 'protect a failing vision of state legitimacy built around technological modernity' with an ambivalent technology that is neither fully indigenous nor able to fully deliver its promises.[40] Abraham's analysis on sedition could be extended to the plethora of other criminal charges. Such a legalistic assault was about re-positioning postcolonial authority and its 'originary crisis of legitimacy'.[41] It was also about reasserting the 'man in the state' to cite Wendy Brown—that is to say, re-establishing the power of the nuclear state and denouncing anyone that threatens its masculinist prowess.[42] On the day of the police and paramilitary assault against protesters outside the nuclear plant walls on 10 September 2012, there were numerous cases of molestation of peacefully protesting women including even a Rapid Action Force officer who pulled out his penis in front of a woman who he chased into the sea.[43] If ever there was a poster of post-colonial patriarchy entrenched by a masculinist state, this would be hard to beat.

Waging Non-violence

The repertoire of Gandhian strategies was both inspiration and the crucible for creative new departures in people's anti-nuclear campaigns. It was largely due to Udayakumar's influence and unwavering commitment to retain the higher moral and political ground that the movement stuck to the tracks of peaceful protest.[44] While he informed, instructed, negotiated, and rallied, others supported him in their collective ventures. Pushparayan commandeered through meetings and in front of the computer with emails and Facebook messages, scouring the news for relevant information while also dutifully sending out regular releases to the media and a growing network of supporters. People became even more motivated to fight for their lives with, in many cases, recalling the anti-colonial movement in the early twentieth century, a 'do or die' mentality.

The body became integral to theatres of resistance. Actions included indefinite hunger strikes or 'fasts to the death' involving a number of committed men and women, and a relay hunger strike where individuals from within and outside the village vowed to fast for a day in the pandal. They also shaved off their hair as a mark of protest, wore black T-shirts or ribbons, and abstained from sex so as strength could be accrued and women avoid pregnancy to fully apply themselves to the cause.[45]

Simultaneously, before March 2012, PMANE representatives visited a range of district officials, politicians, civil servants, lawyers, and judges across the country to solicit their support. While campaigners petitioned the relevant authorities, they refused to accept any benefits of government schemes and barred entry to officials in the village. To keep up the pressure, residents periodically closed shops, boycotted fishing, schools, and elections, and prompted MPs to resign so as to initiate by-elections. Elsewhere in the region, in a symphony of solidarity, people prevented workers' entry to labour sites, blocked rail and road transport, and refused to sell food or provide other services to nuclear plant employees.

Activities took on a more strident turn after the March 2012 crackdown. Supporters returned their voter identity cards en masse to the Idinthakarai parish office to register their disaffection. Their actions vindicated the fallacy of a representative government. They complained: 'we do not want to vote for these selfish and anti-people mainstream politicians and political parties'.[46] Reportedly, about 24,000 cards were collected in the parish.[47]

Campaigners set up their own public hearings.[48] National conventions were repeatedly overturned. The Indian flag and effigies of visiting dignitaries associated with the power plant were ritually burned. Black flags across the village and models of nuclear plants and 'nuclear coffins' with people playing dead in them were created to signal people's oppression.[49] Empty coffins were also paraded through the streets, and burnt on the outskirts of the village on days of symbolic importance to mark the death of democracy.

Remarkable new stunts were designed to regain their popular mandate as a mouthpiece for people's power while garnering further media attention. In the process, they gave dramatic form to the 'slow violence' of environmental toxicity as well as political ruses to contain and oppress them.[50] Men, women, and children formed human chains on land and in the sea activating new sites of civil disobedience. Described as *jal satyagraha*—literally water-based truth-force—rows of people held on to a rope while they immersed themselves in the sea across the coast protesting against the environmental damage of the nuclear power plant (refer to Figure 1.2 of this book). This is to add to a solidarity sea-based siege with the blocking of ships at the Thoothukudi Port and scores of boats with their black flags to form a fisherman's armada around the plant on National Fishermans' Day on 21 November 2012 (Figure 8.7).

The more adventurous buried themselves in the beach sand with only their heads sticking out, appearing to live in what were constructed as mock cemeteries. The stunt was a polysemous marker of radiation burdens that ranged from death through radiation to the death of democracy. Elsewhere, campaigners lay like corpses on the roads and pavements in a die-in. This came about as 'a national

Figure 8.7   A Solidarity Sea-based Siege in Thoothukudi Port, 2012.
Source: Amirtharaj Stephen/Pep Collective.

Figure 8.8   A Postcard Campaign to Send to Political Officials, Idinthakarai, 2012.
Source: Amirtharaj Stephen/Pep Collective.

negligence to our right to live' as the women of the village recalled: 'It was a new experience to lie as if dead because for us the opening of the Nuclear Power Plant is like issuing a Death Warrant.'[51]

Throughout, they initiated postcard, email and signature campaigns to send letters to various individuals (Figure 8.8). All-night vigils, prayer meetings, and candlelight processions were held. Flower throwing ceremonies were made to respect those who had lost their lives in the anti-nuclear struggle along with, on 17 May, those Tamils who had lost their lives in the Sri Lankan civil war that ended in 2009. On another occasion around the midnight hour on New Year's Day after a Mass ceremony, sky lanterns, or mini hot air balloons, were released aimed at the nuclear power plant with hopes towards 'a new year without nuclear', as one participant put it—an 'emotional energy of hope' that literally lifted the spirits up.[52]

On 8 May 2012, the Struggle Committee launched 'Respect India', drawing upon the 'Quit India' call that Gandhi had made to the British in 1942. They explained:

Just as the freedom fighters asked the colonial rulers to *Quit India*, we, the People's Movement Against Nuclear Energy fighters, request the corrupt and communal ruling class in India to *Respect India,* respect the Indian citizens' lives, rights and entitlements.[53]

The call for respect was imbricated in a moral obligation to people, land, and legally enshrined individual and collective rights backed up by an international grammar of millennium development goals. The political theorist John S. Moolakkattu adds that respect involved paying heed to 'their basic necessities' such as:

> ... safe drinking water; nutrition to all mothers and children; toilets and public amenities; health care and insurance; land reform; and protection of farmers and farming, indigenous communities, fisher folks, other depressed castes and minorities. In addition, it called for avoidance of development projects that destroy people's resources and livelihood.[54]

Elsewhere, motorcycle rallies were held as was a *Kudankulam Challo!* (Onward to Kudankulam!) March for supporters based outside of the village. The slogan reworked the freedom fighter Subhash Chandra Bose's call to arms against the colonial government centred in the capital, *Delhi Challo.*

Such was the distrust with authorities that protesters did not put it past them were they to artificially create a power crisis by periodically switching off supplies.[55] Campaigners pressed on, playing devil's advocates with the intent to demonstrate the limits of democratic dialogue as it stood.

In light of how the pillow of the legislature was used to suffocate justice, Udayakumar's campaign almost took on a satirical tone. As the man described by officials as 'the prime accused', Udayakumar called the state's bluff. He responded to the charges by saying that if sedition was one of them, then they had been in meetings with the 'Prime Minister of India, the Chief Minister of Tamil Nadu, the Tirunelveli District Collector, the Tirunelveli Superintendent of Police (SP) and many other officials' who too should be levelled with these accusations.[56] He sent legal notices to the then prime minister Manmohan Singh, and minister of state V. Narayansamy, demanding an apology for unwarranted accusations for being foreign-funded.[57] Several postcard campaigns against 'false cases' of sedition were also launched directed at the Chief Justice of the Madras High Court and

Chief Justice of India in the Supreme Court to add to ongoing judicial activism through supportive representatives from organizations such as the Chennai Solidarity Group and the People's Union of Civil Liberties.[58]

From his open-air jail, Udayakumar made parallels with the Emergency imposed upon Indian civilians in 1975 under Indira Gandhi's Congress government.[59] He stated online:

> ... remember the 'Emergency' and MISA (Maintenance of Internal Security Act) days [a law passed in 1971 that gave inordinate powers to Indira Gandhi's government and law enforcement agencies]. Yes, there is a silent emergency prevailing in India today. The State that accuses us of waging war against it is indeed waging a war against its own people. Also we have to ask what is seditious today in India. The Manmohan Singh government has scores of ministers who are accused of serious corruption and fraudulence charges but it is the common people like us who struggle for the safety and betterment of our people stand accused of sedition.[60]

Continuing resistance to the nuclear development was met with fury in some official quarters and incomprehension in others. They could not understand why residents could be worried about living next to a power plant. They even considered enlisting psychiatrists from the National Institute of Mental Health and Neuro Sciences (NIMHANS) in Bengaluru to 'allay their fears' to which residents responded with howls of laughter, saying it was in fact the nuclear officials that needed psychiatrists. 'The government is getting sicker and sillier' denounced Udayakumar.[61]

The theatre of absurdity vindicates Achille Mbembe's 'mutual zombification' of politics.[62] But not content to drown in baroque brutalization, campaigners continued to idealize and struggle for a transparent, efficient, and just democracy located in a future polity—in the words of the philosopher Jacques Derrida a 'democracy-to-come' that in effect could never come.[63]

A University without Walls

As PMANE's political philosophy was sharpened and honed through collective practice and resistance against a common adversary, bonds of conviction and solidarity eroded religious, caste–class, and rural–coastal–urban divides.[64] While anti-nuclearism dominated their thinking, their goal at heart became the deepening of democracy for *demos*-oriented rather than state-orchestrated development, summarily dismissing out of hand state aspersions of them as anti-development anti-nationals.

The principle of non-violent struggle and desire for deep democracy expanded well beyond fishing communities—designated (More/Most) Backward Classes or Scheduled Castes—to the hinterlands to accommodate more of the farming and business communities bulwarked by supporters from a broad alliance of leftist, environmentalist, and human rights contingents. As they did so, PMANE demands increased to include:

> ... position papers on the selection of the site for reactors 1 and 2, the VVER' reactors' design and engineering, their performance and safety, fuel procurement for the KKNPP and the mode of transportation, nuclear waste disposal and management, plans for reprocessing and the plant for it at the KKNPP site, the Russian and the Indian liability issues [with reference to an Inter-Governmental Agreement secretly signed in February 2008, deeming Russian agencies outside the purview of India's Civil Liability for Nuclear Damage Act, 2010], waste transportation and setting up of a possible weapons facility at the KKNPP.[65]

While they criticized nuclear plans, they constructed others that fused the personal with the political, the everyday with the exceptional. They promoted self and social upliftment through discussions on non-violence, democracy, and development. They showed their solidarity for other environmental and social justice campaigns up and down the country. They organized blood and food donations to help the needy and set up anti-alcoholism campaigns to improve the moral fibre of the collective fabric. They advocated conservation measures for the individual to save energy such as replacing round bulbs with more energy efficient compact fluorescent lamps (CFL). They supported the development of more solar and wind farms to meet India's energy needs. As their slogan went: '*Kattal Yundu, Kadalala Yundu, Varsham Muzhuvan surya oliya ennathikinth aannulai inge? [sic]*' (We have wind and tides. Solar the year long. Why then this nuclear plant here?)[66] They suggested that the nuclear power plant be transformed into a science park or a solar energy production site that could lend itself to India's development plans but without a high cost to the environment, people's health and not forgetting the national exchequer. They developed parallel scientific committees that sought to gather evidence on the safety, feasibility, and alternatives to the plant, and present it in a less biased way than the committees sponsored by the government.[67]

Throughout, they learnt much about the science and technology of nuclear reactors. Apart from persistent control rod problems with the VVER-1000 design, independent experts recorded other flaws.[68] From comparing earlier designs and reports available in the public domain, they learnt that the two reactor presser vessels installed at the Koodankulam sites were not the ones that were initially agreed by the Indian and Russian governments. The installed VVER reactors had a beltline weld that was not in the original plans. This weld among a cocktail of ferocious chain reactions could increase the chances of neutron embrittlement that could damage the vessel and lead to an accident. It was concluded that the equipment was either counterfeit and/or obsolete, and not as resilient as the models originally planned.[69] Such analyses only underlined what residents already felt and knew about the substandard equipment and procedures at the plant.

Problems were also noted with the cabling in the plant. Former Chairman of India's AERB, A. Gopalakrishnan, states that 'the cable system, as completed today, has not conformed to the norms and standards of cable selection, EMI [Electro-Magnetic Interference] shielding, or layout as per Russian, Indian or any other standards'.[70] Even though for the most part, the AERB was compliant with other authorities, peopled as it was by those recommended by the Department of Atomic Energy, Gopalakrishnan was one former employee who stuck his neck out to air his concerns in public.

The efforts to acquire more information and understanding effectively led to the founding of a university without walls. As one young man in Idinthakarai put it:

We sit on the beach and wonder the real meaning of Democracy. All these deep thoughts came to us because we are now able to think and analyse.... The struggle has taught us to listen and relate. It has made us realize the real worth of knowledge and information. In that sense it has been a school of learning.[71]

Through their fast-track learning at the blunt end of nuclear politics, they became expert analysts and spokespersons on several subjects. They began to see for themselves what it felt like to be pawns in the hands of what they saw as a discriminatory and deceitful democracy. They made demands that state practice should stick to the letter of the law and not decry their movement as lawless.

Holding to the letter of the law and its sentences of procedures was itself an impossible ask. As Aradhana Sharma and Akhil Gupta observe, 'excessive devotion to proceduralism itself either creates the possibility of actions that exploit mutually contradictory rules of procedure, or forces bureaucrats and their clients to skirt the rules'.[72] This is indeed how PMANE's demands were met.

Nevertheless, ideals still stood and PMANE supporters' experience of oppression made them even more pertinent. Many visitors noted how the struggle acted like a 'public education program'.[73] Moolakkattu remarked upon:

> ... the outstanding qualities of collective leadership, organizational skills, dedication, and commitment. They have demonstrated their capabilities in analysis of complex questions relating to energy policy, dangers of nuclear power, and radiation and ecological impact of nuclear plants on their livelihoods, on the oceans and the marine biology.[74]

Through their collective endeavours concentrated in the village, they had overcome many of the challenges around 'methods of collective and participatory learning that can deepen processes of conscientization' that Alf Gunvald Nilsen observed for the Narmada Bachao Andolan.[75]

When compared to earlier years, the vagaries of radiation, cancer, and other illnesses and genetic deformations were no longer a subject of incomprehension or disregarded as the workings of fate. Rather they became subjects of grasp and mastery, to the point of obsession. Through their connections with a wider network of anti-nuclear campaigns, people had come across dreadful imagery of deformed children that literally sent shivers down their spine: images of people with bodily defects in the aftermath of the Hiroshima and Nagasaki atomic bombing, those affected by the Chernobyl nuclear disaster and, closer to home, those marred by uranium tailings in the villages around Jadugoda in north eastern India.

PMANE educational programmes were tied in with a profound morality and urgency. Parents and teachers had no option but to share their views with the young, a situation that led to the radicalization of children as young as four and five who were actively involved in marches and political and cultural gatherings: 'Amma, Amma, we called you. You have made us orphans. Abdul Kalam, Abdul Kalam, who are you to speak about nuclear safety?'; Amma being a reference to Mother, often used to describe the then chief minister Jayalalithaa, and Kalam being the president of India.[76] They shouted, sang, presented, danced, performed, and drew satirical graffiti on public walls about pro-nuclear figures of the establishment such as Manmohan Singh and V. Narayansamy. They embodied both the vulnerability and the sanctity of the future.

Time itself became toxic and had a virulent legacy.[77] Understanding that long-term residence in the midst of the mundane release of radioactivity could amount to doses equivalent to those received through a disaster, a woman mourned: 'Our children will be born with defects'. Another elaborated:

We know that this [the nuclear plant] will make our wombs unfit for growing a healthy baby. Will we be able to give birth to deformed babies? Can we bring them up with our tears? Is it not better we do not give birth to such babies?[78]

They viewed their selves, cultures, and futures in a 'world haunted by invisible but intensely noxious waves', a haunting that would persist well into the future as the anthropologist Françoise Zonabend describes another nuclear peninsular in France.[79] One fisherman in Idinthakarai resolved: 'We have to sacrifice ourselves for the nation. The greater good is for the nation, and we are forced to suffer'. 'Cruel optimism' according to Lauren Berlant, is not just cruel for what it promises and cannot deliver, but also in terms of the cruelty enacted on those who *must be sacrificed* in the attempt to create this modernist chimera—the ones who enable the manufacture of these promises even while they cannot be delivered, the futility of which many in the village began to see for themselves.[80]

The Backbone

With their alternative visions and multi-focal chains of coordination and support, came a shaking up of social orthodoxy. It led to the opening up of new spaces for reconceived relationships and the growth of 'rising stars'. Prominent here was the ascent of female organic intellectuals. Earthbound, they grew wings that made them fly, as they directly engaged in processes and decisions about nuclear *realpolitik*. During the heyday of the movement, Idinthakarai had indeed become the site for the largest peaceful protest of women against nuclear power.[81]

From the outset, a Women's Committee of about 85 members was formed with Sundari Pendanpush at the head. They coordinated actions with hundreds of other women inside and outside the village. Udayakumar himself favoured women taking on leading roles in organizing the community for action and liaising with other members including media representatives to present their cause. He admitted that women were the spinal strength of the movement for 'holding the remote control of Kudankulam struggle'.[82] Remote or close control, women were the ones who powered the proceedings even though they may not always be the main spokespeople.

Through their energetic and formidable part in the movement, women's activism was justified in view of their proven credentials to ensure sustained and non-violent struggle against the nuclear plant. They took to the art of non-violence almost like fish to water, and much more readily than many of the men who had to take to the water to fish. They outnumbered men at all agitations, some wearing black T-shirts over their saris loudly proclaiming 'Shut KKNPP'.

Few women were formally educated beyond the age of 16 at school. Tamil, who lived in the CASA hamlet, reflected:

> I know that many decision makers believe we are poor and illiterate. Poor we may be in comparison to the crores of rupees they handle and swindle in the name of people and development.... Illiterate—most of

us have gone to school and can read and write. We have understood all the facts regarding the safety and impact of nuclear power plant. Since a year, some of the best people connected to this issue in our country have visited and shared their knowledge and experiences with us. We love the speeches and imbibe all that is being said.[83]

While the authorities dismissed them as part of the 'ignorant masses' who could not possibly understand the ins and outs of a nuclear reactor, women challenged and confronted their paternalistic and prejudicial views, claiming that it was not their ignorance that was at fault, but the ignobility of the nuclear state.

Rather than be kowtowed by state conduct, women became emboldened through their participation in the struggle. They swiftly learnt to stand firm in the face of allegations from the police. When Selvi spoke from Trichy prison where she was taken after her part in the act of civil disobedience in September 2012, she said:

We have been here for over a month. We get charged with new cases every time we are taken to the Court. It is only after being part of the struggle that we realized that trying to establish one's right to live as one wishes, pursuing traditional livelihoods and also questioning activities that are being implemented without consulting the people is equivalent to crime and sedition.[84]

They inverted the terms of reference, and instead levelled charges of crime and sedition at the state, and not at what they, the people, were doing: acting in accordance with their democratic right to peaceably protest, and to request mandatory procedures and reports for the construction of a nuclear power plant in the name of transparency and accountability.

In the space of only a few months, the horizon of possibilities became more lucid for women. They were engaged in not just a local or national matter but one that was about saving the planet from exploitation and attrition. Sentiments about the need for intergenerational responsibility and equity were expressed in a strongly worded letter in October 2012 addressed to the 'Heads of State/Government c/o The Embassies (or) High Commissions' in New Delhi'. The PMANE Struggle Committee reiterated this cleavage between current and future generations, the former making myopic decisions that would irrevocably harm the young:

> We have no moral legitimacy whatsoever to produce electricity for our present needs and endanger the futures of our children and the unborn generations with the dangerous booty of nuclear waste, contaminated sites and deadly radiation. It is not only immoral but also illegal to help the profiteering MNCs [multinational corporations], corrupt politicians, bureaucrats and technocrats make money at the cost of the Earth, the future inhabitants and their common futures.[85]

The vulnerability of children caught in the political machinations spearheaded by adults sharpened the import of deep time to deep democracy—a temporality that was not just mired in self-interested presentism but a long-term vision of the collective good and the planet we leave behind to the next generation. Sundari emphasized:

> This is not about a personal vengeance or anger. We are here for a common cause—we are here for the world. So I do not feel sad or angry. I know that in the Samara pandal [in front of Lourdes Matha church] at Idinthakarai, all our sisters, children and brothers are keeping the torch of resistance ignited.[86]

Where and when they could, and with their torches of resistance to hand, several women travelled to rallies and other events in distant cities such as New Delhi, Kolkata, Bengaluru, Coimbatore, and Madurai, among other places in India to spread their message. Through their travels and media reportage, they became renowned as campaigners for social justice. Women mobilized other women, leading marches, and at several rallies, went on to become Tamil spokespersons for their plight. Inevitably, in December 2012, Idinthakarai women won a Chingari Award for 'women against corporate crime' set up in the aftermath of the 1984 Union Carbide gas disaster in Bhopal.[87]

With the rise to prominence of particular women, male partners would settle into the background, some even taking on some if not all the daily chores and cooking so as their wives could continue with the struggle. Women also noticed how, since the escalation of the movement, men tended to drink alcohol less, and instead join in organized activities whenever they could.[88] It was almost as if the power of the matriarch was resurrected for a political agenda. Remarks such as the 'wife always wins' or that 'she keeps all the money, not the men' already indicated their decision-making powers in the household.[89] This domestic dynamite was catapulted into the outside world.

Gender equality appeared to be bubbling. However, it did not come to the fore as an express mission for gender justice.[90] Women's rise to political heights did not come without a social cost as there were several problems that they had to abide with as part of the enduring force of patriarchy. Journalist and multimedia producer Karat observed:

> According to the women, their interest in the anti-nuclear struggle is frowned upon by their mothers-in-law. Husbands are often insecure about empowered wives. What if they flout the men's authority? "In the village, someone even said, 'Look at her, she left her husband and children alone to go to jail,'" says Sundari.[91]

This strain between social expectations and political commitment was also evident from our encounter with Savitri in Chapter 5: while new avenues of public engagement opened up, women still had to navigate the undergrowth of tight-lipped convention and loose gossip. Aspersions were cast about their time away from home. Some got jealous of other women over what they might see as their preferential treatment. Simultaneously, husbands might be socially emasculated, mocked as men who 'could not control their wives'. This was a theme that police among others against the movement pounced upon with punishing abandon, trying to coerce women to return to their homes with an array of verbal, visual, and physical abuse. Their affront was not just that they were anti-nuclear activists, but that they had over-stepped their place as women in a patriarchal orthodoxy riven with sexist taunts.[92]

Sundari was one woman who went on to publish her own book in Tamil, *The Fiery Struggle of Idinthakarai*, relaying her experiences as an organic activist. She talks about how she lived with the constant challenges and threats laid against her and her family, and her time in prison after her arrest in September 2012. She had been named in about 350 FIRs, 24 of which were for sedition. But the allegations did not deter her. To journalist and filmmaker Minnie Vaid she recounted her numerous arrests and 'how the police used to tell her to go back

home' and 'be a good wife'.[93] Her experiences reflected several others'. Rather than buckling, she maintained: 'Every time they arrested
me or intimidated me, it just made my will to go back to the protest
stronger'.[94]

Cracking Concrete

While living in Idinthakarai was an open-air jail for the PMANE
Struggle Committee and many of their supporters, people found
surreptitious ways of entering and leaving the village. Despite the
blockade on their main access road in 2012, being located on the
coast proved to be a saving grace. Fishermen and women from neighbouring villages came via boat to deliver supplies and participate in
solidarity events. People also found another route into the village
that meant taking a 30 km detour further up the national highway
and then backtracking through coastal roads to the village. As Karat
reported for one activist cut off from his family:

> Fifty-seven-year-old V. Rajalingam had been holed up in Idinthakarai
> for almost a year. He wasn't sure he would be able to attend his daugh-
> ter's engagement and wedding, which was to happen in May 2013. The
> police would regularly visit his house in Kudankulam to enquire about
> his whereabouts. Stepping out of Idinthakarai would mean imminent
> arrest.[95]

As it transpired, ajalingam managed to attend his daughter's wedding without getting arrested during the 'silent emergency'. It was
a small but no less significant victory in a massive mêlée of minds,
bodies, and future visions.

Worldwide, it would seem that the nuclear lobby has 'a special dispensation' for curtailing people's rights of mobility, assembly, speech
and non-violent protest.[96] This special dispensation has, nevertheless,
been contested on all fronts. In the south Indian case, it led to an ethical and political grammar of peace that drew its source and strength
from Gandhian civil disobedience adapted to suit the contemporary
Tamil Nadu context. The phenomenon also reveals the key role of
women in the struggle as determined advocates of demos-centric
development that is complementary to their lives and environment.

This is in the context of a constitution that promises freedom for peaceful dissent and environmental, human and civil rights. It is with the help of such levers, the mobilization of other laws, and with the capitalization of opportune moments that interstitial cracks had opened up in the concreted layers of power, cracks that at one point in 2011 even threatened to destabilize the nuclear state. Against all the odds, campaigners had done the barely unthinkable: they had stalled the construction of nuclear reactors not just with a passionate protest against the fear of displacement, radiation and nuclear catastrophe, but through reasoned argument and the collective power of ordinary people joining hands in a struggle for a common goal. While they resisted the nuclear plant, they proposed other plans. For them, ecology outweighed economy, and people over policy in debates on the progress of the nation. In the long shadows of non-procedural practices adopted by the nuclear state, they could quite legitimately rise to the role of the rightful custodian of democratic development. It was as if Idinthakarai had become the site of a people's lighthouse that shone across the world.

Notes

1. 'With Love from Idinthakarai', 8 August 2012, compiled by Anitha S. Available on http://www.dianuke.org/with-love-from-idinthakarai/, accessed 10 November 2016.
2. Estimates of fishing households are around the 60–80 per cent mark. See Bhawna. 'Nuclear Energy, Development and Indian Democracy: The Study of Anti Nuclear Movement in Koodankulam', *International Research Journal of Management Sociology and Humanity* 7, no. 6 (2016): 219–29, p. 223.
3. Muhaimin, Tariq Abdul. 'India—The Burden of being a Dissent in Democracy—Story of J. Roslin [sic]'. *Newzfirst*, 2012. Available on https://kractivist.wordpress.com/2012/12/29/india-the-burden-of-being-a-dissent-in-democracy-story-of-j-roslin-rip/; Nichenametla, Prasad. 'Govt Fails to Woo Kudankulam'. *Hindustan Times*, 25 September 2012. Available on http://www.hindustantimes.com/india/govt-fails-to-woo-kudankulam/story-m2lVKucmtjQmHTxbNRkD6L.html; 'Protest against Nuclear Plant', http://www.ndtv.com/topic/protest-against-nuclear-plant, accessed 10 January 2017.

4. This phenomena bear comparison with self-rule instituted in villages along the Narmada river heralded by Narmada Bachao Andolan (NBA) flags along with their micro-hydel projects as a riposte to the large dams being built across the river that threatened to displace residents. See Nilsen, Alf Gunvald. *Dispossession and Resistance in India: The River and the Rage*. New Delhi: Routledge, 2012, pp. 172–3. On the multitude as a differentiated group across divisions and hierarchies in the contemporary global order, see Hardt, Michael and Antonio Negri. *Multitude: War and Democracy in the Age of Empire*. London: Penguin, 2005.

5. There are also comparisons to be made with other states of criticality as with flashpoints that became a sustained movement, community, and occupation over a period of time. In the global north, for instance, Occupy played a fairly recent role in protest camps that re-imagine politics, citizenship, ethics, and ways of living and being away from the logic of global capitalism. See Badiou, Alain. *The Rebirth of History: Times of Riots and Uprisings*, translated by Gregory Elliott. London: Verso Books, 2012, p. 87.

6. See Bornstein, Erica and Aradhana Sharma. 'The Righteous and the Rightful: The Technomoral Politics of NGOs, Social Movements, and the State in India'. *American Ethnologist* 43, no. 1 (2016): 76–90.

7. Srikant, Patibandla. *Koodankulam Anti-Nuclear Movement: A Struggle for Alternative Development?* Working Paper 232, Bangalore: Institute for Social and Economic Change, 2009, p. 5.

8. 'Fact Finding Report on the Suppression of Democratic Dissent in Anti-Nuclear Protests by Government of Tamil Nadu', p. 5; Bhawna, 'Nuclear Energy, Development and Indian Democracy, p. 224.

9. See Paliwal, Ankur, Arnab Pratim Dutta, and Latha Jishnu. 'Kudankulam Meltdown'. *Down to Earth*, 7 June 2015. Available on https://www.downtoearth.org.in/coverage/kudankulam-meltdown--37876, accessed 21 January 2019.

10. Srikant, *Koodankulam Anti-Nuclear Movement*, pp. 7–8.

11. See Prabhu, Napthalin. 'Protest Camp as Repertoire for Antinuclear Protest', forthcoming, 2019. On the significance of claiming symbolic space, see Melucci, Alberto. 'A Strange Kind of Newness: What's "New" in New Social Movements?' In *New Social Movements: From Ideology to Identity*, edited by Enrique Lanana, Hank Johnston, and Joseph Gusfield, pp. 101–130. Philadelphia: Temple University Press, 1994.

12. 'With Love from Idinthakaraï', compiled by Anitha S.

13. Ramana, M.V. 'Nuclear Policy Responses to Fukushima: Exit, Voice, and Loyalty'. *Bulletin of the Atomic Scientists* 69, no. 2 (2013): 66–76.

14. Ramana, M.V. and Ashwin Kumar. '"One in Infinity": Failing to Learn from Accidents and Implications for Nuclear Safety in India'. *Journal of Risk Research* 17, no. 1 (2014): 23–42, p. 27.

15. 'No "Nuclear" Accident in Fukushima, say Indian N-Experts'. *India Today*, 14 March 2011. Available on http://indiatoday.intoday.in/story/no-nuclear-accident-in-fukushima-indian-nuclear-experts/1/132416.html, accessed 10 November 2016.

16. 'No "Nuclear" Accident in Fukushima'.

17. See Ramana, M.V. *The Power of Promise: Examining Nuclear Energy in India.* New Delhi: Penguin Books, 2013, pp. 216–18.

18. On an analysis of post-Fukushima activism in Taiwan, see Ho, Ming-sho. 'The Fukushima Effect: Explaining the Resurgence of the Anti-nuclear Movement in Taiwan'. *Environmental Politics* 23, no. 6 (2014): 965–83.

19. Nilsen, *Dispossession and Resistance in India*, p. 123.

20. See commentary, 'People's Preparedness for Nuclear Emergency'. *Economic and Political Weekly* 28, no. 10 (1993): 377–80.

21. Paliwal, Ankur, Arnab Pratim Dutta, and Latha Jishnu. 'Kudankulam Meltdown', 2005. Available on https://www.downtoearth.org.in/coverage/kudankulam-meltdown--37876, accessed 21 January 2019.

22. Narayanan, Sumana. 'Anti-nuke Protests in Tamil Nadu'. *Down to Earth*, 4 July 2015. Available on https://www.downtoearth.org.in/news/antinuke-protests-in-tamil-nadu-33907, accessed 18 November 2018.

23. Admiral Ramdas and Lalita Ramdas. 'Koodankulam Diary' *Countercurrents*, 9 January 2012. Available on https://www.countercurrents.org/ramdas090112.htm, accessed 21 January 2019.

24. Narayanan, 'Anti-nuke Protests in Tamil Nadu'.

25. 'Koodankulam: Crackdown on Anti-nuclear Activists and NGO's'. *Nuclear Monitor* 744, 16 March 2012, pp. 1–3. Available on https://wiseinternational.org/sites/default/files/744.pdf.

26. 'Thirteen Reasons Why We Do Not Want the Koodankulam Nuclear Power Project'. *DiaNuke*. Available on http://www.dianuke.org/thirteen-reasons-against-the-koodankulam-nuclear-power-project/, accessed 10 January 2019.

27. 'Cabinet Adopts Resolution on Kudankulam Project'. *The Hindu*, 23 September 2011. Available on https://www.thehindu.com/news/cities/chennai/cabinet-adopts-resolution-on-kudankulam-project/article2477355.ece, accessed 10 January 2019.

28. Dix, Benjamin and Raminder Kaur. 'Drawing–Writing Culture: The Truth–Fiction Spectrum of an Ethno-Graphic Novel on the Sri Lankan Civil War and Migration'. *Visual Anthropology Review* 35, no. 1 (2019): 76–111.

29. Ponni. 'Kudankulam: A Brief History and a Recent Update'. *Kafila*, 21 March 2012. Available on https://kafila.org/2012/03/21/kudankulam-a-brief-history-and-a-recent-update/, accessed 10 January 2019.

30. Ponni. 'Kudankulam'.

31. Udayakumar, S.P. 'Kudankulam: The Silent and Telling Emergency In India'. 27 June 2012.Available on http://www.countercurrents.org/udayakumar270612.htm, accessed 10 November 2016.

32. 'After the Mayhem'. 11 September 2012, compiled by Anitha S., *Countercurrents*. Available on http://www.countercurrents.org/kebinston120912.htm, accessed 10 November 2016.

33. 'Kejriwal joins Koodankulam Anti-nuke Stir, High Drama on Site'. *India Today*, 12 September 2012. Available on https://www.indiatoday.in/india/story/kejriwal-joins-koodankulam-anti-nuke-stir-high-drama-on-site-115899-2012-09-12, accessed 10 January 2019.

34. It was later learnt that Jayalalithaa had chaired a state cabinet meeting to reverse her September resolution and proceed to commissioning the reactor. This came with a public statement of appeasement that the cabinet had considered expert reports on safety issues, and had committed a 500 rupees crore [5,000,000,000] development package for the fishing communities. Quite how this package was dispersed was left to the drift of rumour mills. Subramanian, '"Full Steam Ahead": Kudankulam Update'.

35. 'Koodankulam: Idinthakarai Turned into an Open-air Jail, Govt Playing Divide-and-rule', *DiaNuke*, 10 April 2012. Available on https://www.dianuke.org/koodankulam-idinthakarai-turned-into-an-open-air-jail-govt-playing-divide-and-rule/, accessed 7 January 2019.

36. Everyone knew that such charges were exaggerated ploys: 'A senior police officer who doesn't want to be named says this was for two reasons—to "scare people" as any one of them could be among the unnamed; and because naming "all the hundred or thousand was impractical at a time when we were registering multiple cases implicating over 2,000 or 5,000 people in each"'. Janardhanan, Arun. '8,856 "Enemies of State": An Entire Village in Tamil Nadu lives under Shadow of Sedition'. *Indian Express*, 12 September 2012. Available on http://indianexpress.com/article/india/india-news-india/kudankulam-nuclear-plant-protest-sedition-supreme-court-of-india-section-124a-3024655/, accessed 10 January 2019.

37. Cited in S. Senthalir, S. 'Violence against the Non-Violent Struggle of Koodankulam'., *Economic and Political Weekly*, 47, no. 39 (2012): 13–15, p. 14.

38. Senthalir. 'Violence against the Non-Violent Struggle'.

39. Abraham, Itty. 'The Violence of Postcolonial Spaces: Kudankulam'. In *Violence Studies,* edited by Kalpana Kannabiran. New Delhi: Oxford University Press, 2016, p. 325.

40. Abraham, 'The Violence of Postcolonial Spaces: Kudankulam', p. 336.

41. Abraham, 'The Violence of Postcolonial Spaces: Kudankulam', p. 331.

42. Brown, Wendy. 'Finding the Man in the State'. *Feminist Studies* 8, no. 1 (1992): 7–34.

43. '"Report of the Fact-Finding Team's Visit to Idinthakarai and other villages on September 20–21, 2012" by Mr B.G. Kolse Patil, Ms Kalpana Sharma, and Mr R.N. Joe D'Cruz, Chennai'. *Countercurrents,* p. 7. Available on https://www.countercurrents.org/koodankulam260912.pdf, accessed 10 January 2019.

44. '"Report of the Fact-Finding Team's Visit"'.

45. See Alter, Joseph S. *Gandhi's Body: Sex, Diet and the Politics of Nationalism.* Philadelphia: University of Pennsylvania Press, 2000.

46. PMANE Struggle Committee. *PMANE Letters and Press Releases.* Idinthakarai: People's Movement against Nuclear Energy, 2016, p. 6.

47. PMANE Struggle Committee, *PMANE Letters and Press Releases.*

48. 'Report of the Jury on the Public Hearing on Koodankulam and State Suppression of Democratic Rights'. Chennai: Chennai Solidarity Group for Koodankulam Struggle, 2012. Available on https://www.thehindu.com/migration_catalog/article12818959.ece/BINARY/Report%20of%20the%20Jury%20on%20the%20Public%20Hearing%20on%20Kudankulam, accessed 12 January 2019.

49. 'K-ommunity hoists black flags'. *The New Indian Express,* 16 August 2012. Available on http://www.newindianexpress.com/states/tamil_nadu/article590164.ece, accessed 12 October 2015.

50. See Chapter 10 and Nixon, Rob. *Slow Violence and the Environmentalism of the Poor.* London: Harvard University Press, 2011.

51. 'Letter from the Women of Idinthakarai. By Sisters of Idinthakarai'. *Countercurrents,* 19 July 2013. Available on http://www.countercurrents.org/soi190713.htm, accessed 12 October 2015.

52. Hurd, Madeleine. 'Introduction—Social Movements: Ritual Space and Media, Culture Unbound'. *Journal of Current Cultural Research* 6, no. 2 (2014): 287–303, p. 288.

53. Pramod Kumar, G. 'Kudankulam Activists try New Combos to Stop Nuke Plant'. *Firstpost,* 10 May 2012. Available on https://www.firstpost.com/india/kudankulam-activists-try-new-combos-to-stop-nuke-plant-305267.html

54. Moolakkattu, John S. 'Nonviolent Resistance to Nuclear Power Plants in South India'. *Peace Review* 26, no. 3 (2004): 420–6, pp. 421–2.

55. 'Koodankulam: Crackdown on Anti-nuclear Activists and NGO's'.
56. Udayakumar, 'Kudankulam'.
57. 'Kudankulam Protester Udayakumar sends Legal Notice to Manmohan Singh over "Foreign Fund Remark"'. *The Times of India*, 28 February 2012. Available on https://timesofindia.indiatimes.com/india/Kudankulam-protester-Udayakumar-sends-legal-notice-to-Manmohan-Singh-over-foreign-fund-remark/articleshow/12072186.cms, accessed 24 January 2019.
58. 'PMANE Does a Postcard Campaign against False Cases', 24 May 2012. Available on https://www.dianuke.org/pmane-does-a-postcard-campaign-against-false-cases/, accessed 24 January 2019.
59. See Hewitt, Vernon. *Political Mobilisation and Democracy in India: States of Emergency*. New Delhi: Routledge, 2007.
60. Udayakumar, 'Kudankulam'.
61. Udayakumar, 'Kudankulam'.
62. Mbembe, Achille. 'Provisional Notes on the Postcolony'. *Africa: Journal of the International African Institute* 62, no. 1 (1992): 3–37, p. 4.
63. On an elaboration and critique of this term, see Güven, Ferit. *Decolonizing Democracy: Intersections of Philosophy and Postcolonial Theory*. London: Lexington Books, pp. 2–3.
64. See Gramsci, Antonio. *Selections from the Prison Notebooks*. New York: International Publishers, 1971.
65. Bhawna. 'Nuclear Energy, Development and Indian Democracy', p. 224; Subramanian, 'Full Steam Ahead', p. 119; 'Kudankulam Update: Arrests of Peaceful Protesters under the Sedition Law in Tamil Nadu, India', 20 March 2012. See *The Civil Liability for Nuclear Damage Bill 2010* for details on where liability for any nuclear damage and compensation lies. Available on http://www.prsindia.org/billtrack/the-civil-liability-for-nuclear-damage-bill-2010-1042/, accessed 19 January 2018.
66. Bhawna, 'Nuclear Energy, Development and Indian Democracy'.
67. *Report of the People's Movement Against Nuclear Energy (PMANE) Expert Committee Report on Safety, Feasibility, and Alternatives to the Kudankulam Nuclear Power Plant*. 2011. Available on https://www.dianuke.org/wp-content/uploads/2011/12/PMANE_Expert_Committee_Report_Dec_2011.pdf, accessed 20 January 2019.
68. Ramana, *The Power of Promise*, p. 87.
69. Padmanabhan, V.T., R. Ramesh, V. Pugazhendi, K. Sahadevan, Raminder Kaur, Christopher Busby, M. Sabir, and Joseph Makkolil. 'Counterfeit/ Obsolete Equipment and Nuclear Safety Issues of VVER-1000 Reactors at Kudankulam, India'. *Nuclear and Atomic Physics*, 2013. Available on http://vixra.org/abs/1306.0062

70. Gopalakrishnan, A. 'Flaws in Koodankulam Plant'. *The New Indian Express*, 19 June 2013. Available on http://www.newindianexpress. com/opinions/2013/jun/19/Flaws-in-Koodankulam-plant-488169.html, accessed 20 November 2018.

71. Cited in S., Anitha. *No! Echoes Koodankulam*. Trivandrum: Fingerprint Series, 2012, p. 28.

72. Sharma, Aradhana and Akhil Gupta. 'Introduction: Rethinking Theories of the State in an Age of Globalization'. In *Anthropology of the State: A Reader*, edited by Aradhana Sharma and Akhil Gupta. Oxford: Blackwell, 2006, p. 12.

73. Moolakkattu, 'Nonviolent Resistance to Nuclear Power Plants in South India', p. 424.

74. Moolakkattu, 'Nonviolent Resistance to Nuclear Power Plants in South India'.

75. Nilsen, *Dispossession and Resistance in India*, p. 187.

76. Rajappa, Sam, Gladston Xavier, Mahadevan, Rajan, Porkodi for Chennai Solidarity Group for Koodankulam Struggle. 'Fact Finding Report on the Suppression of Democratic Dissent in Anti-Nuclear Protests by Government of Tamil Nadu'. *DiaNuke*, 2012. Available on https://www. dianuke.org/wp-content/uploads/2012/04/Fact_Finding_Report_Sam_ Rajappa_English.pdf, accessed 28 February 2019.

77. See Davies, Thom. 'Toxic Space and Time: Slow Violence, Necropolitics, and Petrochemical Pollution'. *Annals of the American Association of Geographers* 108, no. 6 (2018): 1537–53.

78. S., Anitha, *No!* p. 54.

79. Zonabend, Françoise. *The Nuclear Peninsula*. Cambridge: Cambridge University Press, 1993, p. 107.

80. Berlant, Lauren. *Cruel Optimism*. Durham: Duke University Press, 2011.

81. S., Anitha. 'Call for A Women's Support Group in Kerala for Kudankulam Struggle'. *DiaNuke*, 12 May 2012. Available on http://www.dianuke.org/ call-for-a-womens-support-group-in-kerala-for-koodankulam-struggle/, accessed 10 January 2013. On the significance of women and children to the anti-nuclear weapons movement, see Gusterson, Hugh. *Nuclear Rites: A Weapons Laboratory at the End of the Cold War*. Berkeley: University of California Press, 1998, pp. 209–14; on the role of women in other anti-nuclear movements, see *Women of Plogoff*, trans. Vincent Caillou. France: Editions La Digitale, 2010.

82. Jacob, 'The Movement gathered Momentum only after Women took an Interest'. Women's engagement is extensively covered in S., No, and Vaid, Minnie. *The Ant in the Ear of an Elephant*. New Delhi: Rajpal and Sons, 2016. Therefore, only a summary of key points is presented here.

83. 'After the Mayhem', compiled by Anitha S., *Countercurrents*, 11 September 2012. Available at http://www.countercurrents.org/kebinston120912.htm, accessed 10 November 2016.

84. 'Kudankulam Women from Prison: Tell Everyone We Are Still Here!', Sundari, Xavieramma and Selvi in conversation with Anitha S. on 12 September 2012 at Trichy Women's Prison, *DiaNuke*, 16 October 2012. Availabe on http://www.dianuke.org/Kudankulam-women-from-prison-tell-everyone-we-are-still-here/, accessed 10 November 2016.

85. 'Kudankulam Speaks'. Available on https://chaikadai.wordpress.com/Kudankulam-speaks/, accessed 10 January 2019.

86. 'Kudankulam Speaks'.

87. Available on https://www.transcend.org/tms/2012/12/idinthakarai-women-win-chingari-award-for-nonviolent-struggle/, accessed 16 November 2019.

88. Babu Jayakumar, G. 'Idinthakarai: Women Cheer as Men Sober up'. *The New Indian Express*, 28 December 2011. Available on http://www.newindianexpress.com/states/tamil_nadu/article312544.ece, accessed 10 November 2016.

89. See Ram, Kalpana. *Mukkuvar Women: Gender, Hegemony and Capitalist Transformation in a South Indian Fishing Community*. London: Zed Books. 1991.

90. Parallels may be noted with women who worked in factories, farms, and other places during World War I and II, *demonstrating* rather than *articulating* women's prowess and ability to play equal and vital roles. The aim for gender equality had to be politically intensified after the wars with the suffragette movement.

91. Karat, 'Kudankulam'.

92. See *Women of Plogoff*.

93. Vaid, Minnie. *The Ant in the Ear of an Elephant*. New Delhi: Rajpal and Sons, 2016.

94. 'Every Time They Arrested Me, It Just Made My Will to Protest Stronger'. *The Hindu*, 22 March 2016. Available on http://www.thehindu.com/news/cities/mumbai/news/every-time-they-arrested-me-it-just-made-my-will-to-protest-stronger/article8383692.ece

95. Karat, Jyothy. 'Kudankulam'. Available on http://jyothykarat.com/project/Kudankulam/

96. Bradbury, David. 'David Bradbury on Idinthakarai's Anti-nuclear Frontline'. Available on https://independentaustralia.net/politics/politics-display/david-bradbury-on-idinthakarais-anti-nuclear-front-line,5105, accessed 19 October 2019.

9

Digitalia

By March 2012, events had escalated around the Kudankulam Nuclear Power Plant so much that many outside the village rose to the occasion and organized meetings, rallies, petitions, open letters of appeal, and the like.[1] One of these letters was written collaboratively by environmentalists and activists in Britain, and agreed to and signed off by supportive members of parliament including, at the time of writing, the leader of the Labour party in Britain, Jeremy Corbyn, and former leader of the Green Party, Carolyn Lucas. The letter was sent digitally to the office of the then prime minister of India Manmohan Singh and the chief minister of Tamil Nadu Jayaram Jayalalithaa alongside the Indian media in May 2012.

In view of the sensitive nature of the nuclear topic, a carefully worded letter was composed. Embedded in references to corroboratory sources it listed a series of human rights violations along with the lapse in mandatory procedures for the construction of the plant including the public accessibility of reports on safety analysis, emergency preparedness, and site evaluation in a tsunami and earthquake zone with limited emergency water supplies. The petitioners were keen to strike a conciliatory note, wary of stoking colonial memories or seeming to undermine postcolonial India's sovereignty and development ambitions. An abridged form of the digital letter went as follows:

We write to express our deep concerns regarding human rights and environmental violations around the Kudankulam Nuclear Power Plant in south India.

More than 300 people have been on hunger strike in protest against the construction since May 1st, and a relay strike continues since last year. Several hunger strikers have been admitted to hospital, and other protesters threatened with arrests. The Indian government has reacted to the protests by deploying thousands of police and paramilitary forces in order to commission the reactor in a military style operation. This will have serious consequences for the life and ecology of the whole of peninsular India, as well as the international reputation of India as the world's largest democracy ...

We believe that these draconian measures are not in the interests of a democratic country such as India, and immediate amends should be made to drop these charges and impositions. In a country that has led the world as to non-violent protest which has had enormous impact in South Asia as well as countries such as the USA, South Africa amongst others, we appeal to you to not denigrate the legitimacy of such movements ...

We urge you to:

- Withdraw sedition and 'war against the state' cases amongst other false charges filed against members of Kudankulam People's Movement against Nuclear Energy who have been protesting non-violently and that Section 144 be lifted from the region.
- Ensure that international safety regulations with respect to the above points [on adequate emergency water supplies, safety analysis and site evaluation] are stringently followed.
- Ensure that all reports, reviews and information related to the nuclear plant are made transparent and accessible to the public.

People's basic human rights and environmental safety procedures in the construction of a major nuclear power plant have been abused. Paying short shrift to these mandatory procedures can have mammoth consequences of global proportions.

We fully appreciate that energy needs and security are of a priority for a growing economy and implore that alternative measures be investigated and invested in, rather than resorting to nuclear power plants in ecologically unsound circumstances.[2]

The dispatch accompanied a protest outside the High Commission of India on the Strand in central London (Figure 9.1), and was

Figure 9.1 Anti-nuclear Rally Outisde the Indian Embassy, London, 2012.
Source: Raminder Kaur.

reproduced in part or in full by the Indian media, including the English language press, *The Hindu* and *The Times of India* as well as Tamil dailies such as *Dinakaran* and *Dinamalar*. When doing a Google search on Kudankulam and London the next day, the news reportage amounted to some eight pages of separate entries and played a sizeable part at the time in reigniting and lifting the struggle on to another level with boomerang velocity.[3]

The phenomenal take-up owed largely to increased concerns about safety and the viability of nuclear power in the fallout of the Fukushima Daiichi disaster in March 2011 as it did to the increasing use of social media in 'cybercultural politics' and the imaginings of a virtual global community.[4] This is not to forget other media interventions that might be noted with respect to print, photography, radio, down to more recent innovations as with Blackberry messaging services, 'hacktivism', and Wikileaks.[5] Many modern movements have been triggered by and/or enhanced by communicative potentials through social media.[6] With the de-centring of traditional broadcasting hubs, this new media is characterized by reciprocal one-to-many lines of instantaneous communications in a phenomena described as 'panmediation'.[7] Spanning vast geographies, digital platforms have enabled vigorous and spirited exchange.[8] Not without their inherent problems, the new media has catapulted the collective imagination to motivate social change on a globally conscientized scale.[9]

Email and Facebook communication became even more essential to campaigners based in Idinthakarai whose exit outside the village in the years between 2012 and the year of the national elections in 2014 came with the risk of police harassment and arrest. While the roads to Idinthakarai may have been blocked and the village subjected to surveillance and scrutiny, one of the few advantages of living next to a nuclear power station was that a clampdown of electricity and Wi-Fi transmissions was not entirely possible. Hidden in plain sight, so to speak, virtual communications and coordination continued in quite open ways, protagonists often arguing that there was no need to conceal anything even if they were under surveillance for, after all, what they were doing in the name of non-violent dissent was legal. As one fisherman put it at the time: 'Why should we worry about the state when it is us who are lawful, and they who are lawless?'

An authorized response to the digital letter from Indian officials did not materialize. Nevertheless, plentiful discussion was set alight as is clear in newsreaders' comments and subsequent actions. It is the latter that we then go on to consider with a view to highlighting how the Kudankulam struggle was picked up abroad, and how in turn, those responses were perceived by people based in India and communicated back to Britain. In the process, we will consider the potentials and limits to forge transnational solidarity on the nuclear issue across the global North and South, and probe spaces of criticality across digitalia as they interlace physical with virtual mediation.

Such a transnational enquiry might entail a multi-sited anthropological enquiry entailing a focus on several interconnected field sites as the anthropologist George E. Marcus has proposed.[10] However, the enquiry is more *multi-situated* in that it moves beyond multiple sites as *supplementary contexts* to the life flows of people, materials, and ideas, to consider multi-layered *situations constitutive* of such movements in the making.[11] Despite Marcus' latter-day modifications, multi-sited anthropology continues to subscribe to what the sociologist John Urry has identified as the 'metaphysics of presence' that focuses on 'more or less face-to-face social interactions'.[12] Following decades of work on the decoupling of space, place, and culture along with global flows to do with people and media, sociality needs be decoupled from conventional understandings of being in the same locality.[13] More complex considerations need to be taken on board that can accommodate the materiality of message as well as trace its para/social effects across a diverse range of recipients and sites for creating sociality nearby as well as sociality at a distance.

A succinct example of the synchronous and asynchronous effects of information flows and how they interact in multiple ways, times, and environments, on- and offline, is exemplified by considering a person with a smartphone with an inbuilt GPS navigation unit. The person is 'simultaneously in specific physical places (due to the physicality of their bodies and the planet) and "everywhere" (network space)' as noted by Sally A. Applin and Michael Fischer.[14] This phenomenon is part of a constantly shifting set of configurations. Such a 'multi-plexing of ... mutual blended realities', as the anthropologists put it, transcends the immediate physical space of the local locale to take on board the geo-local context—that is, a locality created alongside a constellation of physical places elsewhere, an else-here perhaps. As goes the title of the book by the sociologists Jennifer Earl and Katrina Kimport, 'digitally enabled social change' is most vividly displayed in activist movements where people move fluidly between on- and offline worlds.[15] This transpired most markedly during the 'Arab Spring' demands for democracy in the Middle East and the Occupy global justice movements in Europe and North America—phenomena that while not simply instigated by technology could not have reached the extents that it did without the affordance of pan-mediation platforms such as Twitter and Facebook.[16]

Our being-in-the-world is not through the corporeal body alone, but through how the body interfaces with other regimes of what Bonnie Nardi calls virtuality.[17] Acknowledging that there remain areas of the world without electricity, Wi-Fi, or access to Web 2.0, there is a need to encompass aspects of digital anthropology not as a discrete *sub*-discipline, but as an integral part of core anthropological focus and method.[18] This is especially important as Nardi notes:

> Huge swathes of human activity have migrated to digital venues where we work, play, study, love, rear children, form relationships, take care of ourselves, and, essentially, exist through digital technology.[19]

The migration to new media is also vested with more significance when it is compelled by socio-political oppression and/or limited financial means to physically travel. Accordingly, as Nardi endorses, the task is to produce 'rich accounts of the affordances of technologies of the virtual and how they are mingled with, and affect, human activity'.[20] It is in this way that I refer to multi-situated ethnography of 'onlife' entanglements-to use the philosopher Luciano Floridi's neologism-to account for the multiplex synergies between computational and on-the-ground complexes.[21]

The Un/holy Trinity

We have already established in earlier chapters that the nuclear is firmly coded as national development; that the national has virtually commandeered the political horizon of possibilities; and that the roping of civilian nuclear plans under national security is sharper in the post-colony sensitized to its sovereign entitlement.[22] Consequently, transnational alliances and grassroots collaborations become a very fraught endeavour in India. The grip of 'nuclear nationalism' or the idea that nuclear power is essential to the growth and sovereignty of the postcolonial nation-state has led to defensive reactions and allegations of the 'foreign hand' when anyone begs to differ.

The foreign hand is a spectral force, both actual and invisible. It connotes alien spectres of interference and interventionism. As these absent-present threats are located outside the national body, it assumes that no self-respecting person in India could, on their own, be able to form a critical or resistive stance to state-backed operations. At a meeting in 2012, S.P. Udayakumar stated:

The authorities cannot believe that someone does something because it's a life and death matter, about the future of their children. They always suspect the foreign hand. It seemed like a convenient way of not identifying any faults in yourself, in your own system, and diverting attention away from this.

Ambiguous yet sturdy, the foreign hand is attached to a chameleon beast that refuses to budge. The discourse came to prominence in the 1970s when the then prime minister, Indira Gandhi, denounced anti-Congress activities as funded or propelled by the USA.[23] Such allegations gathered mileage with the Cold War imprint of US-bashing in the subcontinent where the Western superpower represented the supreme symbol of capitalistic imperialism. Recently, in very different geopolitical and national political economies, the foreign hand discourse has received a boost with the former government led by the United Progressive Alliance to undermine anti-nuclear movements in India.[24] Conveniently overlooking Russia's involvement in the construction of the Kudankulam Nuclear Power Plant, accusations were fired at foreign institutions, charities, NGOs and the CIA (USA's Central Intelligence Agency) for interfering in India's development plans. The digital letter from Britain was submitted, fully mindful of treading such choppy waters.

Individual scholars and other notable personalities based in different regions of the world had also written to the Indian prime minister, with similar points of redress.[25] Amnesty International and Human Rights Watch added to the chorus of criticisms, stressing human rights abuses with reference to the ruthless measures taken against non-violent protesters.[26] In solidarity, a demonstration was organized in May 2012 outside the High Commission of India on the Strand in London. Momentarily, this geo-local patch of transnational India became the place for a combination of visiting Indians, those of Indian descent, students, and social justice and anti-nuclear organizations based in Britain.[27] They included the South Asia Solidarity Group, the Campaign for Nuclear Disarmament, and other activist and Indian diasporic groups such as Tamil Solidarity, Foil Vedanta, Campaign against Criminalising Communities (CACC), Stop the War Coalition, Globalise Resistance, South West Against Nuclear (SWAN), Kick Nuclear, and

Stop Hinkley referring to the EDF nuclear project earmarked for Somerset in west England.

A Digital Letter by a 'Foreign Hand'

The letter was written carefully and collectively through group email including the input of activists and the approval of politicians concerned about their public personas. Campaigners were keen to involve British politicians, not because they particularly revered them, but because it was hoped that this strategy would ensure widespread media attention and possibly constitute a serious engagement with officials in the subcontinent. In addition to Jeremy Corbyn and Carolyn Lucas, other politicians who signed the letter included Labour's John McDonnell (the shadow chancellor at the time of writing), Mary Glindon, Paul Flynn as well as a member of European parliament, Keith Taylor and a former Australian Greens Senator, Scott Ludlum. Digital discussions emphasized the need to focus on the abuse of human rights and mandatory procedures in the building of a nuclear power plant as opposed to an explicit critique of India's development plans. There was an especial need felt to corroborate each statement with footnoted references to other reports and media sources.[28] Once all were satisfied with the letter's content, it was electronically signed and transmitted through the architecture of network space. The next day the letter either appeared in its entirety on activist websites or was fluidly edited as part of newspaper reportage where it became something else in the eyes of Indian readers.[29] Their comments on the letter indicated 'a differentiated genre' to cite the educator and scholar Charles Bazerman.[30] In contrast to the collective and conscientiously composed tone of the digital letter, which aspired to be rational, cautionary, and precise, and the Indian reporters' coverage that aimed for objectivity while trying to hit a sensational note as has been noted in news media in general, readers' responses were much more idiosyncratic, dialogic, dashed, and truncated.[31] The latter presented more of a flickering springboard for self-expression, sometimes with the use of very colourful language when alluding to wider historical and socio-political grievances.

With typographical errors intact, examples of Indian readers' comments to the newspaper articles in the vein of the former sentiment included the following: 'Dr. Singh should shred it [the letter] and Mail back [to] Mr. Cameron [the then Conservative leader and British prime minister]. It is India's internal affair which no [one] should meddle in'.[32] The talismanic presence of a paper letter remained, where the digital letter appears to have morphed into material only fit for the shredder.[33] 'The spooky "foreign hands" that our prime minister saw behind the protest against KKNPP has now come in full public view' wrote another, giving ominous form to suspect interventions.[34]

The attendant discourse of national sovereignty was closely tied with a suspicion of any detractor as reiterated in the following statement:

> ... to all foreigners (anti-nuclear) it is my humble submission that we are sovereign state, and we will decide the energy alternative keeping our resources and demand in mind. i [sic] think that you have funded those anti nuclear NGOs, people of India know about their needs. let [sic] me tell all anti nuclear agents that it is the best nuclear power plant in the world.[35]

Perceived attacks were met with a proud defence of India's techno-developments. Content did not matter as much as the foreign genesis of the letter.[36] Several commentators retreated into a self-justifying nationalism ending their opinion with '*Jai Hind!*' (Glory to India!') The counter-shot manifest itself in a critique of anything British that drew upon age-old colonial arguments as legacies of yore were sparked off by the digital letter.

Noteworthy also is the fact that even though protests were supported by British (or South) Asian people, anti-nuclear organizations and non-Conservative politicians based in Britain, the letter was largely imagined as coming from people in a singularly mono-cultural country—that is, a nation of white, Christian 'Cameronites'. Monocultural assumptions about national identities eclipsed any evidence to the contrary.

One reader wrote mocking British colonial-cockney expressions: 'Manmohan must send a strong message saying "Mind your own business". There ain't british raj in india any more [sic].'[37] Allegations of neo-colonialism became tied to other atrocities and disasters in which Britain's role was impugned:

.... They may better look into their archives and learn more about the autrocites commited [sic] in India when they colonized it. All those treasures looted from us including the Kohinnor, Tippu's sword etc are to be returned to India. We did not see this [sic] MPs reacting when Bhopal caused thousands to die.[38] Their own prime minister vouched for the presence of wepons [sic] of mass destruction in Iraq. Spoiled the future of thousands of Iraquies [sic] by telling lies. India can be taken care of by Indians. Whether it is due to the influence of church or genuine wish to keep Indians in perpetual poverty, your getting involved in this is not correct. [sic][39]

Despite its careful construction, the letter stoked the ghosts of colonial history that fed into a resurgent (H)Indian nationalism accusing Britain of a number of misdemeanours, not immediately linked to the nuclear topic at hand. For some, the letter smacked of a 'Christian conspiracy' where signatories, viewed as all Christians, were deemed to have dubious connections with fishing communities around the Kudankulam Nuclear Power Plant. This alleged transnational alliance in some cases triggered another set of more home-grown Hindu–Christian antagonisms that took on their own rhizomatic narrative arcs quite incidental to the contents of the letter, almost as if the letter was populated by hyperlink triggers that led to other pages.[40] The letter served more as a mutable field for other points of departure through historical recollections, geopolitical reflections and intercommunal tensions.

The tune of powerful and emerging powers indicated by the BRICS nations (Brazil, Russia, India, China, South Africa) as against the shadow of an economically flagging Europe along with US interventionism added to the letter's discordant effects:

Seems to be a desperate attempt of U.S. motivated protests to choke the country which is already facing lots of problems on economic front.... The true mischief makers behind the protests have now emerged out in the open, nations troubled by the rise of the BRICS nations.[41]

This time the line of contention was not simply the colonial legacy of Britain, but the rise of neo-colonial 'mischief makers', the US, and the threat posed to its supremacy by countries amalgamated as BRICS.

Yet there were other registers of reception too including the imagining of impartial and open democracies where dissent could be safely and confidently pursued. With this came a cutting critique of irresponsible Indian politicians with regards to their cavalier plans for nuclear power plants:

> The people who don't even belong to our country can understand the consequences faced by people if something happens at kundakulam nuclear power project ... but why dont our politicians understand the problems occurs [sic] due to these plants...please look at the situation, other alternative resources can be utilised apart from this dangerous plants which are causing many effects to human and environment.[42]

Faced with views from abroad, readers were forced to be globally reflexive from which a certain alternative 'wisdom' against the hegemonic view on nuclear developments emerged. Another commentator wrote:

> Even though Britain has not given up nuclear power, she has come forward to recognise the gross violations of our government in suppressing peaceful protests and refusing to part with vital information. With nuclear power programmes the world over from year to year being only gigantic consumers of high quality energy, with none forthcoming to society outside the nuke industry, it is legitimate to pause and reexamine the wisdom of nuclearisation.[43]

Colonial continuities were highlighted—this time not with reference to British colonial despotism but to the 'gross violations' against Indian citizens by the machinations of its own government. Ironically, contemporary Britain began to stand for just and impartial governance with respect to its citizens, and India as the perpetrator of 'internal colonialism'.[44] A striking vindication of this perhaps rose-tinted view is how several acts that were imposed by the British in the colonial era continue to be used by Indian law enforcement agencies, while those very acts have been repealed, dissolved, or replaced in Britain—a case in point being a section on sedition in the Indian Penal Code that dates back to its implementation by the British in 1860 that had been repealed in Britain in 2009.[45]

Indeed, notions of colonialism, old and new, external and internal, became subject to intense and divisive debates. Support of the letter came in the form of recognizing the global spread of contemporary forms of corporate colonialism:

> If Bhopal culprit Dow chemicals [linked to the 1984 Bhopal gas disaster] is sponsoring the london [sic] [2012 Olympic] games, Chernobyl culprit ASE [Atomstroyexport] is vending the kudankulam nukes. Both mean only business and nothing else. Both are loggerwoods of the same pond. Before criticizing the protest, Can [sic] anyone show a concrete proof that it is powered by foreign funds?.... Christianity triggering, Foreign Funds, Naxalism link [militant Maoist groups in India], False sedition charges. All these managed to remain only at levels of allegations....
> Thanks for your condemns. Protests just got bigger. GOD IS GREAT!!!!!!!!!!![46]

In contrast to the venality of local and national politicians, foreign politicians almost seemed to be cast with the transcendent hand of god.

While the letter writers were careful not to stoke the specters of interventionism or (neo-)colonialism the digital dispatch was recontextualized, reinterpreted, and displaced so as the interests of related actors were aligned with the reader's own.[47] Its responses instantiated an assemblage of contestatory, ambiguous, and consensual opinion and imaginaries—a letter that was either lambasted by some readers for being interfering and suspect, or on the other end of the spectrum, welcomed by others for its supposed attack or 'condemns' of the Indian nuclear authorities and occasionally received almost like a divine intervention to further embolden the anti-nuclear struggle.

Such discursive and discordant effects of the letter are of course familiar. Some of the comments are conceivably part of the stock-in-trade of state-sponsored networks to promulgate what the anthropologist Sahana Udupa reports as 'online nationalism' so as to dominate social media as the content and tone of the responses to the letter appear similar to several other sites.[48] Others registered a very particular strain specific to the Kudankulam issue that demonstrates a more local understanding of Christian communities, transnationalism, and treachery. What makes this episode unique is the way comments were exchanged on a public letter from overseas on what is otherwise cast as a 'national security' concern for the Indian government alone. Debate on an Indian civilian nuclear issue had not entered national, let alone international fora, and it was partly the recent memory of the Fukushima Daiichi disaster of 2011 that enlivened it. Despite the enduring weight of colonial legacies and nationalist discourse, 'outer-national' nodes of solidarity opened up previously closed parameters between the nuclear and the nation.[49] This is not a mean feat considering that, prior to the widespread attention that PMANE agendas received from 2011, while some people may have been unsure or critical about nuclear weapons, they were in the main pro-nuclear energy in view of India's electricity needs.[50] The PMANE struggle publicized the anti-democratic measures that are adopted for a nuclear power plant in the name of the national. Indeed, without the security blanket of nationalism, nuclearism has no legs to stand on in any 'namesake democracy' to cite the words of one Nagercoil-based activist.

A day after the media coverage, a local environmentalist wrote to me in a Facebook message:

> i am very happy and hats of to you ppl [people]. when even our local MP's have disowned us...the MP's signature from london is god send ... pls let them know that i thank them with whole heart.[sic]

When the compound walls began to crumble, London flowed through the journeys of a digital letter into Idinthakarai and back again. This was not as a former colonial or interventionist capital, but as a networked site of transnational solidarity. Indeed, when those from other parts of the world took up the cause, the peculiar historical relationship between Britain and India could no longer provide an anvil to grind an anti-colonial axe. The letter from those based in Britain was further de-territorialized and fluidly adopted by 60 people based mainly in Australia and Canada under the lead of the environmentalist, John Seed, the founder and director of Australia's Rainforest Information Centre.[51] This was shortly followed by another similarly worded letter of support signed by 168 people from 21 countries including many from Japan and other countries from the south such as Indonesia and Taiwan who themselves had their own nuclear anxieties.[52] Foreign bodies built a multilateral bridge with which to forge common grounds against nuclear irregularities and oppression.

While the shady grips of national security, nuclear nationalism, and the foreign hand sum up the un/holy trinity against antinuclear protest in India, the discourses have also been slapped and shaped into other forms in the subcontinent itself. In response to the Indian government's accusation of foreign hand involvement in the Kudankulam protests, the political analyst and journalist Praful Bidwai stated:

> But the real 'foreign hand' is [Manmohan] Singh himself, who is hitching India's energy trajectory to imported reactors, including French reactors at Jaitapur in Maharashtra, and American reactors at Mithi Virdi in Gujarat and Kovvada in Andhra [Pradesh].... As physicist Alvin Weinberg said: 'A nuclear accident anywhere is a nuclear accident everywhere'.[53]

The futility of national borders against nuclear radiation has been acutely felt in the wake of disasters in Chernobyl and Fukushima. Moreover, the neo-liberal Indian state's active engagements with foreign governments, agencies, and corporations in declared or clandestine operations and trade gives the lie to foreign hand allegations against anti-nuclear protestors. This is something that state officials repeatedly understate for, as Itty Abraham observes, any engagement with the foreign for them is 'glossed as encounters with global modern expressed in local idiom'.[54]

Encouraged by transnational interest, Indian journalists such as Sam Daniel from the news channel NDTV in Chennai could confidently write:

> Now the world is watching how India, the largest democracy would tackle this logical and understandable opposition being expressed by local communities. It's for the leaders now to establish that India is a democracy in letter and spirit.... Democracy is precious and priceless. Let's not allow a nuclear plant [to] kill it.[55]

People took what they wished from the coverage and feedback, and the detractive messages in the comments did not deter activists and even a few foreign politicians from pressing ahead with their campaign. Indeed, due to the sheer coverage of the issue, and despite the fact that less than a hundred people turned up to the protests outside the High Commission in London, the protest was seen as 'a success' as one participating activist put it. Indicative of symbiotic relationships in multi-situated spaces, more weight was given to virtual plateaus of intensification rather than just corporeal commitment to the cause, and anti-nuclear campaigners in both Britain and south India were lifted to pick up the trails with more 'onlife' engagements.

Those in Britain set out to organize a parliamentary committee meeting to discuss the Kudankulam issue along with several other plans for nuclear plants. This was set up five months later in a committee room in the Houses of Commons in London at the behest of Caroline Lucas as the Green Party MP, Kate Hudson as the General Secretary of the Campaign for Nuclear Disarmament, and Amrit Wilson from the South Asia Solidarity Group. At the meeting, Lucas began proceedings by stating:

> Global solidarity between those who are resisting nuclear power is absolutely essential. We can all benefit from the strength of knowing that we're part of a wider global movement, to actually sharing expertise and information in how to campaign against multinational corporations who, let's face it, are out there intent on making as much money as they can out of whatever technologies that they can without bothering about paying much attention to the safety, health and effects on the environment of what they're doing.[56]

The convenors of PMANE could not of course be present at the meeting when they resided in an open-air prison in south India. So activists requested Udayakumar to send an audio-visual message from Idinthakarai in view of possible technological problems with a Skype connection on the day. In the transmission to the packed room in parliament, Udayakumar stated:

> We live in a kind of military camp, totally isolated from the rest of the world. Police come and knock on our doors of our houses, ask obscene questions to our women, arrest anyone indiscriminately, and we are not able to resist this. Many political parties and politicians support this ruse ... the normal life in our area has been completely paralysed ... we are living in a total banana republic. We need the solidarity and support of the international community.... We also request the people

of Britain to put pressure on the British government to not engage in nuclear deals with our country.

For the first time on terra firma, voices from a south Indian coastal fishing village could be heard in British parliament (Figure 9.2). Even though mediated through a podcast, the room was charged with a sense of purpose. Despite the militarization and surveillance of the village that curtailed their activities and movement, residents in Idinthakarai forged a contingent political bond through network space across geopolitical locations. Indian petitioners called for global solidarity, not aid or intervention as might be the post-colonial concern. They negotiated treacherous terrains with an attempt to rewrite the familiar script of handouts and foreign hand shadow puppetry. Udayakumar's videographed transmission was then followed by a speech in the committee room by Amrit Wilson from the South Asia Solidarity Group who read out a paper note from Melrit that too had been initially transmitted digitally (Figure 9.3). A

Figure 9.2 Parliamentary Committee Meeting on Nuclear Power, London, 2012.
Source: Rosa Marvell.

Figure 9.3 Melrit Parodies a Nuclear Scientist, Idinthakarai, 2012.
Source: Raminder Kaur.

former smalltime clothes retailer who lived in Idinthakarai, Melrit had joined a large group in boats that surrounded the nuclear power plant on an anti-nuclear day of action after the police descended upon them and ransacked their village in September 2012. Vividly recalling the day, she stated:

> I forgot the desecration of the village church by the police. I forgot the anguish of the wives and mothers of all who were taken away brutally that day. I forgot thirst, hunger.... As I stood on the boat, I remember the demands we have put across ... to let go of the sisters and brothers locked up on unfair charges since September 10 [the day of the camp outside the Kudankulam Nuclear Power Plant] to withdraw all police forces from the villages and reinstate normal life, to close Kudankulam nuclear power plant and convert it into a nature and people friendly energy production plant.[57]

While committee room attendants listened, they were urged to put themselves in the position of those struggling with a nuclear development on their doorstep in the global south. It was as if the sounds of struggles of a south Indian coastal community against a nuclear power plant were crashing around the Victorian edifice of British politics. Social imaginaries went beyond categories of place, ethnicity, race, or nationality as passionate debate on like-minded issues was vented. The Kudankulam Nuclear Power Plant issue was connected to others such as the Fukushima Daiichi reactors in Japan and the reprocessing plant planned for Hinckley Point in Britain. The debate registered a move away from the historical-particular and the national-nuclear, to couching the construction of a nuclear power plant in international human rights discourse and concerns about environmental violations in recursive mirrors of transnational reflections. For a moment at least, it seemed as if the particularities of distant places, and the weight of nuclear-nationalism, national security borders and the foreign hand could be unravelled.

Planetary Forces

One of the main lessons of this chapter is how crucial it is for nuclear authorities to tightly enwrap themselves in discourses of national sovereignty, security, and development so as any connection with the outer-national is viewed as a threat not just for them but for everyone in the country.[58] The transnational digital letter irked this contained vision. While the appeal drew on universalist ideas about the environment, human rights, and democratic principles, it tried to steer clear of national sovereignty, security, and development in a bid to fend off any accusations of interventionism. Nevertheless, its circulation and reception disturbed any tight hold on neat nuclear-nationalisms and parcelled-off visions. As the tenacity of territory was displaced to emphasize de-territorialized connections, the letter provoked, on the one hand, ardent nationalistic defensiveness, and on the other, a means with which to forge transnational solidarity against the abusive turn to nuclear power.

By following the public letter, multi-situated matrices flow and flicker, both physical and virtual. The site is not then a single spatial type, magnified into multi-sites in India and Britain, for instance. Rather it is a complex of several kinds of interlocking Euclidean, networked, fluid and fiery topologies—different kinds of spatialities as proposed by the social scientists John Law and Annemarie Mol.[59] It is around and through these topologies that new discourses took shape and activities took place along sliding global—national—local scales. They then informed stakes to capitalize on the grounds gained in order to set up more physical gatherings that continued to interface with the translocal potentials and limits of online media. Digitalia was energized through the engagement of non-proximate people—students and others from India living in Britain, to those with diasporic connections to the subcontinent, to those who expressed concerns about nuclear irregularities as well as disasters in places like Britain and Japan, to those who fought for environmental and social justice and the right to dissent across the democratic world, to those who lived in an open prison in a coastal village, and to others who tried to rally around on land and on sea in order to register their objections.[60]

Hurdles are not hindrances and anti-nuclear campaigners continue to circumvent obstacles with the conviction of the parallelism of their struggles worldwide. Through a process of what Arjun Appadurai has called 'grassroots globalization', they advocate for an ecological consciousness without borders akin to the oceanic consciousness among fishing communities explored in Chapters 3 and 5.[61] Whether they be located in India, Britain, Japan, Indonesia, Taiwan, Australia, Canada, Germany, France, or other sites of nuclear antagonism, they represent a nascent sense of planetary citizenship against the pressures and limits placed on them by national-nuclear and multi/transnational structures and strictures.[62] Instead, they rewrite nuclear nationalism as trans/national nuclear capitalism where the trans/national indicates, first, between states in terms of the exchange of skills, agreements and trade; for nuclear installations and, second, the way this exchange is then put under erasure indigenously when skills, trade and agreements are masqueraded as in the national interest of everyone.[63]

Moreover, nation-state border mentalities fall apart in view of the environmental effects of nuclear operations, accidents and the inevitable seepage of radioactive waste that make the internationality of the struggle even more vital for the sake of a common earth. When we shift our perspective from land to ocean our horizon alters with radical possibilities of a planetary consciousness. 'Lands, waters, space, people, and time' can indeed be fluidly connected.[64] As a fisherwoman said in

Idinthakarai: 'Why can't the government give us the right to live?.... The wealth of the oceans is ours. The land, the soil, the earth that we live on is ours'.[65] Deep democracy has its deep dimensions in ecology as well as in time—in other words, to add depth is to add to the strength and endurance of claims to what we have inherited and what we leave behind.

Notes

1. For instance, acclaimed filmmaker, David Bradbury, appealed to Australian readers: 'You have to take the information and run with it, and find ways of supporting the people of Idinthakarai by hassling the Indian Government through the local High Commissions in Sydney, Brisbane, and Canberra. Let both the Australian and Indian governments know you don't appreciate this brave community being put through outrageous and anti-democratic actions any more than you appreciate our governments opening up our local area to CSG [Coal Seam Gas] mining and so spoiling the aquifers forever'. 'David Bradbury on Idinthakarai's Anti-nuclear Frontline'. Available on https://independentaustralia.net/politics/politics-display/david-bradbury-on-idinthakarais-anti-nuclear-front-line,5105, accessed 12 January 2019.

2. 'TEXT: Letter to Indian PM and Tamil Nadu CM from British MPs and Leaders', 19 May 2012. Available on http://www.dianuke.org/text-letter-to-indian-pm-and-tamil-nadu-cm-from-british-mps-and-leaders/, accessed 23 October 2012.

3. See Keck, Margaret E. and Kathryn Sikkink. *Activists beyond Borders: Advocacy Networks in International Politics.* Ithaca: Cornell University Press, 1998.

4. Riberio, Gustavo Lins. 'Cybercultural Politics: Political Activism at a Distance in a Transnational World'. In *Cultures of Politics, Politics of Culture*, edited by Sonia Alvarez, Evelina Dagnino, and Arturo Escobar. Boulder: Westview Press, 1996.

5. This is not to forget other media interventions that might be noted with respect to print, photography, radio, down to more recent innovations as with Blackberry messaging services, 'hacktivism', and Wikileaks.

6. See Juris, Jeffrey S. *Networking Futures: The Movements against Corporate Globalization.* Durham: Duke University Press, 2008

7. See DeLuca, Kevin M., Sean Lawson, and Ye Sun. 'Occupy Wall Street on the Public Screens of Social Media: The Many Framings of the Birth of a Protest Movement'. *Communication, Culture and Critique* 5, no. 4

(2012): 483–509; Castells, Manuel. *Communication Power*. Oxford: Oxford University Press, 2009.

8. See Van Aelst, Peter and Stefaan Walgrave. 'New Media, New Movements? The Role of the Internet in Mapping the "Anti-Globalization" Movement'. *Information, Communication and Society* 5, no. 4 (2002): 465–93.

9. Hurd, Madeleine. 'Introduction—Social Movements: Ritual Space and Media, Culture Unbound'. *Journal of Current Cultural Research* 6, no. 2 (2014): 287–303, pp. 298–9.

10. Marcus, George E. 'Ethnography in/of the World System: The Emergence of Multi-Sited Ethnography'. *Annual Review of Anthropology* 24 (1995): 95–117.

11. On a more theoretical elaboration, see Kaur, Raminder. 'The Digitalia of Everyday Life: Multi-Situated Anthropology of a Virtual Letter by a "Foreign Hand"'. *HAU: Journal of Ethnographic Theory* 92, no 2 (2019): 299–319.

12. Marcus, George E. 'Multi-sited Ethnography: Five or Six Things I Know About it Now', 2005. Available on http://eprints.ncrm.ac.uk/64/1/george-marcus.pdf, accessed 30 April 2018. Urry, John. *Mobilities*. Cambridge: Polity Press, 2007, p. 47.

13. See Appadurai, Arjun. 'Disjuncture and Difference in the Cultural Economy'. In *Modernity at Large*. Minneapolis: University of Minnesota Press, 1996, pp. 27–47; and Gupta, Akhil and James Ferguson. *Anthropological Locations: Boundaries and Grounds of a Field Science*. Berkeley: University of California Press, 1997.

    On diasporic communication, see Anderson, Benedict. *Long-Distance Nationalism: World Capitalism and the Rise of Identity Politics*. Amsterdam: Centre for Asian Studies, 1992; Brinkerhoff, Jennifer M. *Digital Diasporas: Identity and Transnational Engagement*. Cambridge: Cambridge University Press, 2009; Conversi, Daniele. 'Irresponsible Radicalisation: Diasporas, Globalization and Long-Distance Nationalism in the Digital Age'. *Journal of Ethnic and Migration Studies* 38, no. 9 (2012): 1357–79.

    On sociality through new media, see Green, Sarah, Penny Harvey, and Hannah Knox. 'Scales of Place and Networks: An Ethnography of the Imperative to Connect through Information and Communications Technologies'. *Current Anthropology* 46, no. 5 (2005): 805–26; Hansen, Mark B.N. *Bodies in Code: Interfaces with New Media*. London: Routledge, 2006; and Miller, Daniel and Don Slater. *The Internet: An Ethnographic Approach*. Oxford: Berg, 2006; and Coleman, G.E. 'Ethnographic Approaches to Digital Media'. *Annual Review of Anthropology* 39, no. 1 (2010): 487–505.

14. Applin, Sally A. and Michael Fischer. 'PolySocial Reality', 2011. Available on http://www.wired.com/2011/05/polysocial-reality-posr/, accessed 30 April 2018.

15. Earl, Jennifer and Katrina Kimport. *Digitally Enabled Social Change.* Cambridge: MIT Press, 2011.

16. Werbner, Pnina, Martin Webb and Kathryn Spellman-Poots eds. *The Political Aesthetics of Global Protest: The Arab Spring and Beyond.* Edinburgh: Edinburgh University Press. 2014; Tufekci, Zeynep and Christopher Wilson. 'Social Media and the Decision to Participate in Political Protest: Observations from Tahrir Square'. *Journal of Communication* 62, no. 2 (2012): 363–79; Fuchs, Christian. *OccupyMedia! The Occupy Movement and Social Media in Crisis Capitalism.* Winchester: Zero Books, 2014.

17. Nardi, Bonnie. 'Virtuality'. *Annual Review of Anthropology* (2015) 44: 15–31,

18. See Miller, Daniel and Heather Horst. 'The Digital and the Human: A Prospectus for Digital Anthropology'. In *Digital Anthropology,* edited by Daniel Miller and Heather Horst. London: Berg, 2012.

19. Nardi, 'Virtuality', p. 16.

20. Nardi, 'Virtuality', p. 25. The webs of significance that Clifford Geertz famously wrote about may be recast so as they do not reside in holistic or systematic notions of culture alone as has been widely critiqued, nor in humanist understandings of social realities. On counts of both diverse cultures, and through our entanglement in technological mediation, we live in an era of *trans-cultural webs of significance.* The trans- in the trans-cultural refers to both a revisitation of classical theories of the field, culture, and the human, as much as it does to technological mediation that inform and co-produce the frame, mode, and 'object' of anthropological enquiry. See Geertz, Clifford. *Deep Play: Notes on the Balinese Cockfight.* Indianapolis: Bobbs-Merrill, 1972.

21. Floridi, Luciano (ed.). *The Onlife Manifesto: Being Human in a Hyperconnected Era.* Cham, Switzerland: Springer International Publishing, 2015. Modifying Marcus' proposals for multi-sited fieldwork, therefore, multi-situated refers to: first, temporally localized and inter-linked situations, which themselves are not exclusively physical sites, in the attempt to learn about how moments effect and transform social realities; second, the reflexive relations set into play by composite onlife entanglements on sliding scales that might be nearby and at a distance; and third, the 'paraethnographic' where the clockwork or 'native point of view', whether it be physically grounded or digitally mediated, leads to new kinds of dialogues and dynamics. Marcus, George E. 'Multi-sited

Ethnography: Five or Six Things I Know About It Now', 2005. Available on http://eprints.ncrm.ac.uk/64/1/georgemarcus.pdf, pp. 23–4.

22. See Abraham, Itty. *The Making of the Indian Atomic Bomb: Science, Secrecy and the Postcolonial State*. London: Zed Books, 1998; and Ramana, M.V. 'India's Nuclear Enclave and the Practice of Secrecy'. In *South Asian Cultures of the Bomb: Atomic Publics and the State in India and Pakistan*, edited by Itty Abraham. Bloomington: Indiana University Press.

23. See Kalugin, Oleg. *Spymaster: My Thirty-Two Years in Intelligence and Espionage against the West*. New Year: Basic Books, 2009.

24. Sharma, Dinesh S. 'Foreign Hand Nuking Tamil Nadu Nuclear Power Project, says Prime Minister Manmohan Singh'. *India Today*, 24 February 2012. Available on http://indiatoday.intoday.in/story/kudankulam-row-prime-minister-blames-foreign-countries/1/175007.html, accessed 23 September 2012.

25. 'Koodankulam: Open Letter from International Scholars to the Indian PM'. *DiaNuke*, 4 May 2012. Available on http://www.dianuke.org/Kudankulam-open-letter-from-international-scholars-to-the-indian-pm/. See also Wankhede, Anuj. 'Solidarity Pours in from Across the World for Anti-nuclear Protests in India'. *DiaNuke*, 30 May 2012. Available on http://www.dianuke.org/solidarity-pours-in-from-across-the-world-for-the-indian-antinuclear-protests/. 'Koodankulam: German Anti-Nuclear Groups Express Solidarity at Chernobyl Vigil in Frankfurt'. *DiaNuke*, 4 May 2012. Available on http://www.dianuke.org/Kudankulam-german-anti-nuclear-groups-express-solidarity-at-chernobyl-vigil-in-frankfurt/, accessed 23 October 2012.

26. 'Koodankulam: Letter to IAEA, Nuclear Regulators and Human Rights Organisations'. *DiaNuke*, 25 March 2012. Available on http://www.dianuke.org/Kudankulam-letter-to-iaea-nuclear-regulators-and-human-rights-organisations/; 'Latest Updates from Koodankulam'. *DiaNuke*, http://www.dianuke.org/latest-updates-from-koodankulam/; 'India: End Intimidation of Peaceful Protesters at Nuclear Site'. *Human Rights Watch*, 11 May 2012. Available on https://www.hrw.org/news/2012/05/11/india-end-intimidation-peaceful-protesters-nuclear-site, accessed 23 October 2012.

27. Kohli, Rashmi. 'Protest in London to Stop Kudankulam: Report and Pictures'. *DiaNuke*, 19 May 2012. Available on http://www.dianuke.org/protest-in-london-to-stop-Kudankulam-report-and-pictures/, accessed 23 October 2012.

28. See Maurer, Bill. 'Due Diligence and Reasonable Man, Offshore'. *Cultural Anthropology* 20, no. 4 (2005): 474–505; Hull, Matthew S. 'Documents and Bureaucracy'. *Annual Review of Anthropology* 41 (2012): 251–67.

29. See Mol, Annemarie and John Law. 'Regions, Networks and Fluids: Anaemia and Social Topology'. *Social Studies of Science* 24, no. 4 (1994): 641–71.

30. Bazerman, Charles. 'Letters and the Social Grounding of Differentiated Genres'. In *Letter Writing as a Social Practice*, edited by David Barton and Nigel Hall. Amsterdam: John Benjamins Publishing, 2000, pp. 17–29.

31. This is not to overlook the fact that, in the shape of a petition, the final digital letter oscillated between the representation of 'facts' to an element of affect ('we believe...we implore...we fully appreciate...'). On newspaper reportage in India, see Udupa, Sahana. *Making News in Global India: Media, Publics, Politics*. Cambridge: Cambridge University Press, 2015.

32. Suroor, Hasan. 'U.K. MPs Sign Anti-Kudankulam Letter to Manmohan'. *The Hindu*, 17 May 2012. Available on http://www.thehindu.com/news/article3429248.ece. Unless otherwise indicated, subsequent quotes in response to news of the letter's contents are taken from the 46 comments on this web page and the 91 comments on this site: 'British MPs Write to PM Manmohan Singh against Kudankulam Nuke Plant'. *The Times of India*, 18 May 2012. Available on http://timesofindia.indiatimes.com/india/British-MPs-write-to-PM-Manmohan-Singh-against-Kudankulam-nuke-plant/articleshow/13262940.cms, accessed 5 June 2015.

33. The phenomenon recalls Kevin Hetherington's discussion about *praesentia* or presence of mind, something absent that can attain a presence. Hetherington, Kevin. 'Spatial Textures: Place, Touch, Praesentia'. *Environment and Planning A*, 35, no. 11 (2003): 1933–44, p. 1941.

34. 'British MPs Write to PM Manmohan Singh against Kudankulam Nuke Plant'.

35. Suroor, 'U.K. MPs Sign Anti-Kudankulam Letter to Manmohan'.

36. See Hetherington, 'Spatial Textures: Place, Touch, Praesentia'; Gitelman, Lisa. *Paper Knowledge: Toward a Media History of Documents*. Durham: Duke University Press, 2014.

37. 'British MPs Write to PM Manmohan Singh against Kudankulam Nuke Plant'.

38. This is a reference to Union Carbide, an American multinational corporation that was running the plant in Bhopal when the disaster happened. It is now re-named the Dow Chemical Company.

39. Suroor, 'U.K. MPs Sign Anti-Kudankulam Letter to Manmohan'.

40. 'British MPs Write to PM Manmohan Singh against Kudankulam Nuke Plant'.

41. Suroor, 'U.K. MPs Sign Anti-Kudankulam Letter to Manmohan'.

42. Suroor, 'U.K. MPs Sign Anti-Kudankulam Letter to Manmohan'.

43. Suroor, 'U.K. MPs Sign Anti-Kudankulam Letter to Manmohan'.

44. See Churchill, Ward. *Acts of Rebellion: The Ward Churchill Reader*. New York: Routledge, 2003.

45. Sonwalker, Prasun. 'Sedition Law in UK Abolished in 2009, Continues in India'. 16 February 2016. Available on https://www.hindustantimes.com/world/sedition-law-in-uk-abolished-in-2009-continues-in-india/story-Pkrvylv6J0T3ddY8uqvKsO.html, accessed 6 January 2020.

46. Sonwalker, 'Sedition Law in UK Abolished'.

47. Barton, David and Nigel Hall (eds). *Letter Writing as a Social Practice*. Amsterdam: John Benjamins Publishing, 2000, p. 9. See also Law, John. 'Notes on the Theory of the Actor-network: Ordering, Strategy, and Heterogeneity'. *Systems Practice* 5, no. 4 (1992): 379–93.

48. Udupa, Sahana. 'Middle Class on Steroids: Digital Media Politics in Urban India'. *India in Transition*, 2016. Available on https://casi.sas.upenn.edu/iit/sudupa

49. See Gilroy, Paul. The *Black Atlantic Modernity and Double Consciousness*. Cambridge, MA: Harvard University Press, 1993, pp. 16, 17.

50. Kaur, Raminder. *Atomic Mumbai: Living with the Radiance of a Thousand Suns*, New Delhi: Routledge, 2013.

51. Krishnan, Jyothi, Anitha Sharma, and Santhi. 'Stop Koodankulam Project: It Involves Abuse of Democracy, Ecology and Human Rights'. *DiaNuke*, 29 May 2012. Available on http://www.dianuke.org/stop-Kudankulam-project-it-involves-abuse-of-democracy-ecology-and-human-rights/, accessed 23 October 2012.

52. 'Koodankulam: 168 International Organizations from 21 Countries Urge the Indian Govt to Stop the Reactor'. *DiaNuke*, 1 June 2012. Available on http://www.dianuke.org/Kudankulam-168-international-organizations-from-21-countries-urge-the-indian-govt-to-stop-the-reactor/, accessed 23 October 2012.

53. Bidwai, Praful. 'Nuclear Power and Democracy'. *The News*, 9 June 2012. Available on http://www.thenews.com.pk/Todays-News-9-113435-Nuclear-power-&-democracy, accessed 23 October 2012.

54. Abraham, Itty. 'The Violence of Postcolonial Spaces: Kudankulam'. In *Violence Studies*, edited by Kalpana Kannabiran. New Delhi: Oxford University Press, p. 334.

55. Daniel, Sam. 'Blog: Democracy in Peril at Kudankulam'. 6 June 2012. Available on http://www.ndtv.com/article/south/blog-democracy-in-peril-at-kudankulam-228251, accessed 23 October 2012.

56. See Marvell, Rosa and Raminder Kaur. 'Civilian Nuclear Proliferation and Deep Democracy'. *Countercurrents*, 23 February 2013. Available on http://www.countercurrents.org/kaur230213.htm, accessed 23 October 2013.

57. On women's narratives in Idinthikarai, see S., Anitha. *No! Echoes Kudankulam*. Trivandrum: Fingerprint Series, 2012; and Vaid, Minnie. *The Ant in the Ear of an Elephant*. New Delhi: Rajpal and Sons, 2016.

58. See Shampa, Biswas. 'Patriotism in the Peace Movement: The Limits of Nationalist Resistance to Global Imperialism'. In *Interrogating Imperialism: Conversations on Gender, Race and War,* edited by Naeem Inayatullah and Robin Riley. New York: Palgrave McMillan, 2006, pp. 63–99.

59. Law, John and Annemarie Mol. 'Situating Technoscience: An Inquiry into Spatialities'. *Environment and Planning D: Society and Space* 19, no. 5 (2001): 609–21.

60. Markedly, the focus on the digital letter highlights how resistance does not just lie along horizontal alliances outside of or peripheral to dominant institutions and networks as noted in much of the literature on direct action and global justice movements, and on resistance in general. 60 Instead, we see the utilization of state representatives where required, pointing to the inherent critique, dissolution, and dissent even within the political establishment as was also the case with PMANE and a handful of supportive politicians in India.

    On direct action and global justice movements, see Juris, Networking Futures; Maeckelbergh, Marian. *The Will of the Many: How the Alterglobalisation Movement is Changing the Face of Democracy*. London: Pluto Books, 2009; Graeber, David. *Direct Action: An Ethnography*. ReadHowYouWant.com, 2009; Berkovitch, Nitza and Sara Helman. 'Global Social Movements'. In *A Companion to Gender Studies*, edited by Philomena Essed, David Theo Goldberg, and Audrey Kobayashi. London: John Wiley, 2009, pp. 266–78.

61. Appadurai, Arjun. 'Deep Democracy: Urban Governmentality and the Horizon of Politics'. *Environment and Urbanization* 13, no. 2 (2001): 23–43, p. 42.

62. See, for instance, the campaign in Germany 'Anti-Jaitapur Activists Go Global, Seek Support in Germany'. *Indian Express*, 20 September 2013. Available on http://archive.indianexpress.com/news/antijaitapur-activists-go-global-seek-support-in-germany/1171542/, accessed 23 October 2013.

63. See Vanaik, Achin. 'National Interest: A Flawed Notion'. *Economic and Political Weekly* 41, no. 49 (2006): 5045–9.

64. Todd, Zoe. 'Protecting Life below Water: Tending to Relationality and Expanding Oceanic Consciousness beyond Coastal Zones'. *American Anthropologist* website, 17 October 2017. Available on http://www.americananthropologist.org/2017/10/17/protecting-life-below-water-

by-zoe-todd-de-provincializing-development-series/, accessed 19 January 2019.

65. Cited in the documentary, *Voices from the Daughters of the Sea, Koodamkulam'* part 2. Available on https://www.youtube.com/watch?time_continue=31&v=muGfMaIyWZg, accessed 27 February 2019.

10

Do We Exist?

Repeatedly we were told that people did not know how to seek justice when doing so would mean going to the very police that had attacked their homes and arrested their people. (B.G. Kolse Patil, Kalpana Sharma and R.N. Joe D'Cruz, 2012)[1]

It is so strange and unfortunate that the central and state governments treat us, nonviolent and democratic Gandhian activists as some kind of dangerous extremists. (People's Movement against Nuclear Energy, 2016)[2]

The more the people in Idinthakarai protested against the nuclear power plant, the more their claim to legitimate citizenry was revoked. It was as if they had become reduced to brutalized and ostracized 'non-persons'. On 23 October 2012, while thousands languished in the 'open-air jail' for fear of arrest outside the village, a group calling themselves the 'Women of Idinthakarai' made a poignant appeal.[3] They wrote in a digital letter:

Do we exist? Do we live within the exclusivity or sterilization zone?... We understand that none of our representations or appeals have been considered.... We do not want to be relocated or rehabilitated. We want to be here by the seashore in our own birthplace. We want to pursue our livelihoods linked to the sea and its bounties. We want good food, water and access to resources here in these villages. We do not want money that is so ephemeral. We are willing to work hard, earn and live well. This is the only representation that we want to make. We have no complaints other than dissent about the way in which the concerned authorities are unwilling to come to us and allay our fears and doubts. We want them to assure us that the KKNPP [Kudankulam Nuclear Power Plant] will not be allowed to attain criticality at the cost of our lives and dreams. We want our sisters and brothers languishing in the jails to be released. We want our peaceful resistance to be dealt with decently and humanely.[4]

For exercising their right to lives and peaceful dissent, village residents were delivered dire blows. In the hands of a punitive state, PMANE supporters were forced to live either in prison or a 'sterilization zone' that is a designated five kilometer zone around the nuclear reactors as stipulated by India's Atomic Energy Regulatory Board. Both places were empty of life-potentials to do with their health and livelihoods.[5] Not integral to or compliant with the state of nuclear biopolitics, they were decreed disposable and subjected to a shadow existence in a case of nuclear necropolitics.

While according to Michel Foucault, the modern state organizes and affirms the lives of populations in biopolitics—'to make live and to let die'—following Achille Mbembe, necropolitics departs from biopower to emphasize the centrality of death to the organization of socio-political life—'to make die and let live'.[6] With a focus on conditions of colonization, slavery and apartheid in which people are subjected to a status of 'living death' under technologies of destruction, Mbembe reminds us that 'death-worlds' are not an archaic form of sovereign government for they persist into the modern era along with biopower.[7] One does not supersede the other. Rather, as the historian Kathleen Siddick contends there is a 'living death in the "make die"' of the modernist political project.[8]

Taken into the realm of energy, the anthropologist Dominic Boyer adapts the Foucauldian strand of biopower: 'Electropolitics infuse governance' he affirms when discussing the grids of electricity—in a neologism, energopower.[9] He elaborates: 'modalities of "biopower" (the management of life and population) today depend in crucial respects upon modalities of energopower (the harnessing of electricity and fuel) and vice versa'.[10] The modern subject cannot envisage a life without electricity especially when it is smoothly and invisibly integrated into the minutiae of their social world.

Energy *necro*power as opposed to *bio*power reverses the optics. By focusing on the deathly underside, alter-worlds are emphasized. From this perspective, death worlds may emerge by way of technologies that, on the one hand, claim to be life-enhancing but, on the other, can be revealed to be life-destroying. The supposed smooth operations of nuclear technologies to produce power, when viewed from another position, can create death conditions for those who have little to gain from them other than environmental harm, health hazards and harassment. The project of life-enhancement by generating electricity and other goods through large-scale installations for metropolitan and industrial hubs comes at huge risks to those living around and objecting to nuclear power stations, a deathly biopolitics.[11]

In the Idinthakarai case, the lives of those who dissented were death-dispensable to the life-power of urban and industrial needs located miles away. Such death conditions were generated by a combination of subtle, surreptitious, and blatant forces, and in this case, compounded by the opacity and autocracy of the nuclear state. First was the ecological condition created by way of a silent and encroaching death where nuclear industries subjected marginalized communities and casual labourers to a life of uncertainty, exploitation, and health hazards. Second, there was the more direct and quick fire policy of state violence that was exacted on- and offline in order to contain and extinguish dissent as with targeting people who protest against nuclear plans or through the hiring of henchmen to enact violence against people and property.[12] Third were social deaths through the more indirect politics of delay, denial, and dismissal through overt and covert practices designed to malign and outcaste protagonists.[13] This third modality came with a wider web of complicity and consent—the benev(i)olent lure of money and other incentives attendant with a fear of denied services and provisions and upsetting the barrel should one desist.[14]

Modalities connote both discursive and figurative elements—processes that (re)produce necropower combined with more authored strategies.[15] The distributed 'authors' might come to the fore as politicians, civil servants, police, intelligence officers, henchmen, and allied media operators that surround the core of nuclear authorities protected as they are under the dark umbrella of national security.

With these combined ecocidal, political and social death conditions, victims, suspects, and/or targets were created by a number of syncretic subjugations to do with 'let die' and 'make die'. In this chapter, we examine in more detail nuclear necropower, and its effects on those who dare to defy the nuclear power plant. In so doing, we can begin to scrutinize what Judith Verweijen and Alexander Dunlap describe as 'political (re)actions "from above"'—a combination of state, corporate, and allied endeavours that range from 'managing' dissent to 'manufacturing' consent so as to prevent and pre-empt discord emerging in the first place.[16] The (re)actions are labyrinthine and often covert as we now attempt to bring into the light.

Silent, Quick, and Slow

The first modality of nuclear necropolitics was vindicated with sphinx-like silence through an encroaching death that is more a case of 'let die' rather than 'make die'.[17] This more ecocidal death condition chimes with other regional studies on racial and class hierarchies described by Rob Nixon and Gabrielle Hecht's proposals as 'slow violence'—a violence that might accompany contact with radioactive material, waste or through the radionuclide emissions from reactors in general that enter into the food chain and the DNA of nearby residents, an increasing percentage of whom will succumb to genetic mutations, disease, and death over the decades.[18] From these perspectives, nuclear plants are little more than emblematic tombs: death reaped by a grim reactor that, aside from accidents, intensifies what Nixon calls 'long dyings' over time (Figure 10.1).[19] The radionuclides isotopes and low-level liquid waste whose dangerous half-life spans days to decades is akin to 'attritional violence' from the nuclear reactors. It is 'a violence that is neither spectacular nor instantaneous, but rather incremental and accretive, its calamitous repercussions playing out across a range of temporal scales' as Nixon elaborates on the politics of toxicity. The ecological meets with the economical as the death condition is compounded by structural violence to do with neo-Brahmanic hierarchies highlighted by a lack of health and safety provisions for low-caste casual labourers in the nuclear industries, a phenomenon that we have already covered in Chapters 2, 4, and 7.

Regarding the second modality to do with the nuclear state's more punitive re(actions) to 'make die': the most glaring are homicides of anti-nuclear protesters who rise against their subjugation. These go

Figure 10.1 Poster Highlighting the Deathly Effects of Nuclear Power Stations, Nagercoil, 2006.
Source: Raminder Kaur.

back to the 1980s when a fisherman known as Ignatius was killed in police fire at a rally in 1988. The toll picked up with the resurgence of resistance and the police firing at Anthony John in a solidarity rally in 2012, along with the detrimental dive of an Indian Coast Guard surveillance plane on a crowd of people engaged in *jal satyagraha* (water-based truth-force struggle) that led to the demise of Sahayam Francis (refer to Figure 1.2 of this book.) (in the same year).[20] Other deaths were less direct, emerging from the (re)production of death conditions through the combined forces of make die and let die.

The collective commitment to advance towards the nuclear plant walls on the coast in September 2012 with which we opened this book was the result of careful organizing in order to publicly register peoples' exasperation with the nuclear state. The classifying of the administrative territory as a prohibited zone for assembly under the colonial hangover of Section 144 discernibly shifted the state's stance from March that year (Figure 10.2). This shift and peoples' affront to the shift became the pretext for a violent assault on those who defied the order with *lathi* beatings and tear gas attacks. Simultaneously, while people gathered outside the compound walls of the nuclear power plant, another batch of paramilitary soldiers had entered the near empty village of Idinthakarai, vandalized and looted people's

Figure 10.2 Police and Paramilitary Cordon Idinthakarai Village and the Kudankulam Nuclear Power Plant, 2012.
Source: Amirtharaj Stephens/Pep Collective.

homes, bikes, and boats, threw sand into household sacks of grain, destroyed the sacred statue of Mother (Matha) Mary, urinated in the church, and threatened and harassed anyone that they encountered.[21]

The multipronged assault left a big dent in the morale and purses of families and communities.[22] Those who were arrested were subjected to isolation and intimidation, and treated like criminals rather than the political protesters that they were. While some were detained for a few hours or days, others remained in prison for months before being conditionally released with severe limitations on their movements and freedom of expression.[23] With trumped up police charges, the primary aim was to create a prolonged psychosis of discipline and distress for anyone who chose to dissent.[24]

Most people survived their ordeal, but 63-year-old J. Roslin who was arrested during September's civil disobedience, had a severe ailment and was denied treatment. It led to her further deterioration in prison. She was released on bail and asked to sign into a Madurai police station every week, an overland journey that took five to six hours one way. Shortly after being conditionally released, she succumbed to her illness. People were both saddened and incensed, convinced that her death was yet another outcome of state vengeance. They declared:

Roslin is a victim of neglect, and the vengeance of a state that views the very holding of a contrary opinion on nuclear power as a crime warranting imprisonment under harsh sections.[25]

The death toll around Kudankulam is on top of one other in protests around the Jaitapur nuclear power plant in Maharashtra when a police officer driving a four-wheel drive Sumo SUV rammed into a 40-year-old activist, Irfan Qazi, on his scooter.[26] There are also three elderly farmers, Ishwar Singh Siwach, Bhagu Ram, and Ram Kumar, who died due to the accumulative stress of a sit-in (dharna), a long-term land-based occupation that had been going on since 2010 against acquisition plans for the Gorakhpur nuclear power plant in the blazing heat of Haryana state.[27] In all cases, the deaths were spurred by necropolitics produced by the nuclear state.

Unsurprisingly, authorities have denied culpability, a familiar pattern that has characterized state (re)actions to people's protests against large-scale developments across the field. These are just the tip of the iceberg for there are plenty of 'accidents' that have led to critical illnesses among those employed in nuclear industries, particularly among those subcontracted by middle-men who are not on any labour or occupational health register, descending even further from death by intent and neglect to the silent or rather, silenced death condition of the first modality.[28]

Violence might itself be 'outsourced', distantly orchestrated through a series of middlemen who have no official standing. While the siege and standoff was ongoing in Idinthakarai from March 2012, assaults were foisted upon anyone or anything outside the village associated with PMANE. Henchmen rumoured to be connected to local political groups, Hindu Munnani and Congress, took to damaging the properties of the lead spokesperson with two attacks on the South Asian Community Centre for Education and Research (SACCER) Matriculation School led by S.P. Udayakumar's wife, Meera, on the outskirts of Nagercoil. This is just another vindication of, as the political economist Prabhat Patnaik describes it, 'mosaic fascism' with its dispersed local power centres that characterize Indian politics.[29] Aggression came with systematic and not-so-systematic attempts to intimidate Udayakumar's extended family through threats, rumours, and other measures to undermine the school as a 'den for terrorism' so as to put off parents from sending their children to the school. Despite formal complaints to the Rajakkamangalam Police Station officers, the Deputy Superintendent of Police in Kanyakumari Town and to the Superintendent of Police in Kanyakumari District followed by a letter to the state's chief minister, no one was taken to task.[30] The silence and inaction pointed to a state-wide complicity that spanned official and not-so-official elements.

In the aftermath of police charges, many named protesters were financially and psychologically wrecked by legal fees and constant trips to either attend court hearings or, if on bail, periodically report to police stations as far away as Madurai over 200 km away from their homes. Often, to turn up at a hearing was simply to hear that the case has been adjourned and a date set for a new hearing. Stupefying to say the least, court cases drawn out over months and years were like an albatross around their necks, reined into policing prerogatives to clamp down on dissent. Defendants became puppets to larger machinations as lawyers—if they showed up at all—debated right from wrong in an excruciatingly slow Indian judiciary. The process itself was the punishment.[31]

To not attend court in what many considered to be unjust and distorted charges in the first place was to risk further punishment. R.S. Mugilan, a member of the PMANE Struggle Committee, was himself arrested in September 2017 for ignoring a succession of court summons. A determined environmentalist to the core, Mugilan continues to be harassed to this day.[32] To add misery to the outrage, those who are in pre- or post-trial custody atrophy with unwarranted isolation, beatings, substandard food, unhygienic conditions, and swarms of mosquitoes as if nature too is conscripted for mercenaries.[33] Women in prison face the added burden of being vulnerable to sexual assault. Its suspicion alone was a merciless poison if and when they are released.[34]

Udayakumar reflected that the struggle collapsed because repeated court appearances and police threats 'broke them mentally. The government wanted just that'.[35] It was part of a grand project of nuclear social engineering—to create a coerced consensus for the nuclear plant. Even while some people persisted, socio-psychological breakdowns encouraged a situation of 'coerced neutrality' for the rest, to cite the anthropologist David Stoll.[36] Along with this came strong-armed silence as people chose to keep quiet and turn away from the politics of resistance even if they might have played an active part in the recent past.

The politics of indifference came also with the official snubbing of further hunger strikes against KKNPP. Following the September 2012 camp around the KKNPP compound wall, another 'indefinite fast to death' was organized in the Lourdes Matha (Mother) Mary church in Idinthakarai. More than three hundred people mainly women embarked upon the act demanding the release of protestors that had been arrested across Tamil Nadu along with the reiteration of their earlier demands (Figure 10.3).[37] As part of a ritualized repertoire of resistance, the hunger strike signalled a meaningful act of self-destruction. As Ferit Güven elaborates: 'It is an act where the "oppressed" assumes the role of the "oppressor" against his or her own body in order to undermine the oppressor'.[38] In the process, it dramatised the struggle for recognition in an effort to reclaim power over death conditions on behalf of their community and humanity at large.

All hunger strikes from the previous year in August 2011 had been called off with meetings with officials. Worries about the public

Figure 10.3   Mass Hunger Strike at Lourdes Matha Church. Idinthakarai, 2012.
Source: Amirtharaj Stephens/Pep Collective.

attention a political death might garner compelled authorities to act. So began a prolonged series of actions and reactions with officials making promises to PMANE representatives that in the end turned out to be empty.

A year later, the hunger strike was no longer effective or disruptive for the body of the striker had been stripped of any association of citizenship or indeed humanity. Along with growing disinterest amongst mainstream media, the protest was evacuated of its political potency. Authoritarian desires to discipline the protesters were replaced by desires to disappear them. Debased in the eyes of officials and mainstream media, an indefinite fast would just complete their political and physical degeneration.[39]

No stranger to political activism himself, the anthropologist Akshay Khanna comments on the contemporary legacy of the hunger strike more generally:

> ... when Gandhi first developed the idea of the hunger-strike, or the fast-unto-death in the face of atrocities of the colonial state, it was new, disruptive and 'unruly'. Today the fast-unto-death has emerged as one of the best recognised idioms of political action in India, which while continuing to be effective in some circumstances, is very much tame and draws its power from its reference to the legacy of Gandhi rather than to the conditions it attempts to address.[40]

Weighed down by history, the hunger strike cannot on its own address political injustice in the contemporary era. As the political idiom of peaceful resistance became commonplace, it became almost redundant. In the Idinthakarai case, rather than addressing political injustice, it imploded into a death condition. The body of the hunger striker was cheapened to the bare life of radiation burdens.[41] A dalliance with death made little political impact if the body itself was deemed death-dispensable. The verdict might have already been there for (More/Most) Backward and Scheduled Caste fishing communities on the margins of the nation, but was extended to others such as S.P. Udayakumar through his political allegiance and activities as we shall see below.

Inevitably, the corpo-political act also lost its sensationalist currency—yesterday's news. The fast until death drew hardly any sympathy, concern or even interest outside of supporters' circles. Eventually those in Idinthakarai decided to abandon the indefinite fast while continuing the relay strike where many joined in to take part in a fast for a day or a limited period of time. This way they could signal their collective protest without it taking a serious toll on their health.

Progressively, the relay fast took on a more internalized rather than an external orientation. As the fast was always enacted in front of Matha Mary in the church pavilion, and as the unlikeliness of genuine dialogue and favourable outcomes set in, the act of political penance took on more spiritual significance. For believers, divinity provided a sanctuary of solace against state violence. Mother Mary provided recourse at times of desperation, a figure that in the eyes of many of the religiously observant was more likely to listen to them than any nuclear official. A septuagenarian woman, Celine, asserted, 'Not a single government, not a single political party is willing to take up our cause [against the Kudankulam Nuclear Power Plant].... Only Mother Mary can save us now'.[42] It was as if divinity was the only reliable agent firmly on their side, the side of the struggling subaltern. Several village residents mentioned that the reactors' commissioning had been delayed for so long because of her intervention, and that 'Mother Mary will never bless the plant'. The hunger strike was geared less towards political redress and more an appeal for continuing spiritual solace, health and justice from a divine source. The fast might also be conceived as a penance to exact more individual and collective powers, a power that would, it was hoped and believed by many, destabilize the power of nuclear authorities in another manner. As one woman advised me in the village:

Think of your problems, but put them to the side and focus on Mother Mary, God and Jesus. They will sort out your problems, and make them go away. They are bigger than any nuclear reactor, and wiser than any scientist or technocrat.

A miracle was the last chance saloon for disempowered communities against a brutal behemoth. The lyrics of a globally famous song come to mind: 'When I find myself in times of trouble, Mother Mary comes to me' but her words of wisdom, 'let it be', fell short of helping with people's nuclear subjugation.

Crackdown to Meltdown

To recall Foucault's rhetorical question on the subject of governmental oppression or fascism in the modern era: 'How can the power of death, the function of death, be exercised in a political system centered upon biopower?'[43] So, how did we get to this state of affairs in south India? How does the nuclear state reduce citizens to victims, suspects, targets or, as Giorgio Agamben described them in his book, *Homo Sacer*, bare life that might be killed with impunity and cannot be sacrificed for a larger cause?[44]

Gaining the higher moral ground went hand-in-hand with criminalizing or diabolizing anti-nuclear campaigners such that their reputation was marred and they underwent a social death consonant with the third modality of death conditions. Since the escalated struggle in peninsular India from 2012, we have seen an intensification of predatory prosecutions. Extreme and excessive legislature has been deployed for 'protracted lawfare' to use the term by the human rights scholar Sarah Joseph based on her work in Ecuador. This, as she elaborates, becomes the route of choice for trans/multinational 'corporations accountable for environmental and human rights harms when those harms have taken place in a weak developing State in a virtual regulatory vacuum'.[45]

In the case of India, an overdeveloped state in a neo-liberal era entwined with a mosaic of local political powers meant that while checks and balances were present, for the most part, they could be quashed depending upon how much power and influence one wielded. This could also create the conditions for protracted lawfare. In the southern peninsular, more than 55,000 people have been charged in about 380 FIRs, charges of a cognizable offence, from the nearest police station to the nuclear plant in Tirunelveli district, which is among the highest recorded from one place in India's history.[46] Of these, about 9,000 had been accused of 'sedition' and 'waging war against the state', allegations that carry with it the prospect of a life sentence or a death penalty. Typically, a few names would be mentioned in any charge with the added number of anywhere between a few hundred to a few thousand to add to the list of named individuals. The exact figures have vacillated along with the prevarications of officials and some were subsequently dropped after campaigners appealed to the

Supreme Court.[47] In a flurry of FIRs, Udayakumar is 'Accused No. 1' in over 300 charges along with nearly 100 criminal charges.[48]

Even though decried as a 'parody of law', the allegations had a toxic tenor.[49] A Human Rights Watch report on 'stifling dissent' in India notes on accusations of sedition:

> These laws have been misused, in many cases in defiance of Supreme Court rulings or advisories clarifying their scope. For example, in 1962, the Supreme Court ruled that speech or action constitutes sedition only if it incites or tends to incite disorder or violence. Yet various state governments continue to charge people with sedition even when that standard is not met.[50]

Whatever their credentials, protagonists of the state have long realized that the easiest way to crush dissent is to frame their spokes-persons and supporters as suspect even in spaces where democratic dissent is permitted and no violence is incited.[51] For the accuser, whether the allegations are genuine or artificial is beside the point, for it is the performative effects that are deemed more important in creating compliance. For the accused, it is as if they 'fake it until they make it' as goes an all-too-true colloquialism.

Officious means are deployed for vicious ends. On state officials, the legal scholar Usha Ramanathan notes:

> They know even if all the cases fall at the end of the day it doesn't mat-ter, because they've had their purpose served. You can beat people up, you can put them away.[52]

Anthropologists Sharika Thiranagama and Tobias Kelly add: 'To make an accusation of treason is to make a claim to power, to try police the boundary of permissible politics, and to exert authority in the face of constantly shifting affiliations'.[53] Such allegations are rife as India goes through a fever of sedition revivalism. Accusations are banded around against anyone who seeks or seems to go against official narratives on politically sensitive or hot topics. The relative ease of allegations acts like a schoolmaster's stick to discipline the subject, while discharging infectious sprays for anyone that surrounds the accused.

More worryingly, as the political scientist Anushka Singh points out in her research, 'persecution in the name of sedition finds popular acceptance in India'.[54] A political culture has emerged that has suppressed people's voices and credentials for environmentalist, activist, and even charitable work to help the poor, abused, and ill. It is a culture characterized by limited imaginaries beyond the horizon of nationalism entrenched by the state to 'rationalize forms of repression' to cite the anthropologist Kay B. Warren.[55]

Rationalized repression converged with micro-aggression. In and around Idinthakarai, specific sectors of society were targeted so as their social and/or economic standing could be expediently undermined. A fact-finding team consisting of the advocate B.G. Kolse Patil, the former judge of the Bombay High Court, the senior journalist Kalpana Sharma, and the Tamil writer R.N. Joe D'Cruz reported that women were frequently molested and abused when they joined campaigns.[56] The team noted how women were abused by police officers: '"Why all of you are going behind Udayakumar? What has he got that I don't have? Come I'll show" and even worse abuses' [sic].[57] State protagonists also targeted young men seen to be the most pugnacious of protesters: 'We were told that most of the young men had left the village and some had even gone as far as Chennai to escape the police dragnet'.[58] Profiling individuals by way of their age and gender on top of their caste and minority religious identities meant that vulnerabilities could be exploited to maximum effect. There is an intersectionality to bare life where, to add to their generic class-caste subjugation, specific aspects of their identities and lives were further exploited as part of necropolitics—what could be termed necrosectionality.[59]

This 'diffuse violence in daily life' was attendant with pressing on pressure points associated with the intimacies and weaknesses of the individual and his or her chains of relatives, friends, and associates.[60] The politicization of mundane practices might include the blocking of essential permissions, services, or provisions to which people might be entitled. Negotiations with otherwise banal bureaucracies became more apprehensive.[61] As many of the men in the village sought to find work overseas in countries like those in the Gulf, for instance, a passport was mandatory. However, if it was known that the applicant lived in the epicentre of anti-nuclear struggle, Idinthakarai, their passport would invariably not be processed, their home having been branded a pariah place. In other cases, state permissions for essential work might not be granted as happened when an activist was threatened by a police officer in 2012, stating that his son's company in another part of Tamil Nadu would not be cleared for business if he did not stop his support for PMANE.

While the authorities sought to cut off the blood supply of any kind of anti-nuclear funding on grounds of anti-nationalism, they injected funds into their own operations to ensure that they were successful. At the collective level, it is with respect to the creation of 'welfare schemes' that might include the promise of jobs, housing, health benefits, new schools, computers, and scholarships. The incentives promise comfort while co-opting people into the nuclear state's own vision to varying degrees. At the individual level, money might also be offered to pay local leaders such as the sarpanch (village head) or 'local stooges' who, on the one hand, can inform authorities about anything happening on the ground; and on the other, spread rumours against particular individuals and groups while stirring up caste and/or religious tensions. Once snakes are released in the grass, rumour and suspicion abound with viral velocity. With rebound force, village gossip about who was a government stooge itself initiated antagonisms as aspersions were cast on particular individuals lured by bribes or incentives to act as locally embedded 'spies' for the police.

Such examples of manufactured micro-aggression led to a culture of distrust in Idinthakarai and neighbouring villages and towns where one individual and/or community was pitched against another. Often micro-aggression was accompanied by lawfare and criminalization. After India's Independence Day on 15 August 2013 was celebrated as Black Day to mark the death of independence, yet more allegations were made, this time from police lackeys along the coast. The neighbouring Vijayapathi village administrative officer registered a case against the protesters, including the Struggle Committee members, Udayakumar and Pushparayan, under the Prevention of Insults to National Honour Act (1971).[62] It was manifestly clear that the local officer had succumbed to 'state stoogery', with several suspecting that he had been paid off to register the case.

The neighbouring village also had a large number of Muslims, and it was rumoured that some of the stooges were planting seeds of communal division against Christians at the head of the struggle. The classic divide-and-rule formula from the colonial era was reinstated.[63] As Dunlap elaborates:

These strategies are designed to adapt and merge with local interests, which often seek to widen existing social and political divisions as a means to fragment, break and isolate resistance groups, often intentionally blurring the line between counterinsurgency and inter-communal conflict.[64]

Udayakumar's Hindu Nadar identity itself became problematic in terms of him speaking for a largely Christian fishing community. On one occasion, he was delivered a postcard supposedly from 'south Indian fishermen'. It was addressed to him with the opening line, 'Dear idiot Udayakumar'. Then it went on to ask him to concentrate on the plight of fishermen who were being killed by the Sri Lankan navy, rather than the nuclear power plant. With characteristic nonchalance, Udayakumar duly threw it into a pile of numerous other baseless allegations in FIRs issued by the police.

The majority of the local police officers in Kudankulam are themselves of Hindu Nadar background. The more perceptive observers in the Struggle Committee held them accountable for fermenting trouble among coastal communities 'because of their caste and religious prejudices'.[65] Once successful with their tactics, resistance collapsed into caste-communal cleavages.

Careerism too came into the picture when those from the PMANE Struggle Committee who knew the police officers reported on how much they were compelled to impress their seniors: 'They are ... overzealous in their respective tasks as they may receive possible promotions and other pecuniary benefits such as increments in their department'.[66] While there might have been a slackening in Intelligence, as will be made evident below, attractive incentives for containing the anti-nuclear struggle ensured that there was no slackening in policing.

The Stealthy State

Non-violent movements that have a wide support base compel indirect means of control by the state to avoid backfire from the general public and social justice or human rights agencies. To turn James C. Scott's classic anthropological work on its head, 'hidden transcripts' were deployed less as part of the art of resistance for the dominated, and more as part of the craft of control by the dominant.[67] The subordinated were publicly open about the ethics and justice of their nonviolent cause. The subordinator was not and had to rely on strategies comparable to 'infrapolitics' that Scott had hitherto identified in the zone of constant struggle by the peasantry. Such infrapolitics of inversion were instrumental to the slighting of anti-nuclear advocates that meant moving away from direct repression to mobilizing policies and actions that marred their reputations and lifelines in less noticeable ways. The Indian government certainly did not want more international human rights and environmental appeals as elaborated in the previous chapter. They veered away from the dramatically draconian to more in-house strategies.

Indicative of circular and circuitous control, the nuclear state with its congeries of state departments, policy-makers, police, and intelligence and surveillance agencies attempted to reframe nonviolence and dissent as violence and threat to the entire body politic. They did this in a variety of ways including taking recourse to the production of on- and offline paranoia about accusations of anti-nationalism, weaponizing policy and information, and leaking tip-offs and 'Intelligence' to the media to create grey zones of doubt against peaceful dissidents. Backed by economic and political clout, such inverted infrapolitics were an alternative to physical violence, but equally influential in maligning individuals and organizations while generating a psychosis of public paranoia.[68]

As Stephan Feuchtwang terms it, 'politicized paranoia' has been powerfully whipped up through the entrenchment in India of the 'foreign hand bogey'.[69] This has most traction in the post-colonial context with its history of anti-colonial struggle sensitized to anything that disturbs national sovereignty, although it is not altogether absent in other countries with respect to the terrain of national security.[70]

Arguably casting the 'enemy within' with a treasonous charge was even more effective than highlighting the 'enemy outside'. The imagined relationship to what might be called 'Othered selves'—someone that could in fact be you—was highly effective in keeping people disciplined and docile. Allegations mutated realities, setting up new regimes of truth according to which people oriented their thoughts and behaviour, themselves fearful of victimization and public shaming.[71] In such a scenario, mere supposition of treachery led to a state

of 'guilt before innocence' against which the accused then had to operate. This burden of extrication was a heavy one in contexts where the
stakes were high and where there was less than satisfactory scrutiny
of state conduct and the criminal justice system.

Deriding the foreign is a crucial organizational principle of soci-
etal control and stemming dissent. In a disinformation campaign
designed to discredit, the Indian state portrayed specific individuals as funded by foreign powers to bring India down. The rumour
itself was radioactive, mutating realities as it spread.[72] Once moral
and political claims of the accused were undermined, their status as
law-abiding citizens became questionable, and they became the legitimated—rather than legitimate—targets of other strategies to harass
or intimidate.

On top of name-calling, anyone supporting an anti-nuclear movement might be targeted financially. Bank accounts might be frozen
or blocked if individuals are accused of stirring anti-nuclear dissent.
If it was suspected that NGOs were partial to anti-nuclear protest, the
Ministry of Home Affairs might revoke permissions to receive foreign
funds under Section 18 of the Foreign Contributions Regulation Act
(FCRA), an act that was established in 1976 to scrutinize the voluntary
sector during Indira Gandhi's Emergency rule and revised in 2010
under Manmohan Singh's coalition government to tackle terrorism
and money laundering.[73] In fact, the FCRA threat was applied to
any organization working on social, environmental, or human rights
inimical to the government's agenda. It was extended to church-based
NGOs seen as part of a transnational Christian conspiracy supporting coastal Catholic communities. Those that supported the struggle
including the Tuticorin Multipurpose Social Service Society, Tuticorin
Diocese Association, People's Education for Action and Community
Empowerment and Good Vision Charitable Trust.[74] Priests are
answerable to the Roman Catholic Diocese in their respective districts, and even though their laity might support the struggle, most
began to think twice about backing anything that might rankle the
government.

With the election of the Bharatiya Janata Party (BJP) in central
government in 2014, measures against social, environmental, and
human rights organizations saw a step-change with more and more
names placed on a security watch list.[75] After alleging that 'donors'

to the anti-nuclear movement were based in western countries such as the USA, Netherlands, Germany, and 'Scandinavian countries' in a conspiratorial scheme to 'take down' India's development projects, the BJP government threatened to dismantle NGOs that were receiving funds from foreign sources if they supported the campaign against the Kudankulam Nuclear Power Plant.[76] In 2016, up to twenty thousand NGOs were threatened in a FCRA crackdown on bank accounts.[77] This compared with around 4,000 cancellations after the revised FCRA Act was first implemented under the previous United Progressive Alliance government in 2011 for the stated reason of non-submission of annual returns within nine months of the end of the financial year.[78] Even well-established bodies such as Greenpeace India had their funding status revoked in 2015 for about a year before a court decision in their favour reversed the decision.[79] This organization is one of several that had raised awareness about environmental violations with KNPP and, among their other work on health and environmental justice, they have been pulled up for 'prejudicially affecting public interest and economic interest' as reported by the journalist Bharti Jain.[80]

While embroidered charges against anti-Kudankulam campaigners were foisted, the leaner machine of technocracy was put in attack mode. From lawful parody came weaponized policy that, in effect, stemmed dissent. As argued in Chapter 3, the apparently depolitical nature of law, bureaucracy, policy, and punctuality are both rooted in and have their effects in politics. Critics fully recognized that the crackdown on FCRA was to expunge any movement that the government did not approve of, and to shackle opposition in the country where FCRA status was set up as yet another 'tool of repression'.[81]

Another way numbers were weaponized in the neo-liberal era was to enumerate the percentage by which individuals and organizations were alleged to have decreased India's development. This strategy is clear in a report sealed by the institutional signature of the Joint Director of the Ministry of Home Affairs Intelligence Bureau, S.A. Rizvi, that was leaked to the media—a document that is freely available online.[82] Its contents reflect the message that the government pass to staff, trainees and the media in order to monopolize financialized 'common sense'.[83]

In the report, a variety of individuals and NGOs are accused of having a 'negative impact on GDP growth', attributing it a decrease of

'2–3%'.[84] Blame on economic decline in what otherwise is the inevitable consequences of capitalist fluctuation is attributed to specific
people. They include those who stand up to nuclear power plants,
uranium mines, coal-fired power plants, genetically modified organ-
ism enterprises, hydel (hydroelectric) projects, extractive and other
large-scale industries. The document states that economic decline is
encouraged by 'identified foreign donors [who] cleverly disguise their
donations as funding for protection of human rights; "just deal" for
project-affected displaced persons; protection of livelihood of indigenous people; protecting religious freedom, etc' [sic]. It is then alleged
that the 'foreign donors lead local NGOs to provide field reports which
are used to build a record against India and serve as tools for the
strategic foreign policy interests of Western Governments'. Human
rights and social justice charities such as Amnesty International and
Action Aid are claimed to 'dedicate a small portion of their annual
donations to such project under varied veils such as "democratic and
accountable Government", "economic fairness" etc.' The report continues to allege that such tactics are designed to stop India's growth
and keep it in a 'state of under-development'.[85]

Casting a historical view, Erica Bornstein and Aradhana Sharma
remind us that the organizations mentioned in the report 'were the
very civil society bodies that international development organizations
and the Indian state ... had lauded in the 1990s as ideal partners in
development and democracy'.[86] Wrapped in an armoured cloak of
righteousness, however, contemporary ministries and associated
agencies can neither be reflective, nor visionary. Rather than taking time to understand grievances against large-scale operations or
develop alternative plans that are consultative and complementary
to peoples' lives, their attitude comes with a swiping dismissal. They
brand any complaint as the conspiratorial consequence of foreign
support and intervention, thus invalidating the content of the complaint. In the register of paternalistic patriotism, they adopt a 'father
knows best attitude' where any ounce of disrespect or disagreement is
met with indictments of various kinds.[87]

More specifically, the aforementioned Intelligence Bureau report
details how anti-nuclear activists brought down India's Gross
Domestic Product by stalling the construction of the Kudankulam
Nuclear Power Plant 'spear-headed by Ohio State University funded,

SP Udayakumar, and a host of other Western-funded NGOs'.[88] The 'connection with Ohio State University' is made on the basis of two payments to Udayakumar when he was a Research Fellow in the International Program of the Kirwan Institute for the Study of Race and Ethnicity at the University. Worried that he was being singled out with extrapolated allegations, Udayakumar maintained that he was researching and writing on issues to do with 'globalization, racism, minority welfare, BRICS etc.' and not 'India's development or India's nuclear program'.[89] At any rate, payments to an academic who had studied abroad in the 1990s and since researched, taught and presented in several foreign countries is quite routine for many scholars. However, this is an aspect of his career that state protagonists rarely prioritize. In the numerous FIRs and reports against him, hardly ever is his name attached to the prefix of Dr as a testimony to his postdoctoral scholarship.

Intellectual achievements hold little ground when Intelligence reports are attributed more aura, importance, and integrity. Backed by such reports, the then Union Home Minister P. Chidambaram can declare that there was 'information to show that [a German national] was associated with semi-political and protest organisations who were involved in the anti-Kudankulam stir'.[90] Indeed, the intelligence in such reports thrives on stemming critical intelligence of another order. While there is no evidence to point to financial transactions on this point, the allegation now is premised on a marked map. It is held to be proof of a 'larger conspiracy' that was 'unravelled when a German national provided Udayakumar a scanned map of all nuclear plant and uranium mining locations in India'. Elsewhere in the report the number comes down to six proposed nuclear power plants and five uranium mines that were 'marked prominently' on the map with hand-written name slips on 50 Indian anti-nuclear activities along with a 'Blackbery PIN graph'. The handwritten material is held 'to avoid possible detection by text search algorithms said to be installed at e-gateways'. Altogether, the piecemeal information is purported to reveal 'the involvement of an organized agency and/or a highly professional well-paid entity' and 'an intricate "Network"' designed to 'take down India's nuclear programme through NGO activism'.

Parading themselves as national saviours who had discovered a dangerous and intricate conspiracy with what they call a 'network analysis', the Intelligence Bureau identify a "Superior Network" (prominently driven by Greenpeace and renowned activists)' along with five 'Territorial Networks'. The five networks are deemed to be based in Idinthakarai (Tamil Nadu), Trivandrum (Kerala), Hyderabad (Andhra Pradesh), Ahmedabad (Gujarat), and Meghalaya (Shillong) attached to other named individuals, eleven of which are 'frequent foreign travellers' in flights bound to USA and Germany.[91]

As it transpired, the German with the perfidious map in question, Sonntag Reiner Hermann, turned out to be a 'hippy'. In February 2012, while the nuclear power plant's construction was stalled due to the anti-nuclear struggle, he was summarily deported from the sub-continent without any media or legal access. In a rare online interview after he was extradited, Hermann denied funding the movement. He stated that he lived on $10 a day, and that 'I'm unemployed and don't receive orders from any group in Germany or anywhere else'.[92]

Hermann was hardly the picture of an affluent funder to take down India's nuclear programme. An environmentalist who got to know some of the people in the Indian movement, he shared an anti-nuclear agenda as is widely prevalent in Germany and intensified after the Fukushima Daiichi disaster.[93] Moreover, the scanned map of nuclear plants and uranium mining that he stood accused of circulating among what is presented as an insidious and dangerous network is easily accessible on the internet. Shreds of evidence such as this are included in the Intelligence report without any means of ascertaining whether or not it was Hermann's writing that marked the map so prominently.

Such reports rest on the maxim that 'the proof is in the pudding' when the pudding is overcooked and consumed in secret. The peddling of Intelligence combines pieces of prominent information that might be verifiable, immersed in motivated extrapolation, rumour and inevitable error, witting or unwitting. The reports make for revolving doors of partial evidence and half-baked enquiries.[94] With the tailoring of allegations, the Intelligence Bureau ends up belying their title. However, intelligent or non-intelligent is beside the point. The imperative of the surveillance state is to performatively (re)produce penal power where the exact content and context of the Intelligence does not matter so much as the *effects* of the allegations.[95] As goes the irregularity of state secrecy and disclosure, such powerfully affective reports are for the most part kept behind the scenes but released when advantageous so that media and interested parties can milk them for at-once crowd-pleasing and crowd-controlling aims.

The jouissance of exposure is not without the anxiety of self-interested survival. The creation of fear, terror or anger before the facts is well established as an organizing principle of social control.[96] Intelligence gathering is enlisted to this project of control. P. Thirumavelan, the editor of the *Junior Vikatan* magazine, who himself faces several criminal cases, testifies to this lack of 'hard facts' behind allegations:

> The government is not interested in pursuing a case. The intention of the government is only to create a fear psychosis among journalists and newspapers. Because if the government were really serious, they would counter with evidence in a court of law.[97]

If there is any evidence, it is fragmentary and forcefully extractive under conditions the observer cannot always or easily ascertain. The more one looks into these matters the more one realizes that the agencies capitalize upon their high status, a plethora of secrecy clauses, the weaponizing of law and policy, an exceedingly slow criminal justice system, and the rhetoric of nationalism and national security to veil proceedings rather than produce rigorous research and (untampered) evidence that could stand up to a fair trial in court. However, as is demonstrated below, the conjunction of forces can certainly lend itself to a post-truth trial by media.

Such capricious and manipulative measures bring to mind the controversial case of Iftikhar Gilani, a Jammu-based journalist who was imprisoned without bail under the Official Secrets Act in 2002: 'His crime—possessing out-of-date information on Indian troop deployments in "Indian-held Kashmir" culled from a widely circulated monograph published by a Pakistani research institute', comments a senior editor, Siddharth Varadarajan.[98] The document was available on the internet and in some public libraries. After a media frenzy, the case fell apart after seven months due to the inconsistencies of fabricated reports by the Union Ministry of Home Affairs, Delhi Police, and Intelligence Bureau. The curried statements reveal a political culture of cooking up tenuous scraps of information to feed to the media. The glare of publicity in this case, however, does not deter from the utility value of causing trouble and setting off political contagion for the indicted.

The subsequent exposure of an injustice does not right the wrong. Even while assertions lack foundation and are sorely insufficient, they can enact powerful damaging effects on an individual's psychological, physical and social standing that linger well after the accused might be cleared: and these are the lucky ones. Such dubious practices therefore continue unabated.

Onlife Ondeath

'Onlife operations' are another means to attack and harass antinuclear protagonists in a combination of the second and third modalities of death conditions.[99] With rare exception, their intent can only be sensed through effect, or gauged through hearsay.[100] With respect to the people's movement in India, e-mails, blogs, websites, petitions, and Facebook groups amongst other outlets have been subject to the 'silent arms race' of digital surveillance and cyber-attack. This has included bringing down websites, circulating spam mail, hacking accounts, implanting malware and/or spyware, phishing (when the recipient is enticed to enter sensitive information on a fake website), and trolling by posting inflammatory statements to discredit the status of activists who have gained popularity.[101] A Tirunelveli District resident declared on social media:

> ... phones are being tapped and mobile calls are monitored ... emails are being hacked regularly and obscene mails are sent from the email and facebook profiles of key persons ... it is creating a wrong msg [message] among the general public ... and sometimes creating a wrong image on the leaders of the [PMANE] movement as there were phishing of porn mails and post to others FB [Facebook] account from the leaders account.[102]

These are just the more explicit onlife measures. There are also plenty of other setups that are not so easy to pin down as they draw upon now widely prevalent military–industrial 'psychological operations'. These are on- and offline practices that are becoming de rigeur for political and economic elites in cahoots with databank analysis companies to influence digital users, and demonize issues and individuals while exalting others, in the course of which, interfering with due process. As the journalist Carole Cadwalladr explains, they are 'the same methods the military use to effect mass sentiment change. It's what they mean by winning "hearts and minds"'.[103] We are in the midst of the weaponization of information for social and political influence on the hearts and minds of many.

Recent disclosures from the USA have revealed that we have moved from the state of exceptional surveillance to the ordinary state of surveillance of all such that, when we are plugged into electricity and Wi-Fi with our devices, we subscribe to a surveillance state whether we want to or not.[104] The revelations also demonstrate that indiscriminate surveillance is a shared and now standard practice of many governments around the world. Plugged in is also about being cordoned in, a connection that might or might not be activated—a fine formula for the production of paranoia.

Announcements in 2015 during Prime Minister Narendra Modi's launch of the Digital India campaign along with earlier promotion of e-governance have to be seen in the light of not just postcolonial infrastructural ambition but also surveillance expansionism.[105] Utopias of political transparency and immediacy are not without their dystopias of persecution and paranoia. With the slogan 'power to empower', the stated aims of Digital India are to develop secure and stable digital infrastructure covering urban and rural territories, deliver government services digitally, and to enhance universal digital literacy.[106] However, an imagined golden era of communication convenience makes it also a golden era for surveillance convenience. Virtual media has become another crucial site to control, contain and influence. This has in fact come to light in 2018 with revisions to the Information Technology Act (2000) such that 10 central agencies are able to intercept, monitor and decrypt 'any information generated, transmitted, received or stored in any computer resource under the said Act (section 69 of the IT Act, 2000)'.[107] Critics have condemned it as the rise of a 'stalker Sarkar [government]' to add to its relentless witch-hunt of 'anti-nationals'.[108] With respect to the Kudankulam Nuclear Power Plant, technologies that were once aimed at terrorists are now being used to track and contain anyone who does not agree with governmental decisions. Whether they resort to tactics of terror or not is besides the point for they are all deemed 'trouble-makers'.

An all-consuming and hardened response to anti-nuclear movements is reported more widely for the surveillance security state.[109]

On post-9/11 USA, the sociologists Robert Futrell and Barbara G. Brents note how non-violent direct action dissent is criminalized as terrorism with the example of:

> ... a bill before the Pennsylvania State legislature, which has been hailed as a model bill for Congress by antienvironmental lobbies [that] would so broadly define ecoterrorism as to potentially cover acts such as sit-in, blockades, and other forms of nonviolent protest.[110]

What happened in south India reflects a global governmental trend. All kinds of activists are bunched together, decried as part of 'single-issue terrorism' and treated as such by intelligence and law enforcement agencies.[111] The war front moves to the home front as a redefined battlefield becomes everywhere yet visibly nowhere. As with radioactivity, it is a world of in/visibles that has powerful effects in stifling dissent and opposition. In so doing, it ends up keeping administrative regimes in power, further eroding principles of democracy, however they are imagined.

Toxygen

Information weaponization entails multipronged battles on several fronts. They have three main features—diminishment, exaggeration, and the embroilment of 'new news'. Key issues and events might be ignored, on the one hand, or, on the other, selectively capitalized to disparage not just a person but a whole collective or community to underline their illegitimacy. Information warfare might also include flooding the field with 'fake news' and 'sting operations'. While news is continuously curated, stings entail a trap laid out for particular individuals to fall into, which is secretly recorded, while others pose as *agents provocateurs*. Generally conceived to *exact* the truth, they can also be flipped to *manufacture* the truth.[112] For the consumer, even while they may be familiar with formulae for fiction, it is increasingly difficult to detach an exposure of a truth from the manufacture of a truth.

The leaked intelligence report mentioned above rests on an aporia that resonates across the pleats and folds of secrecy and publicity.[113] It is dangerously affective because of its association with spy-like shenanigans, and highly credible because of the weight of Intelligence behind it—a perfect storm for sensation-seeking media to whip up. In view of their inflammatory effects, Intelligence leaks are far from accidental.[114]

In 2017, a curated exposé was announced on the basis of the leaked report on Arnab Goswami's Republic News channel, a TV and digital broadcaster who is renowned for his support for the BJP—conceivably another channel for what Sahana Udupa has called 'enterprise Hindutva'.[115] In what is billed as 'a hard hitting sting exposé', it declares that 'Indian Intelligence says Udayakumar is part of a western conspiracy' leading 'a plot against the nation'. Journalists Schweta Kothari and Sanjeev masquerade as Indian students from Britain who want Udayakumar to admit that he would take donations from abroad even though he does not have an accessible bank account, let alone FCRA clearance. After Udayakumar fell into the trap—albeit by saying that his 'party' could only accept donations through an Indian bank account—the broadcast concluded that he is personally culpable for 'routing money from foreign funders and agencies to destabilize India, floating political parties with foreign funds to spearhead anti-national protests, and create dissatisfaction towards the Indian state'.[116] Described as leading 'activism that compromises India's security', he is presented as 'one man with very deep American connections' who is destabilizing the country.[117]

With aggrandized claims to an investigation, the broadcast brashly asserts that 'now the pieces are coming together'.[118] This disjointed detective work is stitched together into a threadbare patchwork whose holes are filled in with bluster. This becomes the premise for a trial by media and a damning indictment of all NGOs that according to the sting journalist:

> ... curb development all in the name of civic activism. They hunt in packs, these NGOs, they trawl in herds, but we couldn't care less. Republic will bust them again and again and again.[119]

Not unlike the Intelligence Bureau, the station presents itself as the saviour from what they describe as 'double-crossing' Indians. Mileage is gained in a Manichean universe where the self-proclaimed heroes vanquish the villains for the benefit of the nation. True or not, such melodramatic traps end up fulfilling a larger motive to increase viewership.

From the dreary days of state-run Doordarshan television, Indian broadcasting has fallen down the rabbit hole of corporate sensationalism. From a discourse of patriotic paternalism comes one of jingoistic adventurism. In this hyped-up fast lane, material is selected, edited, distorted and commodified to fit a publicity-hungry goal. Confirmation bias means that only select information is needed to validate the prior belief that Udayakumar is already suspect. This aspersion is backed up by tokenistic stock footage that foregrounds the anti-nuclear movement's unruly violence rather than its progressive politics of peace through, for instance, the optic of fishing communities clashing with police in the September 2012 act of civil disobedience. The curated exposé is followed by a gesture to a debate with the wheeling in of senior politicians to lend authority and veracity to their sting operation. Altogether, they endorse the stereotype of pre-modern and anti-development activists intent on bringing the country down as if they would prefer to live in the Stone Age.[120]

Time and again, such public spokespeople fail to appreciate that most activists are not against development per se, but campaign for development that is justly pursued and socially and ecologically complementary to people's lives. While they understand the country's progress, they cannot understand its progressiveness. Instead, a topoi of urgency and threat, barbarism and modernity, villain and saviour, and history and future is promulgated in which the anti-nuclear protagonist is dismissed as illegitimate, inauthentic, and backwards. Another axe is wielded between the consumptive citizen of neo-liberal India attracted to mammon and media, and the constitutional citizen struggling for a democratic India that is demonized and eviscerated.

Through a single yet cleverly layered message, multiple and contradictory audiences are addressed. A worm of doubt is implanted even among supporters who cannot ascertain otherwise. The sting sponsor comes off as redeemer while the dissenting target becomes the threat with the (il)logic of victim/perpetrator reversal. This situation creates its own realities when the politics of denial, blame games and their constant reiterations echo further down the line as happened later when Udayakumar was accused again, this time of supporting secession for the state of Tamil Nadu, another serious anti-national charge in yet another trial by media.[121]

Sensationalist media was both bane and aid. It was fixated with novelties, tainted by cowboy adventurism, and infected by state–corporate

collusions of various kinds. Yet, it was needed for the oxygen of publicity without which a people's movement cannot breathe. However, too
much oxygen left a movement trapped like a beached whale.[122] With
the onslaught of time, public fatigue and indifference set in reflected
in reduced media interest in PMANE peaceful activism while exaggerated interest developed in slanderous associations of its key protagonists. Against such an apathetic yet toxic political environment,
Udayakumar contended that they had no ulterior designs against the
government, that all resources for PMANE had been raised through
the collection of 10 per cent of earnings from fishing communities
when it was based in Idinthakarai, and that their primary interest
lay with the protection of the country and its future. However, such
defence stood little ground in a post-truth world where Udayakumar
was already framed as 'guilty' by a curated media optic.

A critical disposition to the rise of sensationalist ploys, however,
creates the conditions of their demise. A sting might well reek of a
stink, for instance.[123] Quite fittingly, one sting was aimed at former
diplomat and Congress politician Shashi Tharoor who described
Republic News as 'the digital equivalent of a toilet roll'.[124] Taking the
muckraking metaphor further, a Nagercoil-based critic recognized the
dynamics:

> ... sting journalists are like dog shits. They set up scenes to attract flies.
> And if the dog shit can't get what it wants then it stands on the path hop-
> ing that someone would slip on it. Then they will smear the person with
> that shit. This is what they present to their viewers and readers. Most of
> the intelligent ones see through it but others don't. And others like poli-
> ticians jump in the shit themselves making more of it because they're
> very much attracted to dog shit. The sting journalists and the politicians
> who support them are lower than the shit. They are the ones who bring
> humanity down. They are the ones that are bringing the country down.
> Not the flies, or the person who unwittingly slips on the dog shit.[125]

Despite their lack of news credibility, casualties of stings still have
the burden of extrication from the 'dog shit'. Denouncing the curated
exposé makes little mark in a moral field that has moved beyond truth
and falsity. The stung have to lift themselves away from the smears
and retrieve some sense of dignity against the 'toxygen' of penal publicity. Energy is expended and redirected from the issues at hand for

it is a suspicion that is hard to scrub away. The sting creates its own stigmata.

Murk

We have noted how subaltern communities and their allies against the Kudankulam Nuclear Power Plant were subjected to death conditions by way of three overlapping modalities – silent and encroaching, quick and punitive, and dismissive, deflective and demonising. These modalities include ecological toxicity entrenched in social hierarchies that are compounded by the neglect of low caste-class casual labourers working for the nuclear industries; more punitive and direct intentions to suppress and extinguish dissent though the actions of particular agents or agencies; and more covert, demonising and snowballing features that demonstrate an inverted infrapolitics to outcaste anti-nuclear campaigners as anti-national criminals so that they can become socially tabooed and targets of further intimidation.

The enquiry on political (re)actions from above demonstrate that nuclear necropolitics are less from a '"centaur" state—neo-liberal at the top and penal at the bottom' as David Mosse contends.[126] While these obviously penal elements might be identified with the second modality in particular, they are but a spar in a ramshackle architecture of oppression, or to use a more organic metaphor, part of a 'hydra state' with its multiple heads and arms hand-in-hand with other corporate and media actors enhanced by technological advances, special security and enforcement forces, and the recruitment of henchmen and stooges. This has unleashed several lines of dispersed attack that ranged from the aggressive to the stealthy, overt to the covert, and financial to the sensational in which the sponsor's role becomes 'plausibly deniable' even when 'denials lack plausibility' as goes the astute analysis of the international relations scholars Rory Cormac and Richard J. Aldrich.[127]

As Foucault identifies, media has become 'a design of subtle coercion for a society to come'.[128] Public slurs and bruising therefore becomes just as desired an objective in the (re)production of, on the one hand, death conditions for detractors, and on the other, social control of the wider public. Streaked around the three modalities of death conditions along the spectrum of make die to let die are

those that cast targeted persons into a moral impasse. In a context
where 'grey is the new black', as Cormac and Aldrich argue, 'leaders
are embracing implausible deniability and the ambiguity it creates'
to make charismatic and oppositional leadership non-operational.[129]
Satyagraha is made redundant in a post-satya world. In this twilight
scenario, express political interests might be detectible, but they
hardly ever come to laser sharp focus. There can be little of a battle
when the dominant opposition, with its brand of piecemeal peasant
politics, ends up lurking in the murky light.

Notes

1. '"Report of the Fact-Finding Team's Visit to Idinthakarai and other
   Villages on September 20–21, 2012" by Mr. B. G. Kolse Patil, former
   judge of the Bombay High Court, Pune; Ms Kalpana Sharma, senior
   journalist, Mumbai; and Mr. R. N. Joe D'Cruz, Tamil writer, Chennai'.
   Countercurrents. Available on https://www.countercurrents.org/
   koodankulam260912.pdf, accessed 7 January 2019, p. 4.
2. PMANE Struggle Committee. *PMANE Letters and Press Releases (April
   2012–January 2014)*. Idinthakarai: People's Movement against Nuclear
   Energy, 2016, p. 9.
3. 'Koodankulam: Idinthakarai Turned into an Open-air Jail, Govt Playing
   Divide-and-rule'. *DiaNuke*, 10 April 2012. Available on https://www.
   dianuke.org/koodankulam-idinthakarai-turned-into-an-open-air-jail-
   govt-playing-divide-and-rule/, accessed 7 January 2019.
4. 'Do we Exist? Fifth Letter from the Women of Idinthakarai'.
   Countercurrents, 23 October 2012. Available on http://www.countercur-
   rents.org/5thletter231012.htm, accessed 3 October 2016.
5. See Abraham, Itty. 'Geopolitics and Biopolitics in India's High Natural
   Background Radiation Zone'. *Science, Technology and Society* 17, no. 1
   (2012): 105–22, pp. 107–8.
6. The quotations are from Foucault, Michel. *Society Must Be Defended:
   Lectures at the Collège de France, 1975–76*, edited by Alessandro
   Fontana and Mauro Bertani, trans. David Macey. London: Penguin,
   2004, p. 247.
7. Mbembe, Achille. 'Necropolitics'. *Public Culture* 15, no. 1 (2003): 11–40.
8. Biddick, Kathleen. *Make and Let Die: Untimely Sovereignties*. Goleta,
   California: punctum books, 2016, p. 4.
9. Boyer, Dominic. 'Anthropology Electric'. *Cultural Anthropology* 30, no. 4
   (2015): 531–9, p. 534.

10. Boyer, Dominic. 'Energopower: An Introduction'. *Anthropological Quarterly* 87, no. 2 (2014): 309–33, p. 309.

11. See Raminder Kaur. 'Southern Spectrums: The Smooth and the Raw Edges of Energo-Politics around an Indian Nuclear Power Plant'. In *Energopolitics: Citizenship, Governmentality and Violence along the Grid*, edited by Simone Abram, Nathalie Ortar, Tristan Loloum, European Association of Social Anthropologists book series, New York: Berghahn, 2019.

12. On other comparable strategies adopted by energy industries, see Dunlap, Alexander. 'Counterinsurgency for Wind Energy: The Bíi Hioxo Wind Park in Juchitán, Mexico'. *The Journal of Peasant Studies* 45, no. 3 (2018): 630–52, pp. 644–6.

13. This third modality of social deaths is a debateable extrapolation of necropolitics. Jared Sexton's critical reading of Mbembe suggests that it is not possible to apply social death, developed to characterize the violence exercised on enslaved black people, to contexts outside the Caribbean or North American plantation. Others such as Katherine McKittrick suggest otherwise, which is the way I have adapted the term. Sexton, Jared (2010) 'People-of-color-blindness: Notes on the Afterlife of Slavery', *Social Text* 28 (2 (103)): 31–56; McKittrick, Katherine (2016) 'Diachronic Loops/Deadweight Tonnage/Bad made Measure', *Cultural Geographies*, 23 (1): 3–18.

14. Together, these three modalities are specific to nuclear installations in India with parallels in other regional contexts. While they overlap, they are at a tangent to theories on 'hard' and 'soft' techniques of population control. See Dunlap, 'Counterinsurgency for Wind Energy', pp. 635–6; and Dunlap, Alexander and James Fairhead. 'The Militarisation and Marketisation of Nature: An Alternative Lens to "Climate-Conflict"'. *Geopolitics* 19, no. 4 (2014): 937–61.

15. Modalities are comparable to Michel de Certeau's concept of strategies developed as a critique of Foucauldian discourse. de Certeau, Michel. *The Practice of Everyday Life*, trans. Steven Rendall. Berkeley: University of California Press, 1984.

16. Verweijen, Judith and Alexander Dunlap. 'Engineering Extraction and Land Control: Examining Political (Re)actions "From Above"'. *Political Geography*, Special Issue, forthcoming (2019). See also Geenen, Sara and Judith Verweijen. 'Explaining Fragmented and Fluid Mobilization in Gold Mining Concessions in Eastern Democratic Republic of the Congo'. *The Extractive Industries and Society* 4, no. 4 (2017): 758–65.

17. See Li, Tania M. 'To Make Live or Let Die? Rural Dispossession and the Protection of Surplus Populations'. *Antipode* 41, no. 1 (2010): 66–93.

18. Nixon, Rob (2011) *Slow Violence and the Environmentalism of the Poor*, Cambridge: Harvard University Press; Hecht, Gabrielle (2018) 'Interscalar Vehicles for an African Anthropocene: On Waster, Temporality, and Violence' HYPERLINK "https://culanth.org/articles/936-interscalar-vehicles-for-an-african-anthropocene" \l "bck_cuanHecht_bib037" https://culanth.org/articles/936-interscalar-vehicles-for-an-african-anthropocene#bck_cuanHecht_bib037, accessed 20 November 2018. See also Alexis-Martin, Becky and Thom Davies (2017) 'Towards Nuclear Geography: Zones, Bodies, and Communities, *Geography Compass*, 11(9) https://doi.org/10.1111/gec3.12325; and Elizabeth Cardis et al. (2005) 'Risk of Cancer after Low Doses of Ionising Radiation: Retrospective Cohort Study in 15 Countries', *BMJ*, 331 (7508): 77–83, http://www.bmj.com/content/331/7508/77; Churchill, Ward (2003) *Acts of Rebellion: The Ward Churchill Reader*, London: Routledge. Comparisons might also be made with the racialised aspects of 'slow violence' in a toxic town in Louisiana due to petrochemical pollution that Thom Davies elaborates. (2018) 'Toxic Space and Time: Slow Violence, Necropolitics, and Petrochemical Pollution', *Annals of the American Association of Geographers*, 108(6):1537–1553, p. 1540.

19. The seminal article on structural violence is Johan Galtung (1969) 'Violence, Peace and Peace Research', *Journal of Peace Research*, 6(3): 167–191, a theorist who also greatly influenced the PMANE convener, S.P. Udayakumar, as elaborated in Chapter 1.

20. Senthalir, S. 'Violence against the Non-violent Struggle of Koodankulam'. *Economic and Political Weekly* 47, no. 39 (2012): 13–15.

21. See 'Report of the Fact-Finding Team's Visit to Idinthakarai and Other Villages on September 20–21, 2012'; 'The Real Situation in Koodankulam', 12 September 2012. Available on https://geetacharusivam.blogspot.co.uk/2012/09/the-real-situation-in-koodankulam.html, accessed 10 January 2019.

22. While people mourned their ordeal and the state ambush of their village, some kept empty canisters of the tear gas as evidence, pointing out that they were made in USA and had all gone over their expiry dates in 2002. They were thus more hazardous and had left facial sores on young children. According to the US Environmental Protection Agency Resources Conservation and Recovery Act (1987), old and expired tear gas is classed as ignitable waste. Available on https://www.osti.gov/servlets/purl/6651928, accessed 3 March 2019.

23. 'Report of the Fact-Finding Team's Visit to Idinthakarai and Other Villages on September 20–21, 2012'; 'Koodankulam: Letter to IAEA [International Atomic Energy Agency], Nuclear Regulators and Human Rights Organisations'. *DiaNuke*, 4 March 2012. Available on http://

www.dianuke.org/koodankulam-letter-to-iaea-nuclear-regulators-and-human-rights-organisations/.

24. Lovelace, Douglas (ed.). *Hybrid Warfare and the Gray Zone Threat*. Oxford: Oxford University Press, 2016, p. xi. See 'Three Koodankulam Protestors Face Arrest: Amnesty Action Alert'. *DiaNuke*, 18 May 2012. Available on http://www.dianuke.org/three-koodankulam-protestors-face-arrest-amnesty-action-alert/, accessed 23 October 2012.

25. 'Kudankulam: Jailed Idinthakarai Woman Dies for Want of Timely Treatment', 21 December 2012. Available on http://www.indiaresists.com/Kudankulam-jailed-idinthakarai-woman-dies-for-want-of-timely-treatment/, accessed 10 November 2016.

26. Viju, B. 'Jaitapur Boils after Activist's Death'. *The Times of India*, 19 December 2010. Available on https://timesofindia.indiatimes.com/city/mumbai/Jaitapur-boils-after-activists-death/articleshow/7125307.cms, accessed 22 November 2016.

27. Sundaram, P.K. 'Farmers Mourn their Third Martyr in Anti-nuclear Power Struggle, Pledge against Another Fukushima (Fatehabad, India)'. *DiaNuke*, 21 September 2011. Available on http://www.dianuke.org/farmers-anti-nuclear-struggle-fatehabad-fukushim/, accessed 22 November 2016.

28. See Udayakumar, S.P. (ed.). *The Koodankulam Handbook*. Nagercoil: Transcend South Asia, 2004.

29. Patnaik, Prabhat. 'Neo-liberalism and Democracy'. *Economic and Political Weekly* XlIX, no. 15 (2014): 39–44, p. 43.

30. 'Kudankulam: Letter to CM about Attack on School'. *DiaNuke*, 30 April 2012. Available on http://www.dianuke.org/Kudankulam-letter-to-cm-about-attack-on-school/, accessed 10 November 2016.

31. See Human Rights Watch, 'Stifling Dissent'.

32. Ananth, M.K. 'Environmental Activist Mugilan begins Indefinite Fast in Palayamkottai Central Prison'. *The Times of India*, 8 January 2018. Available on http://timesofindia.indiatimes.com/articleshow/62414922.cms?utm_source=contentofinterest&utm_medium=text&utm_campaign=cppst, accessed 11 July 2018.

33. Available on https://www.facebook.com/amirtharaj.stephen/videos/pcb.10155258388737202/10155258380312202/?type=3&theater. On the prevalence of torture, see Kaur, Baljeet. 'India's Silent Acceptance of Torture Has Made It a "Public Secret"'. *EPW Engage*, 6 September 2018. Available on https://www.epw.in/engage/article/indias-silent-acceptance-torture-has, accessed 11 January 2019.

34. See Sen, Atreyee. 'Slaps, Beatings, Laughter, Adda, Puppet Shows: Naxal Women Prisoners in Calcutta and the Art of Happiness in Captivity'. In *Arts and Aesthetics in a Globalizing Worlds*, edited by Raminder Kaur

and Parul Dave-Mukherjee. Bloomsbury, London: Association of Social Anthropologists.

35. Sukumar, Varun. 'Activist S P Udayakumar vs Arnab's Republic TV'. *SIFY*, 23 June 2017. Available on http://www.sify.com/news/activist-s-p-udayakumar-vs-arnab-s-republic-tv-news-columns-rgxk18aeeichc.html, accessed 10 January 2018.

36. Stoll, David. *Between Two Armies in the Ixil Towns of Guatemala*. New York: Columbia University Press, 1993.

37. 'Kudankalum Update: Arrests of Peaceful Protesters under the Sedition Law in Tamil Nadu, India', 20 March 2012. Available on https://indian2006.wordpress.com/2012/03/20/kudankalum-update-arrests-of-peaceful-protesters-under-the-sedition-law-in-tamil-nadu-india/, accessed 10 November 2018.

38. Güven, Ferit. *Decolonizing Democracy: Intersections of Philosophy and Postcolonial Theory*. London: Lexington Books, 2015, p. 81.

39. The hunger strike itself might be denounced as a criminal activity: as occurred in 2006 with the social activist, Medha Patkar, the police could slap a charge of 'attempted suicide' according to Section 309 in the Indian Penal Code. In April 2018, suicide in India was decriminalized. *Human Rights Watch World Report*. New York: Seven Stories Press, p. 267. Available on https://www.hrw.org/sites/default/files/world_report_download/201801world_report_web.pdf, accessed 7 January 2019.

40. Akshay Khanna. 'Seeing Citizen Action through an "Unruly" Lens'. *Development* 55, no. 2 (2012): 162–72, p. 15.

41. On bare life, see Agamben, Giorgio. *Homo Sacer: Sovereign Power and Bare Life*. Stanford: Stanford University Press, 1995.

42. Cited in Doshi, Vidhi. 'The Lonely Struggle of India's Anti-Nuclear Protesters'. *The Guardian*, 6 June 2016. Available on http://www.theguardian.com/global-development/2016/jun/06/lonely-struggle-india-anti-nuclear-protesters-tamil-nadu-kudankulam-idinthakarai, accessed 10 November 2016.

43. Foucault, *Society Must be Defended*, p. 254.

44. Agamben, *Homo Sacer*.

45. Joseph, Sarah. 'Protracted Lawfare: The Tale of Chevron Texaco in the Amazon'. *Journal of Human Rights and the Environment* 3, no. 1 (2012): 70–91. See also Dunlap, Alexander. 'Permanent War: Grids, Boomerangs, and Counterinsurgency'. *Anarchist Studies* 22, no. 2 (2014): 55–79, pp. 64–6.

46. Janardhanan, Arun. '8,856 "Enemies of State": An Entire Village in Tamil Nadu Lives under Shadow', *Indian Express*, 12 September 2012. Available on http://indianexpress.com/article/india/india-news-india/

kudankulam-nuclear-plant-protest-sedition-supreme-court-of-india-section-124a-3024655/, accessed 10 January 2017.

47. Janardhanan, '8,856 "Enemies of State"'.

48. Janardhanan, '8,856 "Enemies of State"'.

49. Rajappa, Sam, Dr. Gladston Xavier, Mahadevan, Rajan, and Adv. Porkodi for Chennai Solidarity Group for Koodankulam Struggle. 'Fact Finding Report on the Suppression of Democratic Dissent in Anti-Nuclear Protests by Government of Tamil Nadu', 2012, p. 5. Available on https://www.dianuke.org/wp-content/uploads/2012/04/Fact_Finding_Report_Sam_Rajappa_English.pdf, accessed 28 February 2019.

50. *Human Rights Watch.* 'Stifling Dissent: The Criminalization of Peaceful Expression in India', 24 May 2016. Available on https://www.hrw.org/report/2016/05/24/stifling-dissent/criminalization-peaceful-expression-india, accessed 23 October 2018.

51. On parallel cases in Germany, see Brock, Andrea and Alexander Dunlap. 'Normalising Corporate Counterinsurgency: Engineering Consent, Managing Resistance and Greening Destruction around the Hambach Coal Mine and Beyond'. *Political Geography* 62 (2018): 33–47; and in Peru, Dunlap, Alexander. 'Agro sí, mina NO!' The Tía Maria Copper Mine, State Terrorism and Social War by Every Means in the Tambo Valley, Peru'. *Political Geography*, 71 (2019): 10–25.

52. Cited in *Human Rights Watch*, 'Stifling Dissent'.

53. 'Introduction: Spectres of Treason'. In *Traitors: Suspicion, Intimacy, and the Ethics of State-Building*, edited by Sharika Thiranagama and Tobias Kelly. Philadelphia: University of Pennsylvania Press, 2012, p. 4.

54. See Singh, Anushka. *Sedition in Liberal Democracies*. New Delhi: Oxford University Press, 2018. '"Persecution in the Name of Sedition Finds Popular Acceptance in India", Says Anushka Singh, Author of the Recent Book, "Sedition in Liberal Democracies"'. *Live Law*, 21 April 2018. Available on http://www.livelaw.in/persecution-name-sedition-finds-popular-acceptance-india-says-anushka-singh-author-recent-book-sedition-liberal-democracies/. See also http://foreignpolicy.com/2016/08/16/amnesty-international-accused-of-sedition-in-india/, accessed 23 October 2018.

55. Warren, Kay B. 'Conclusion: Death Squads and Wider Complicities: Dilemmas for the Anthropology of Violence'. In *Death Squad: The Anthropology of State Terror,* edited by Jeffrey A. Sluka. Philadelphia: University of Pennsylvania Press, 1999, p. 228.

56. On other cases of sexual taunts, harassment, and torture by law enforcement agents, see Arya, Divya. 'Soni Sori: India's Fearless Tribal Activist'. *BBC News,* 22 March 2016. Available on https://www.

bbc.co.uk/news/world-asia-india-35811608, accessed 20 January 2019; Stephen, Lynn. 'The Construction of Indigenous Suspects: Militarization and the Gendered and Ethnic Dynamics of Human Rights Abuses in Southern Mexico'. *American Ethnologist* 26, no. 4 (1999): 822–42; and Brock and Dunlap, 'Normalising Corporate Counterinsurgency', p. 43.

57. 'Report of the Fact-Finding Team's Visit to Idinthakarai and Other Villages on September 20–21, 2012', p. 9.

58. 'Report of the Fact-Finding Team's Visit to Idinthakarai and Other Villages on September 20–21, 2012', p. 12.

59. On the confluence of gender, race and class as part of intersectionality, see Crenshaw, Kimberley (1991) 'Mapping the Margins: Intersectionality, Identity Politics, and Violence against Women of Color', *Stanford Law Review* 43 no. 6: 1241–1299. For perspectives that veer away from the additive, see Brah, Avtar and Ann Phoenix (2004) 'Ain't I a Woman? Revisiting Intersectionality', *Journal of International Women's Studies* 5, no. 3: 75–86. On the contemporary Indian case, see Dingli, Sophia and Navtejj Purewal (eds). 'Gendering (In)security: Post/Neocolonial Security Logics and Feminist Interventions'. *Third World Thematics* 3, no. 2 (2018): 153–63,

60. Warren, 'Conclusion', p. 234.

61. See Sharma, Aradhana and Akhil Gupta. 'Introduction: Rethinking Theories of the State in an Age of Globalization'. *Anthropology of the State: A Reader*, edited by Aradhana Sharma and Akhil Gupta. Oxford: Blackwell, 2006, p. 11.

62. 'Case Registered against Protesters'. *The Hindu*, 17 August 2012. Available on http://www.thehindu.com/todays-paper/tp-national/tp-tamilnadu/case-registered-against-protesters/article3783244.ece, accessed 24 January 2019.

63. Tharoor, Shashi. *Inglorious Empire: What the British Did to India*. London: Christopher Hurst, 2017; and Wilson, Jon. *India Conquered: Britain's Raj and the Chaos of Empire*. London: Simon and Schuster, 2016. Although Jon Wilson argues that colonial divide-and-rule was not as systematic as it appeared to be.

64. Dunlap, 'Counterinsurgency for Wind Energy', p. 649.

65. PMANE Struggle Committee, *PMANE Letters and Press Releases*, p. 17.

66. PMANE Struggle Committee, *PMANE Letters and Press Releases*.

67. Scott, James C. *Domination and the Arts of Resistance: Hidden Transcript*. New Haven: Yale University Press, 1990.

68. See Wodak, Ruth. *The Politics of Fear: What Right-Wing Populist Discourse Means*. Los Angeles: Sage, 2015.

69. Feuchtwang, Stephan. 'Afterword: Question of Judgement'. In *Traitors: Suspicion, Intimacy, and the Ethics of State-Building*, edited by Sharika Thiranagama and Tobias Kelly. Philadelphia: University of Pennsylvania Press, 2009, p. 231. See Venkataramanan, K. 'Kalam for Boosting Local Economy to Allay Nuclear Fears'. *The Hindu*, 8 November 2011. Available on https://www.thehindu.com/news/national/kalam-for-boosting-local-economy-to-allay-nuclear-fears/article2607289.ece; Ittyipe, Minu. 'Doom Under These Domes?' *Outlook India*, 21 November 2011. Available on https://www.outlookindia.com/pwa/magazine/pwa_story_first/278934, accessed 1 March 2019.

70. See Mbembe, Achille. 'Provisional Notes on the Postcolony'. *Africa: Journal of the International African Institute* 62, no. 1 (1992): 3–37, pp. 3–4; and Mazzarella, William. 'Internet X-Ray: E-Governance, Transparency, and the Politics of Immediation in India'. *Public Culture* 18, no. 3 (2006): 473–505, p. 474.

71. Foucault, Michel. *Discipline and Punish: The Birth of the Prison*. Penguin: London, 1977.

72. See Raj, Jayaseelan. 'Rumour and Gossip in a Time of Crisis: Resistance and Accommodation in a South Indian Plantation Frontier'. *Critique of Anthropology* online, 2018. Available online https://journals.sagepub.com/doi/abs/10.1177/0308275X18790803, accessed 6 January 2020.

73. All NGOs who receive foreign funds have to have FCRA approval from the government every five years. With latter-day revisions, approval is conditional on no political or 'anti-national' activity.

74. 'Koodankulam: Crackdown on Anti-nuclear Activists and NGO's'. *Nuclear Monitor*, #744, number 6238, 16 March 2012. Available on https://www.wiseinternational.org/nuclear-monitor/744/koodankulam-crackdown-anti-nuclear-activists-ngos, accessed 21 February 2019.

75. 'Kudankulam Protests, Church and Western NGOs—A Citizen's Probe'. Available on http://ariseasia.blogspot.co.uk/2012/02/Kudankulam-protests-church-and-western.html, accessed 10 November 2016.

76. Rizvi, S.A. 'Concerted Efforts by Select Foreign-funded NGOS to "Take Down" India's Development Projects', Ministry of Home Affairs, Intelligence Bureau. Available on https://kractivist.org/wp-content/uploads/2014/06/NGO-Report_leaked-IB-report.pdf, accessed 6 January 2020.

77. 'FCRA Licenses of 20,000 NGOs Cancelled'. *The Times of India*, 27 December 2016. Available on http://timesofindia.indiatimes.com/india/fcra-licences-of-20000-ngos-cancelled/articleshow/56203438.cms, accessed 12 October 2017.

78. Dubbudu, Rakesh. 'Massive FCRA Cancellation Is Not a New Thing. 4138 NGOs Lost Their FCRA Licenses in 2012'. *Factly*, 29 April 2012.

Available on https://factly.in/massive-fcra-cancellation-is-not-a-new-thing-4138-ngos-lost-their-fcra-licenses-in-2012/, accessed 12 January 2019.

79. 'India Cracks Down on Greenpeace and Foreign NGOs'. *Aljazeera*, 27 May 2015. Available on http://www.aljazeera.com/news/2015/05/india-cracks-greenpeace-foreign-ngo-150526102622208.html, accessed 12 October 2017.

80. Jain, Bharti. 'MHA Goofs Up, Renews Greenpeace's FCRA License'. *The Times of India*, 14 December 2016. Available on http://timesofindia.indiatimes.com/india/MHA-goofs-up-renews-Greenpeaces-FCRA-licence/articleshow/55970187.cms, accessed 12 October 2017.

81. 'FCRA as a Tool of Repression: Greenpeace India Signs Civil Society Statement in Solidarity with Those Denied Registration'. *Greenpeace India*, 14 December 2016. Available on http://www.greenpeace.org/india/en/news/Feature-Stories/FCRA-as-a-tool-of-repression-Greenpeace-India-signs-civil-society-statement-in-solidarity-with-those-denied-registration/, accessed 12 October 2017. Others brought to attention the communalization of allegations against minority communities in what might be called 'Hindia'. It is certainly true that the BJP government has not reproached Hindutva organizations such as the Rashtriya Swayamsevak Sangh (RSS) who are held to be 'getting crores [10,000,000] from migrant Indians living in other countries'. 'Republic Channel Exposed "Activism for a Price" in Anti-nuclear Protests, Funded by the Church. S. P. Udayakumar was in the Sting. What Is Your Take?' *Quora*. Available on https://www.quora.com/Republic-Channel-exposed-Activism-for-a-price-in-anti-nuclear-protests-funded-by-the-church-S-P-Udayakumar-was-in-the-sting-What-is-your-take, accessed 23 October 2018.

82. Rizvi, S.A. 'Concerted Efforts by Select Foreign-funded NGOS to "Take Down" India's Development Projects'.

83. See Crehan, Kate. *Gramsci's Common Sense: Inequality and Its Narratives*. Durham: Duke University Press, 2016.

84. Rizvi, 'Concerted Efforts by Select Foreign-Funded NGOS to "Take Down" India's Development Projects'.

85. Rizvi, 'Concerted Efforts by Select Foreign-Funded NGOS to "Take Down" India's Development Projects'.

86. Bornstein, Erica and Aradhana Sharma. 'The Righteous and the Rightful: The Technomoral Politics of NGOs, Social Movements, and the State in India'. *American Ethnologist* 43, no. 1 (2016): 76–90, p. 76.

87. The expression was used by a liberal Democrat congressman, Pete Stark, as cited in Gusterson, Hugh. *Nuclear Rites: A Weapons Laboratory*

at the End of the Cold War. Berkeley: University of California Press, 1998, p. 183.

88. Rizvi, 'Concerted Efforts by Select Foreign-Funded NGOS to "Take Down" India's Development Projects'.

89. Udayakumar, S.P.) 'India after BJP Victory: I Fear for My Life, Says Anti-Nuclear Activist SP Udayakumar after Intelligence Bureau Report'. *Europe Solidaire sans Frontières*, 14 June 2014. Available on https://www.europe-solidaire.org/spip.php?article32223. See also 'Is the IB Trying to Frame Anti-nuclear Activist Udayakumar?' *Weekend News*, 11 June 2014. Available on http://www.theweekendleader.com/ Headlines/2443/is-the-ib-trying-to-frame-anti-nuclear-activist-udaya-kumar-.html, accessed 13 January 2019.

90. Subramanian, 'Full Steam Ahead', p. 120; 'Kudankulam Update'.

91. Rizvi, 'Concerted Efforts by Select Foreign-Funded NGOS to "Take Down" India's Development Projects'.

92. Jacob, Jeemon. 'Hermann Seems More a Scapegoat than an Instigator'. *Tehelka*, 29 February 2012. Available on http://archive.tehelka.com/ story_main51.asp?filename=Ws290212Koodankulam.asp, accessed 23 October 2012.

93. Milder, Stephen. *Greening Democracy: The Anti-Nuclear Movement and Political Environmentalism in West Germany and Beyond, 1968–1983*. Cambridge: Cambridge University Press, 2017.

94. Allegations were also made that the Nepal-based transnational conglomerate the Sharda Group, had funded the movement to the tune of five crores rupees, according to a joint investigation by the Enforcement Directorate, the Central Bureau of Investigation, the Serious Fraud Investigation Office and the Income Tax Department that was reported in the media. Bhawna. 'Nuclear Energy, Development and Indian Democracy: The Study of Anti Nuclear Movement in Koodankulam'. *International Research Journal of Management Sociology and Humanity* 7, no. 6 (2016): 219–29, pp. 226–7.

95. See Taylor, Diana. *Disappearing Acts: Spectacles of Gender and Nationalism in Argentina's 'Dirty War'*. Durham: Duke University Press, 1997.

96. On its relevance in USA, see Masco, Joseph. *The Theatre of Operations: National Security Affect from the Cold War to the War on Terror*. Durham: Duke University Press, 2014.

97. Cited in *Human Rights Watch*, 'Stifling Dissent'.

98. Varadarajan, Siddharth. 'My Foreword to Iftikhar Gilani's *My Days in Prison*', 1 February 2005. Available on http://svaradarajan.blogspot. com/2005/02/my-foreword-to-iftikhar-gilanis-my.html, accessed 12 January 2019.

99. See Floridi, Luciano (ed.). *The Onlife Manifesto: Being Human in a Hyperconnected Era*. Cham, Switzerland: Springer International Publishing, 2015.

100. On Hindutva 'IT cells' used for tweeting and trolling, see Udupa, Sahana. 'Enterprise Hindutva and Social Media in Urban India'. *Contemporary South Asia* 26, no. 4 (2018): 453–67.

101. Shulsky, Abram N. and Gary James Schmitt. *Silent Warfare: Understanding the World of Intelligence*. Washington, DC: Potomac, 2002. See Udupa, Sahana. 'Middle Class on Steroids: Digital Media Politics in Urban India'. *India in Transition*, 2016. Available on https://casi.sas.upenn.edu/iit/sudupa, 6 January 2020.

102. Facebook correspondence, 17 May 2012.

103. Cadwalladr, Carole. 'The Great British Brexit Robbery: How Our Democracy Was Hijacked', 2017. Available on https://www.theguardian.com/technology/2017/may/07/the-great-british-brexit-robbery-hijacked-democracy, accessed 30 April 2018. See also Wooley, Stanley C. and Philip N. Howard. 'Political Communication, Computational Propaganda, and Autonomous Agents—Introduction'. *International Journal of Communication* 10 (2016): 4882–90.

104. *CitizenFour* (2014, dir. Laura Poitras), Channel 4; *Vice: State of Surveillance*. HBO, 2015. Available on https://plus.google.com/+ArashPayan/posts/a3p3QKLuHUh, accessed 18 November 2018.

105. https://www.digitalindia.gov.in/

106. https://www.digitalindia.gov.in/

107. 'India Accused of Being "Surveillance State" after Allowing 10 Central Agencies to Snoop on any Computer'. *Technology Intelligence*, 21 December 2018. Available on https://www.telegraph.co.uk/technology/2018/12/21/india-accused-surveillance-state-allowing-10-central-agencies/; 'Opposition Slams Cyber Surveillance Order'. *Live Mint*, 21 December 2018. Available on https://www.livemint.com/Politics/fLVGbRUfvYv1pf4YDtPf6M/Computers-in-India-will-now-be-under-government-surveillance.html, accessed 21 January 2019.

108. 'Opposition Slams Cyber Surveillance Order. See also Bornstein and Sharma, 'The Righteous and the Rightful', p. 77.

109. See Choudry, Aziz (ed.). *Activists and the Surveillance State*. London: Pluto Press, 2018.

110. Futrell, Robert and Barbara G. Brents. 'Protest as Terrorism: The Potential for Violent Anti-Nuclear Activism'. *American Behavioral Scientist* 46, no. 6 (2003): 745–65.

111. Brock and Dunlap, 'Normalising Corporate Counterinsurgency', p. 35.

112. This is a slightly different argument to 'revealing something that everybody already knows' about political corruption that Mazzarella concentrates on with respect to the web-based newspaper, *Tehelka's* exposé of members of the defence establishment in negotiations with journalists posing as arms dealers. 'Internet X-Ray', p. 473. Here, it is a case of ensuring political concoctions stick to public opinion as part of political (re)actions 'from above'.

113. On the discrepant nature of rules and practices to do with secrecy and disclosure, see Gusterson, *Nuclear Rites*, pp. 68–100. See also Taussig, Michael. *Defacement: Public Secrecy and the Labor of the Negative*. Stanford: Stanford University Press, 1999; Kaur, Raminder and William Mazzarella. *Censorship in South Asia: Cultural Regulation from Sedition to Seduction*. Bloomington: Indiana University Press, 2009.

114. 'Intelligence Bureau Report on NGOs Comes under Attack from Activists'. *NDTV*, 13 June 2014. Available on https://www.ndtv.com/india-news/intelligence-bureau-report-on-ngos-comes-under-attack-from-activists-577864, accessed 6 January 2020.

115. Udupa, 'Enterprise Hindutva and Social Media in Urban India'; Sukumar, Varun. 'Activist S P Udayakumar vs Arnab's Republic TV', 23 June 2017. Available on http://www.sify.com/news/activist-s-p-udayakumar-vs-arnab-s-republic-tv-news-columns-rgxk18aeeichc.html; 'Were Kudankulam Protests Funded By Foreign Money?', 20 June 2017. Available on https://www.youtube.com/watch?v=B3i4rYnz7Hc, accessed 20 February 2019.

116. 'Were Kudankulam Protests Funded by Foreign Money?'. On how Udayakumar's admission was twisted for a nugget of news, see https://www.facebook.com/watch/?v=1722607367768652, accessed 6 January 2020.

117. 'Were Kudankulam Protests Funded by Foreign Money?'

118. 'Were Kudankulam Protests Funded by Foreign Money?'

119. Sukumar, 'Activist S P Udayakumar vs Arnab's Republic TV'.

120. Sukumar, 'Activist S P Udayakumar vs Arnab's Republic TV'.

121. 'Activist SP Udayakumar Unmasks Himself'. Available on https://www.youtube.com/watch?v=mT6iG-gm_bY, accessed 20 February 2019.

122. See Roy, Arundhati. *The End of Imagination*. Chicago: Haymarket Books, 2005, p. 96.

123. 'Republic Channel Exposed "Activism for a Price" in Anti-nuclear Protests, Funded by the Church'.

124. 'Outrage over Videos that show Shashi Tharoor surrounded and followed by Republic TV Scribes', *The News Minute*, 4 August 2017

https://www.thenewsminute.com/article/outrage-over-videos-show-shashi-tharoor-surrounded-and-followed-republic-tv-scribes-66252, accessed 31 October 2019.

125. See also https://www.facebook.com/watch/?v=1722607367768652, accessed 31 October 2019.

126. Mosse, David. 'The Anthropology of International Development'. *Annual Review of Anthropology* 42 (1993): 227–46, p. 237.

127. Cormac, Rory and Richard J. Aldrich. 'Grey Is the New Black: Covert Action and Implausible Deniability'. *International Affairs* 94, no. 3 (2018): 477–94, p. 477.

128. Foucault, *Discipline and Punish*, p. 171.

129. Cormac and Aldrich, 'Grey Is the New Black', p. 477.

11

The Sparks That Hover

I am fearless, the more I am tortured, the stronger I become. If they are successful in silencing me, they would show me as an example to silence everyone else. (Soni Sori 2016)[1]

Imagine what it might feel like to be at both the surge and ebb of a monumental movement that gained much in terms of raising a public debate about the opaque plans and irregular procedures around a nuclear power plant but inevitably has little to show for it in terms of reaching its main goals. The first Kudankulam reactor went critical in July 2013, the second three years later in 2016, and it looks like the relevant authorities are bulldozing ahead with plans for more. For a long time afterwards, the skeletal pavilion outside Lourdes Matha church in the coastal village of Idinthakarai stood as a fragile reminder of large-scale collective resistance against a nuclear plant. Perhaps it will be reconstructed to host more in the future. Who can say?

To look back at the history of the anti-nuclear movement in peninsular India: we began by considering what part the construction played with regards to discourses of democracy, development, and nationalism in post-colonial India—each term being subject to a variety of perspectives in ever-changing spaces of criticality. We grounded the study in an exploration of the ecological, economic, demographic, and social contours of the region. We then moved on to moments of concerted dissent through a history of local struggle against the nuclear industries in the Kudankulam region as soon as the news was first made public in 1988, plans that were later dropped to be reinstated at the turn of the millennium. A lens on the people's struggle around Kudankulam enabled an overview on changes in Indian social movements over the last four decades in response to the continuing popularity of non-violence, growing environmental awareness, and a culture of entitlement for project-affected people. Whereas in the 1980s, those in the establishment might have taken nuclear concerns seriously and even be converted by them as happened over plans for a nuclear plant in Kerala, by the turn of the century, these anxieties were dismissed out of hand to matters of policy where the right to electricity as development's lynchpin took precedence over all else. Against such forces, we learnt about the gradual working out of politics, practice, and philosophies and the broadening of the anti-nuclear momentum. It is a period that has been much overlooked in the mainstream media coverage of Kudankulam, some of which deemed the anti-nuclear movement as simply a post-Fukushima Daiichi disaster development.

As we explored these spaces, we did not confine our analysis solely to the most visible sparks of resistance even though this remains important. Chapters 4 and 5 therefore took us to an examination of discourses of risk, radiation, and the production of knowledge and ignorance where the latter was a necessary prologue to nuclear developments. While activists presented radioactivity as a pathological life invader, nuclear authorities channelled it through somewhat contradictory registers to do with the dismissal of any dangers to do with radioactivity, its normalization, its technical manageability, and its many virtues to the point of aggrandizing ionized evolution. This was while nuclear authorities promulgated different rationales for radiation to their own contracted nuclear employees and trainees, much of which overlapped with the views of critics and activists but differed in terms of their understanding of 'safe dosage'.

In the subsequent sister chapter, we looked at how and where opinion, resilience, and resistance might emerge with regards to people's responses to the prospect of a nuclear power plant near their homes in terms of adaptation, diversion, diversification, solace, and satirical release. This was followed by a focus on three people, Josef, Savitri, and Rajesh, from different walks of life navigating competing and contradictory discourses of radiation while they steered social and political risks of various kinds. Their stories demonstrate how autochthonous resistance was simmering that was not merely the outcome of outside orchestrations as characterized the dismissive stance in corridors of power.

We moved on to an account of a 'secret' public hearing in order to reflect on the ostensible as well as the more subtle circuits of power, resistance, and its disguises.[2] The event on 6 October 2006 was the first of several attempts for a public hearing on the nuclear plant with NPCIL. It provided an exemplary occasion with which to consider the clash of epistemologies between the nuclear state and peninsular residents with their range of direct, creative, and nuanced challenges.

We also looked at concurrent attempts to collate statistics and create an evidence base of radiation readings and proximal health problems that could inform an anti-nuclear movement. It represented a tactic that emerged around the turn of the new millennium in the region— one that called upon 'hard facts' and a politics of transparency for all public authorities as enshrined in the Right to Information Act (2005) even while such acts decreed nuclear agencies as off-limits.

The next three chapters signalled a turn to the recent past. They honed in on political tensions and ruptures around the village of Idinthakarai that, from 2011, became the heart of the anti-nuclear movement in south India. We considered the village's role in jettisoning PMANE as a powerful force to be reckoned with, and highlighted the prominent role of women and children's protest against the nuclear power plant.

Digitalia in modern social movements—or the significance of social media entangled with various on-the-ground constituencies—was no less significant to Kudankulam. Therefore we moved onto digital activism by focusing on the content and reception of a public letter from British to Indian politicians written collaboratively in May 2012 with culturally diverse activists in Britain. In its fallout, the potentials and hurdles in the way of forging a transnational anti-nuclear movement *across* the global south and north, the formerly colonized and colonizing, were highlighted. The letter's reception evident in news readers' commentaries point to another series of debates that indicate, on one end, colonial legacies, the stranglehold grip of nationalism and the fear of the 'foreign hand' on the postcolony; and on the other, a reignited promise of transparency, accountability, recompense, and the promise of a grassroots planetary consciousness.

Once a progressive vehicle for people's sovereignty against feudalism, elitism, and colonialism, the hot-blooded energies of nationalism have morphed to execute their own oppressions in the subcontinent. The dyad in nation-state that underwrites nationalism has seen a major shift to the aggrandized power of the latter along with the weakening of the former. In the previous chapter we concentrated on the way people's claim to legitimate citizenry and dissent was revoked the more they mobilized against the nuclear state. Such measures were executed through creating 'death conditions' of various modalities from slow to quick, covert to overt, entailing syncretic subjugations from 'let die' to 'make die'. In the process, we considered how victims, suspects, and targets were created out of citizens in direct and indirect ways, whether it be by way of political officials, village-based castigation, law enforcement and surveillance agencies, policy makers, or psychological and media operations to influence the public. However, this did not deter all of the accused from persevering. From bare life were ongoing efforts to return to political life. We now assess PMANE convenors' entry into electoral politics and end with our hands around the embers that remain of the social movement.

Electoral Ventures

In 2014, there was yet another blow to the Idinthakarai-based movement, indicating another phase to its recent history. This time it was largely at the behest of those among its frontline—a 'fission within' as M. Abdul Rabi reports.[3] The main convenors of PMANE, S.P. Udayakumar, Michael Pushparayan, and Father Michael Pandian Jesuraj, made a reluctant decision to partake in May's state elections amidst much argument among their supporters. They were to be single-issue candidates for the Aam Aadmi Party (AAP) in the Lok Sabha polls from the Kanyakumari, Tirunelveli, and Thoothukudi constituencies respectively. AAP was an emergent power with the main aim to eradicate corruption in Indian politics and became a vocal challenge to the mighty weight of the Congress and the BJP. Key AAP figures Arvind Kejriwal and Prashant Bhushan had earlier visited PMANE in the village in 2012. Ongoing discussions between them and the PMANE Struggle Committee led to the launch of political candidacies for the three men.

Several PMANE supporters were disheartened about this tactic, however. They believed that the nuclear struggle could not be won through electoral means, and that they would rather see them court arrest as political martyrs to the cause. Interestingly, bare life was the preferred state when the civic alternative was considered as mired in deceit and corruption. Others worried that they would be sucked into the crooked entrails of official politics. The radicality of the PMANE movement was seen to be compromised with its entry to the formalities of elections. Still others wanted the three men to stay in Idinthikarai, to protect them with their collective strength from what they saw as an inevitable and detrimental arrest that would only break their backs as well as wreck the movement. The PMANE spokespersons would be forgotten—like countless others who languish in prison for years on end without conviction, waiting for their cases to be heard and resolved.[4]

Wracked with such anxieties, the AAP candidates nevertheless began to canvass in the southern districts, early meetings being addressed through video-conferencing from Idinthakarai. However, over time and with the support of other PMANE activists including the women, the candidates ventured out of the village, relatively freely to address seminars and rallies.[5] Udayakumar's wife, Meera, also campaigned on his behalf.

The candidates travelled out of the village with an uneasy confidence that they would not get arrested during the election period. It was a wary re-entry into the realm of citizenry. With the Election Commission of India's and its 'moral code of conduct for parties and candidates', nuclear and enforcement officials had to reluctantly accept the role of the three candidates in public life.[6] However, this was also a time when adversarial authorities believed that they had won the fight, that any threats of stalling nuclear reactor construction had been quashed, and that the first reactor had already been commissioned. They surveilled but they did not strike.

As it transpired, PMANE co-ordinators were unsuccessful in securing AAP seats. Pushparayan and Jayakumar returned to their village, dismayed by the direction that PMANE had taken into electoral politics. But for Udayakumar, the process was a blessing in disguise. The outcome of the elections was almost incidental to Udayakumar's main motive that was to be able to leave the village of Idinthakarai and return to his family in Nagercoil for the first time in over two years. In retrospect, he declared that he had adopted the electoral route as 'the only way to revive the agitation' and a result of the 'insult and intimidation' by the state government he suffered when coordinating PMANE.[7] It was a case of push- rather than pull-factors into *realpolitik*. While Udayakumar complained that 'BJP and Congress are selling our natural resources to foreign corporations', his primary interest was in fact not to become a politician, but to further his anti-nuclearism as a grassroots campaigner.[8] As 'a person of the people', the wider politics of electricity remained more important than the politics of elections.

In October 2014, Udayakumar decided to leave AAP stating that the party lacked 'clarity on the nuclear policy of the nation' and that autonomy was not granted to their unit to 'deal with Tamil Nadu issues, especially like the Tamil fishermen's problem, Sri Lankan Tamil issue, beach mineral sand mining, anti-people's projects, social justice and minority issues [sic]'.[9]

With his base back in Nagercoil living with his wife, children, and elderly parents, Udayakumar carried on, venturing across the country to further his campaign of social justice. This was not without great consternation for Udayakumar remained a prime target of intimidation. Police and intelligence officers followed him everywhere making daily visits or phone calls to check on his whereabouts. His family constantly worried about whether or not he would return that day, either taken away by the authorities or even killed by henchmen, perhaps even those employed by Russians disgruntled by his criticism of their joint venture with the Indian government at Kudankulam. A stubborn cloud of surveillance tied to police and gangster harassment hung over him wherever he went largely fuelled by anxieties about his proven potential to rouse the public.

If Udayakumar's movements were not the subject of long police files, they were curtailed by the police. Travel to the neighbouring country, Nepal, ordinarily does not require a passport for an Indian citizen. But when Udayakumar was at the airport in New Delhi in September 2014 in order to attend a United Nations Special

Rapporteur on Human Rights Defenders meeting at Kathmandu, he was taken away for questioning.[10] He was under a 'lookout notice' from Tamil Nadu police. Soon after in 2015, he contested his right to have a passport that had been earlier impounded on 'humanitarian grounds' by appealing to the External Affairs Minister, Sushma Swaraj.[11] But, as with his other appeals, this was to no avail—silence, indifference, or diversion being a common ruse for state inaction without the oiling of prominent palms.

Any short-term optimism in a newly elected government in 2014 withered fast. The BJP continued similar practices with an added layer of muscular Hindu conservatism. In 2016, central government sanctioned work to begin on the third and fourth reactors at Kudankulam that was initiated by the Indian and Russian premiers, Narendra Modi and Vladimir Putin, through video-conferencing.[12]

Those left in Idinthakarai still respected the legitimacy of non-violence and deep democracy. They welcomed Udayakumar any time he visited. But they were also fully aware that the core of the movement had disintegrated. Some felt abandoned and forlorn, pointing to the way that a brume of rumour and blame had destroyed the struggle's unity. While they talked about how the village had destroyed the movement, others maintained it was the movement that destroyed the village. One man, Sagayaraj, reflected on the magnitude of fighting the nuclear state mired as it is in transnational trade networks:

> I supported the protests at first, but now I've realised that we just need to take what we can get. This is not just about Idinthakarai or Kudankulam—it is an international issue. If they close the plant here, there will be protests to close nuclear plants all around the country. Now people realise the scale of what we're trying to do. It's like we're protesting against Russia—against all the foreign governments. How can our small village take on international powers?[13]

The sheer weight of the struggle rebounded on them. The epicentre of this phenomenal number of people and their commitment to nonviolence was splintered and shattered in all directions by dubious state-endorsed mechanisms and allied forces. The 'broken shore' of Idinthakarai became host to broken people. Lives and livelihoods were trampled by profiteering powers. The a(nta)gonistic state of criticality crumbled.

Looking Forward

The story of the Kudankulam Nuclear Power Plant is a story of many moving parts. Once it was beautiful ... life was calm ... then came the humachines ... they brought with them rivers of fire, dark realms of vast caverns and magisterial gates ... they lurked in the shadows like serpent gods of science...they said they will create a future supernova that will shower everyone with immeasurable joys...instead they broke every bone in their bodies ... they danced on their blood ... they created a death-space where there could be no sadness, pain or anger.... All stories have an ending but this one cannot.

On 15 March 2017, Udayakumar recalled his instruction in Peace Studies when he tweeted 'Never give up, Never stop, Never loose hope—Johan Galtung, my teacher.' [sic]. It was his erudite study of writings on peace activism including the works of sociologist and mathematician Johan Vincent Galtung and the freedom fighter Mohandas Karamchand Gandhi among others that informed much of Udayakumar's thinking. Violence is embedded throughout the hegemonic social order. By people becoming blind to it was yet another effect of violence.[14] Udayakumar resolved to continue raising awareness, linking people's experiences with theories for action. No matter what government is in power, this intellectual and political conscientization cannot be wiped out. As Udayakumar's home image on his Facebook account endorses: 'Advice from a tree. Stand tall and proud'.[15] With such efforts, the lens on nuclear necropolitics is flipped. Do we exist? Yes definitely, but existence on our terms outside of nuclear and neo-liberal compulsions.

It was an existence that was led by an eco-conscious and demos-orientated future vision that others across the subcontinental spectrum began to appreciate. One Nagercoil resident not linked to the movement reproached:

> Yes, let the country develop, but let the environment on which it is built develop as well. Don't destroy it just to make money. We need to have sustainable development, but the politicians do not even know how to spell it.

State officials may not know how to spell, but plenty of others had learnt, despite their lack of formal credentials, through a soaring learning curve on the wings of the social movement.

The emergence, rise, and petering out of anti-nuclear resistance shows many parallels with other movements—the growing awareness of biased decision-making and scientific processes, the peak years of unleashed energy, the volley of attacks and counter-attacks, the passionate and intellectual arguments that powered them on, the mobilization of an assembly of workers, organic intellectuals and their supporters, the solidarity expressed with other social justice movements, the anxiety, pain, and heartbreak of civilian loss, charges, and arrests, the ongoing onslaught by an unsympathetic mainstream media, the fraught engagements with and as politicians, and the gradual dissipation through a combination of direct and indirect techniques and procedures to stem the revolt.[16] As the political philosopher Michael Hardt observes more widely:

> ... in most cases the movements of the recent decades did not fail but were defeated, by ideological and media forces, by the police, and by the ruling institutions. Whereas failure is closed in a dead end, political projects that suffer defeat live on beyond their death and are often reanimated in new form. Recognizing diverse temporalities, then, has benefits not only for scholarship. The movements themselves are enriched by maintaining multiple attachments to the past. Just as important as learning lessons from mistakes, then, is recognizing the need and possibility to continue the projects of past movements and develop them in new ways.[17]

If we were to return to the reactors: the first commissioned reactor was dogged by forced power outages, lay off-grid for 468 days in the space of three years, has not worked to full capacity since, and even had a boiler room accident in the plant after which six workers were clandestinely taken to hospital only a year after the reactor was commissioned.[18] Through independent research by citizen scientists, the reactors have been cast as 'sub-standard, unsafe and unviable' for continuous outages and mishaps.[19] Such disruptions vindicate what PMANE supporters have been saying all along about the problems that beset the story of the nuclear power plant. Two nuclear reactors may have attained criticality but the space of criticality, 'an area of power and anti-power' to modify John Holloway's proposal, continue to erode the 'NPCIL success story'.[20]

Embers

Criticality is an unpredictable ocean across diverse spaces and temporalities. It is about what was, is now, and can or will be. It has been used in the book as a focused yet flexible lens through which to examine multi-layered and multi-situated spaces with which to consider the ever-changing encounters, ruptures, and tensions of socio-political conduct from the everyday to the extraordinary, the ambivalent to the more resistive and outspoken. Despite the crushing of the people's movement, Udayakumar maintains that 'Embers of the protest are still there'.[21] Reasserting his popular authority and moral leadership, he continues to spread awareness against nuclearism along with other industries damaging to the environment and public health.[22]

In May 2016, Udayakumar floated a green party that had its roots in thoughts, conversations and activities from a decade earlier recounted in Chapter 3. Under the name of Pachai Thamizhagam (Green Tamil Nadu), he contested one seat with a campaign that sought to expose the 'inefficient disaster management system in India'.[23] He elaborated on his main agenda: 'Our main motive is ecology and safeguarding our environmental wealth, our natural resources, participatory democracy, feminism, non-violence, future-mindedness and being responsible'.[24] His commitment to derail nuclear plans in favour of alternatives that were complementary to people and habitat went undiminished: 'Struggle still continues against Kudankulam and Kalpakkam. We are persisting with that as in Germany and as a Green party we will continue to advocate anti-nuclear stand [sic]'. His attention was wider than nuclear politics to generate a culture of awareness and critique with a focus on 'creating the next generation of leaders' with alternative pathways for demos-centric development.[25] This is in a country where hardly any mainstream party has adopted an ecologically grounded democratic development agenda. Against all the backlash, only time can tell whether he has effectively sown the seeds of an electoral party in India with its main prerogative being the protection of the environment and its inhabitants.

Do we make history or does history make us? A perennial causality dilemma perhaps, but we can definitely say that it is the people's struggle around the Kudankulam Nuclear Power Plant that has made history for India, and not the construction of the plant itself that has followed by now a fairly standard route for large-scale developments in the subcontinent. Itty Abraham adds: 'What makes the Koodankulam struggle different from other accounts of local resistance to mega-projects are the origins of violence.'[26] This violence emanated from the nuclear state. To counter it, bring it into the spotlight, and raise pertinent questions on democracy, development and nationalism with respect to India's energy future required exceptional strength and bravery, the likes of which has not been seen before against a nuclear power plant in quite the same way. People like Udayakumar, Pushparayan, Jesuraj, Mugilan, Sundari, Melrit, Tamil, Xavieramma, and Celine among many other women and men have risked all to keep the memory and relevance of Mohandas Karamchand Gandhi alive with those like Anthony John, Sahayam Francis, and J. Roslin who are now united in his spirit. Even though concepts and words like ecology and sustainability were not rife at the time of Gandhi's life, PMANE activated his freedom-fighting 'soul-force' for environmental and political justice against a nuclear state that continues to

Figure 11.1   Kudankulam Nuclear Power Plant from Idinthakarai Beach, 2015.
Source: Raminder Kaur.

(re)produce silent and sometimes deafening emergencies on those who defy them. Supporters revived notions of self-rule (*swaraj*) to promote devolved forms of ecologically complementary energy production. They struggled to bring their own perspective on people and ethics to the heart of democratic politics. This is a very different chord to the way Gandhi has been appropriated by latter-day Hindu nationalists to advocate a path of purity while aggressively availing of neoliberal dividends to 'Make in India'. Here, Gandhi's associations with the selfless struggle for justice and freedom against oppression are conveniently cast aside.[27]

While the anti-Kudankulam movement has dissipated, the energy of the people that led to it continues to circulate (Figure 11.1). Based in Idinthakarai, Tamil sums up this hushed mood:

My eyes smart with the salt of tears and my throat hurts with the laments and cries, but somewhere I see a ray of hope ... as always. Be with us and let us make this ray wider and wider till the whole world shines with peace and justice.[28]

A hope against hope perhaps but the rays of peace and justice have not entirely flat-lined. Rather like an ECG heart monitor, the beat goes on.

Notes

1. Arya, Divya. 'Soni Sori: India's Fearless Tribal Activist', *BBC News*, 22 March 2016. Available on https://www.bbc.co.uk/news/world-asia-india-35811608, accessed 20 January 2019.
2. See Gledhill, John. *Power and Its Disguises: Anthropological Perspectives on Politics*. Sterling, VA: Pluto, 2000.
3. Rabi, M. Abdul. 'After Spearheading Struggle against Kudankulam N-Plant, PMANE Now Fights Fission Within'. *The New Indian Express*, 10 February 2016. Available on http://www.newindianexpress.com/states/tamil_nadu/After-Spearheading-Struggle-Against-Kudankulam-N-Plant-PMANE-Now-Fights-Fission-Within/2016/02/10/article3269353.ece, accessed 12 October 2017.
4. See *Human Rights Watch. Prison Conditions in India*, 1991. Available on https://www.hrw.org/sites/default/files/reports/INDIA914.pdf; Home Office. *Policy and Information Note India: Prison Conditions*, 2016. Available

on https://assets.publishing.service.gov.uk/government/uploads/system/ uploads/attachment_data/file/565771/CPIN-India-Prison-Conditions-v2-November-2016.pdf, accessed 15 February 2019.

5. Arockiaraj, J. 'Idinthakarai Womenfolk Take Turns to Work for AAP Candidates'. *The Times of India*, 25 March 2014. Available on http:// timesofindia.indiatimes.com/city/madurai/Idinthakarai-womenfolk-take-turns-to-work-for-AAP-candidates/articleshow/32628784.cms, accessed 12 October 2017. On an earlier history of video-conferencing, see Mazzarella, William. 'Internet X-Ray: E-Governance, Transparency, and the Politics of Immediation in India'. *Public Culture* 18, no. 3 (2006): 473–505, pp. 486–7.

6. 'Election Commission of India Model Code of Conduct for the Guidance of Political Parties and Candidates'. Available on https://eci.gov.in/mcc/, accessed 19 October 2019.

7. Venugopal, Vasudha. 'Electoral Politics Only Way to Revive the Agitation: Anti-Kudankulam Activist Udayakumar'. *The Economic Times*, 2 May 2016. Available on http://articles.economictimes.indiatimes.com/2016-05-02/news/72775693_1_kudankulam-nuclear-power-plant-sp-udayaku-mar-aam-aadmi-party, accessed 12 October 2017.

8. Venugopal, 'Electoral Politics Only Way to Revive the Agitation'.

9. Aruloli, M. 'SP Udayakumar quits Aam Aadmi Party', 19 October 2014. Available on http://www.deccanchronicle.com/141019/nation-current-affairs/article/sp-udayakumar-quits-aam-aadmi-party, accessed 12 October 2017.

10. Sikdar, Shubhomoy and P. Sudhakar. 'Udayakumar Stopped from Flying to Nepal', *The Hindu*, 17 September 2014. Available on http://www. thehindu.com/news/national/tamil-nadu/udayakumar-stopped-from-flying-to-nepal/article6418652.ece, accessed 12 October 2017.

11. 'Kundankulam activist S.P. Udayakumar Demands Passport from Sushma Swaraj on "Humanitarian Grounds"', 15 June 2015. Available on http://www.india.com/news/india/kundankulam-activist-s-p-udaya-kumar-demands-passport-from-sushma-swaraj-on-humanitarian-grounds-423172/, accessed 12 October 2017.

12. 'India Renews Talks on Building Nuclear Power Plants', 25 October 2016. Available on http://economictimes.indiatimes.com/industry/energy/ power/india-renews-talks-on-building-nuclear-power-plants-report/arti-cleshow/55051338.cms. On other nuclear power plant announcements by the BJP, see M.V. Ramana and Suvrat Raju who suggest that such declarations are addressed to the Nuclear Suppliers Group so as India too could become a member. Ramana, M.V. and Suvrat Raju. 'Old Plans, Ongoing Handouts, New Spin: Deciphering the Nuclear Construction

Announcement'. *Economic and Political Weekly* 52, no. 24 (2017). Available on http://www.epw.in/journal/2017/24/web-exclusives/old-plans-ongoing-handouts-new-spin.html, accessed 12 January 2018.

13. Doshi, Vidhi. 'The Lonely Struggle of India's Anti-Nuclear Pro-testers'. *The Guardian,* 6 June 2016. Available on http://www.theguardian.com/global-development/2016/jun/06/lonely-struggle-india-anti-nuclear-protesters-tamil-nadu-kudankulam-idinthakarai, accessed 12 October 2017.

14. See Galtung, Johan. 'Violence, Peace and Peace Research'. *Journal of Peace Research* 6, no. 3 (1969): 167–87.

15. Available on https://www.facebook.com/142885519131001/photos/a.38 0136035405947/380136042072613/?type=1&theater.

16. Abraham, Itty. 'The Violence of Postcolonial Spaces: Kudankulam'. In *Violence Studies,* edited by Kalpana Kannabiran. New Delhi: Oxford University Press, 2016, pp. 318–19. See Tarrow, Sydney G. 'Struggle, Politics, and Reform: Collective Action, Social Movements, and Cycles of Protest'. Occasional Paper No. 21 Centre for International Studies, Cornell University Western Societies Program, 1989. On how 'activists find themselves caught between the celebration-exhaustion, hope-despair, unity-splintering dynamics', see Osella, Caroline. 'Utopia Interrupted: Indian Sex/Gender Dissident Activism and the Everyday Search for a Life worth Living'. In *Urban Utopias,* edited by Tereza Kuldova and Matthew Varghese. London: Palgrave Macmillan, 2017, pp. 227–46, p. 207.

17. Hardt, Michael. 'Multiple Temporalities of the Movements'. *tripleC: Journal for a Global Sustainable Information Society* 15, no. 2 (2017): 390–2, p. 392.

18. 'Six Injured in Kudankulam Nuclear-plant Accident in Tamil Nadu'. *DNA,* 14 May 2014. Available on http://www.dnaindia.com/india/report-six-injured-in-kudankulam-nuclear-plant-accident-in-tamil-nadu-1988224, accessed 12 January 2018. Padmanabhan, V.T. 'Kudankulam Reactor Tripped 20 times, was Off -grid for 468 days, who is Accountable? *News Minute,* 12 February 2016. Available on https://www.thenewsminute.com/article/kudankulam-reactor-tripped-20-times-was-grid-468-days-who-accountable-38901, accessed 6 January 2020; Padmanabhan, V.T., R. Ramesh, V. Pugazhendi, Raminder Kaur, and Joseph Makolil. 'Report of the 14 May 2014 Accident at the Kudankulam Nuclear Power Plant in India'. *Countercurrents,* 2014. Available on https://www.countercurrents.org/kknp150614.pdf, accessed 12 January 2018.

19. 'PMANE's Statement Demanding an Immediate Stop to the Koodankulam NPP Expansion', (nd). Available on http://www.cndpindia.org/stop-nuclear-power-plant-expansion/, accessed 12 January 2018.

20. Holloway, John. *Change the World without Taking Power* https://platypus 1917.org/wp-content/uploads/readings/Holloway_Change_the_World. pdf; Padmanabhan, V.T. 'Kudankalam Reactor Tripped 20 Times Was Off Grid for 468 Days, Who Is Accountable?' *The New Minute*, 12 February 2016. Available on http://www.thenewsminute.com/article/ kudankulam-reactor-tripped-20-times-was-grid-468-days-who-account- able-38901, accessed 12 January 2018.

21. Kolappan, B. 'Kudankulam Issue on the Back Burner This Election'. *The Hindu*, 18 April 2016. Available on https://www.thehindu.com/news/ cities/chennai/kudankulam-issue-on-the-back-burner-this-election/ article8487832.ece, accessed 18 January 2019.

22. This included the campaign against Sterlite Copper to set up a polluting smelting unit in Thoothukudi. See T.T.B. Sivapriyan (2018) One Allowed, the other Party Nurtured Sterlite', *Deccan Herald*, 2 June HYPERLINK "http://www.deccanherald.com/national/sunday-spotlight/one-allowed- other-party-nurtured-sterlite-673067.html"www.deccanherald.com/ national/sunday-spotlight/one-allowed-other-party-nurtured-ster- lite-673067.html, accessed 1 November 2019.

23. Venugopal, 'Electoral Politics Only Way to Revive the Agitation'.

24. Mathew, Pheba. 'Say Hi to Tamil Nadu's Newest Political Party, and They Have One Main Agenda—Environment'. *The News Minute*, 18 January 2016. Available on https://www.thenewsminute.com/article/ say-hi-tamil-nadu%E2%80%99s-newest-political-party-and-they-have- one-main-agenda-environment-37850, accessed 12 January 2018.

25. Venugopal, 'Electoral Politics Only Way to Revive the Agitation'.

26. Abraham, 'The Violence of Postcolonial Spaces', p. 319.

27. Chaturvedi , Rakesh Mohan (2019) 'After Sardar Patel, BJP aims to Claim Gandhi', *Economic Times*, 10 October, https://economictimes.indiatimes. com/news/politics-and-nation/bjp-out-to-claim-mahatma-gandhi/ articleshow/71514592.cms?from=mdr, accessed 1 November 2019.

28. 'After the Mayhem', 11 September 2012, compiled by Anitha S. *Countercurrents*. Available on http://www.countercurrents.org/kebin- ston120912.htm, accessed 12 January 2018.

Epilogue

In the beginning is the scream. We scream.

When we write or when we read, it is easy to forget that the beginning is not the word, but the scream. Faced with the mutilation of human lives by capitalism, a scream of sadness, a scream of horror, a scream of anger, a scream of refusal: NO....

The reality that confronts us reaches into us. What we scream against is not just out there, it is also inside us. It seems to invade all of us, to become us. That is what makes our scream so anguished, so desperate. That too is what makes our scream seem so hopeless. At times it seems that our scream itself is the only fissure of hope. (John Holloway, 2002)[1]

One of the most astounding developments of the twentieth century was how a phenomenally expensive and devastating war-time technology became pacified for national use. As with several other technologies of destruction and surveillance, nuclear power has been domesticated—in this case, in the form of electricity legitimated through consumer interest, energy security and national development. It can be surprising, even shocking to learn of what the nuclear lobby have managed to get away with in terms of human, environmental, financial, and political costs, bulwarked by their special dispensation to act in the national interest.

Although several anti-nuclear social movements have configured world history since at least the 1960s, my main journey was vastly informed by movements such as the PMANE in south India from the 2000s. No matter what side of the political spectrum one may be in, once the external layers are peeled away, and even if one cannot get to a nucleus of truth, the half-truths on nuclear organizations and histories strip away like peeling paintwork. Whether it be India, Pakistan, Britain, USA, Russia, France, Japan—to name but a handful—the conduct of nuclear authorities are sadly all too familiar. Their operations in the name of national and/or energy security, and increasingly now in terms of mitigating global warming, means that vast sectors of the populace are kept onside. Nuclear nationalism becomes yet another baton to beat detractors into something that is contrary to the reality—they are twisted and paraded as a threat to national livelihoods and futures, when it is actually the nuclear state that presents a larger threat. This is the perilous irony of our times—an irony that evaporates when, during political crises, nationalism boils over to jingoism further castigating any critique as unpatriotic or traitorous.

A decade after I had met her, Savitri was having a personal crisis of her own. Six days after her birthday, like a belated bitter gift from an unknown sender, she was diagnosed with cancer. I had gotten close to her after my first meeting with her in a Kanyakumari District village, and the news was distressful all round. At the time of diagnosis, she had been married seven years and had two young children. The moment of declaration was both surreal and a slap in the face as she listened to the consultant relay the details of the tests that she had undergone. It was as if gravity itself had fallen away. On hearing the word cancer, a security siren had sounded. It was the silent scream of evacuation. She felt removed from her flesh, transported into a place where she peered onto her body as if it was an intimate stranger. In a narrative that has become quite a common one, she began to wonder how otherwise she looked, lived, and felt fit: ' ... how on earth could a cancer creep up on me?' It was not an acceptance of fate but one tainted with questions to do with what and why.

Another struggle began. A crash course in new life–death emotions, experiences rallied with knowledge about different kinds of cancer, biopsies, cell grades and stages, MRI and CT scans, and a smorgasbord of radio-, chemo, and hormonal therapy offerings. Her

body became a place of concerted forensics—scanned, probed, prod-
ded, pierced, marked, cut, bandaged, and scarred. Imaging in the
form of X-rays and radionuclide dyes were pumped around her body
to mark out the sites of consternation. Photon and electron beams
were targeted at designated zones on her body. The small amount of
radiation was not life-threatening but life-enhancing, she was told,
as she was directly subjected to radiation's schizophrenic killer-cure
identity. Invader or liberator, it was astounding how much special-
ism, yet at the same time, mysteries, were packed into a gremlin like
cancer.

So, while writing this book, it felt like I had reached another preci-
pice, this time through the body of someone familiar. Hovering at the
edge of a great unknown expanse and the prospect of falling over was
less about awe and more about fear and melancholy. I began to wonder whether she had stayed too long in places with high background
radiation in south India. Was it the nuclear-scape internally tattooing
her body? Or perhaps something she had imbibed or put in her body
over her life span? Environmental factors have been minimized or
denied as one factor in cancer causation despite the intensive industrial and military toxicity that surround us.[2] However, she chose to give
the view some credence for cancer-prone genetic and chronic lifestyle
factors mattered little to her personal history. After all, it has generally
been agreed among the medical community that many cancers are
caused by the environments in which we live, studies attributing as
much as up to 90–95 per cent to lifestyle behaviours and environmental factors.[3] These inevitably have an impact on our mitochondrial
and genetic make-ups. Although a part of downstream effects, it is
biomedical conditions that have received the most funding in terms
of pharma-corporate backed research and treatment options.

As we all know, there can be no definitive answer for the upstream
area of cancer causation. Cancer is one of the effects of a cocktail of
factors. The statements I heard in south India about radiation burdens such as the woman whose narrative opens the book returned
with a vengeance. However, its stochastic character meant that one
could have no causative proof other than a malignant body in pro-
lapses. Nobody knows, not even the constellation of experts Savitri
met on the helter-skelter slides of surgeons, oncologists, radiologists,
and their support staff.

When you are in the business of knowing, not knowing is a most disturbing if not humbling experience. Explanations become mere chimeras in the face of cancer. Comfort betrays you. Words fail you. Even the idea of a cure for cancer is a misnomer for there is no cure—only remission in five year personal milestones. Once invasive malignant cells are inside you, they snitch about like shadowy characters in a film noir, shadows that might stay still, move, grow, spread, or metastasize in other parts of the body. Who knows? Who next?

It was with respect to Savitri's latter-day diagnosis that I began to see my fieldwork experiences in a different light. Looking back at my stay in south India took on the quality of an existential excavation as I recalled the places Savitri and I met and visited together, and remembered those others who worried, contracted, suffered, or had known people who had died from cancer ... those people who chose to live in denial ... those who had cancer on their mind day and night as they looked on at the sight of the nuclear reactors on the beach ... those who tried to keep their children safe from harm with preventative measures ... those mothers in advanced stages of cancer who could not afford further treatment, and lived out their karma on worn-out hospital charpoys with their infants ... [4]

Criticality emphasizes not just the critical condition of a nuclear reactor, but also the changing responses of our emotional, physical, circumstantial and political disposition towards it: how the same situation may alter with the introduction of new entrants to the lives of social actors, how the corners of our visions fold in on themselves to take centre stage. Lochlann S. Jain elaborates on the creepiness of cancer: 'After it shows up one realizes that it must have been there for a while, growing, dispersing, scattering, sending out feelers and fragments'.[5] It unleashes 'prognostic time [that] demands that we adopt its viewpoint, one in which the conclusion haunts the story itself'.[6]

Suffice to say that Savitri was diagnosed with a 'good cancer': a verdict that was not about whether it would be terminal or not. Rather, that it was/is treatable. It was not a death sentence but a sentence on life. It was/is a 'sneaky cancer', she was told by the oncologist. Nevertheless, it behaves well and can, in the majority of cases, be managed. Good cancer was not then a judgement on the malignant cells, but indicating a set of relationships with medical experts and treatments that were not outside the realm of knowability that could white-cloak this enigma of malignance.

I began to reflect on the idea of a 'good cancer'—to have a potentially fatal condition that is nevertheless controllable. Cancer has swayed from terminal illness to chronic or light burden to a question of luck. It was/is the same that could be said for government mandates for nuclear reactors—a chronic burden that nevertheless was decreed as good, almost lucky, for the nation. They could be sneaky but they are manageable. They could be potentially fatal but not always ordained. They too are a good cancer, the emphasis being on the good.

In a very hefty sense, we all live in a potential state of prognosis in what have become carcinogenic cultures of the contemporary era, a silent epidemic. If as recent studies show, one in two will experience a cancer diagnosis over their lifetime, we must ask who is the winner and loser in this lottery.[7] The industrial-military toxicity that makes for cultures of cancer are all around us in virtually every country. No border can militate against air-, water-borne, and increasingly terrestrial toxins. Where the main regional differences lie are to do with historical legacy, global positionality and the servicing of health provisions.

The story of KKNPP cannot be seen in isolation but as part of larger trails of distributed carcinogens that amount to a global apartheid of toxicity. Products that are avoided or banned in western countries invariably end up having a new lease of life in countries to the global south where life is seen as cheap.[8] A telling example of the global political economy of toxicity is pesticide. Some are produced in industrialized countries for export even where they may be proscribed in the country of manufacture. Aggressive marketing, influencing policy makers with lucre and incentives, and contesting scientific evidence becomes part of this campaign. As cited in a report to the United Nations Human Rights Council: 'The burden of the negative effects of pesticides is felt by poor and vulnerable communities in countries that have less stringent enforcement mechanisms'.[9] Discriminatory camp mentalities with respect to bare life exceed the nation.

Yet, even if about 200,000 people die from pesticide toxicity every year, agrochemicals industries persistently deny the scale of the dangers of their pesticides. In a convoluted logic, health becomes determined by the *right to food* however it is produced subject to the dictates of agricultural capitalism, which then takes precedent over the *right to lives* through a healthy environment. While a 'disease of the developed world' associated with the industrialized north, carcinogenic cultures are showing an exponential rise in industrializing nations to the "south".[10]

Similarly, the right to electricity in modern lives is overshadowing the right to healthy lives and habitation for many others. Electricity lords over our environments. Electricity also hides a lot of atrocities. Apart from death conditions, inferior reactor technologies might be produced in industrialized countries where they fall short of fulfilling regulations, but then sold to complicit countries with lower safety mechanisms, procedures, and regulations who might bend laws to strike the desired deals. The environmental scientist V.T. Padmanabhan concludes with his team of researchers: 'The civilian nuclear establishment of Russia has decided, at the highest level to dumb the surplus inventories of obsolete equipment to Asian–African countries, where regulations are lax or non-existent'.[11] Despite accusations of corruption and the sale of 'shoddy equipment', the Russian Atomstroyexport has expanded its operations in Finland, Slovakia, Bulgaria, Turkey, Iran, Morocco, Vietnam, China, and for the first time in a trilateral pact with Indian nuclear subcontractors operating abroad, Bangladesh.[12] With discounts and offers to take back the nuclear waste, Atomstroyexport's bullish approach surpasses other trans/multinational contenders such as EDF, Toshiba, or General Electric, a task that is now pledged to be taken over in the subcontinent by multinationals such as Reliance Infrastructure Limited, themselves donors to the BJP administration in central government from 2014.

How has this state of affairs in the subcontinent emerged? In a line: through the dictates of development, democracy and nationalism. Development is the mantra waved with abandon to make up for lost time under colonialism and compete in a neo-liberal world. Dubious practices hide behind and within the charming masks of democracy. Nationalism wraps everything up and gifts it to the people in shiny packaging. Yet, due to a civic space that marginalizes and discriminates, this does not bode well for minorities nor those seeking a fair debate on burning issues in a space that is clamped down by the stranglehold of national and energy security. The practice of sedition charges against any detractor comes with a sweeping turn to the right. Aggressive nationalism has seen an upsurge that has affected not just India, but much of the world. It is deeply fed into the very veins of body politics. Although only about a century old in its modern inception, hardly any politician or decision-maker would care to question the legitimacy of the nation-state for fear of losing favour. It has now become a Möbius strip in which citizens also partake, seeing a nationality and national culture as naturalized as one's race, ethnicity, gender, or sexuality. Indeed, democracy too has colonized our thinking as Ferit Güven contends: 'a conceptual, theoretical anchorage that determines the limits of what is thinkable'.[13] As a post-Enlightenment call for sovereignty and suffrage, we are unable to think through any other kinds of political possibilities other than its continual rebooting.

'The timeless and timely narratives upon which expressive language rests, narratives so ingrained and pervasive they seem inextricable from "reality", require identification', writes the author Toni Morrisson.[14] If people could look beyond the discursive effects of educational, cultural, and political initiatives, then perhaps they can begin to see how some of the biggest lies of the twentieth century— nuclear electricity 'too cheap to meter' and 'nuclear deterrence' against 'enemy' nations—sit in hollow yet treacherous thrones of truth in the twenty-first. If this was possible, and if nuclear-nationalisms could be unravelled, then we might see the emperor's clothes for what they actually are—a noxious rhetoric become real, based on little else but hot air. This is a somewhat belittling metaphor: if only it was but hot air. An ecological crisis has emerged out of an egological surplus driven by neo-liberal agendas. Achille Mbembe adds a cautionary note, 'The world will not survive unless humanity devotes itself to the task of sustaining what can be called the reservoirs of life'.[15]

With the intensifying collaborations of public and private enterprise, we have reached what the philosopher John McMurtry has described as the 'cancer-stage of capitalism' with its globalizing and mutating market paradigms.[16] This stage has informed a governing rationality of our times that has multiplied into several carcinogenic eruptions and metastases that may even be decoupled from commercial logics such that alternative visions and autonomous lives become hard to realize. Who could live without citizenship, governments, and national borders, for instance? Who could live without electricity and the comforts it provides? Many residents in the embattled vicinity of Kudankulam, however, could see through the veiled rationalities, mutations, and offshoots. Through being subject to the construction of an Indo–Russian nuclear power plant near their homes, the economic, the socio-political, and the physical were all too worryingly entwined—it could quite literally manifest itself in somatic pathologies in all too critical carcinogenic conditions, a silent emergency that comes with a silent epidemic.

Anthony explained that despite the fact that people are being duped, displaced, and marginalized by the nuclear state without due protocol, consultation, or compensation, meeting violence with violence will only mean that nuclear capitalism will prosper. Indeed, he preferred to see it as a case of nuclear colonialism. As with cancer, he could not immediately see an antidote or cure, only remission. With the switch of a light bulb, he added: 'The last chapter will be written by the person with real civilisation.'

What is this real civilization? Can it be dislocated from the scream that both surrounds us and seeps into our bodies? Can this scream be about a return to life conditions rather than merely a response to death conditions? Can it be about development with a genuinely pro-people rather than a destructive pro-profit agenda? This book cannot provide the answers, only to note that they are being worked out.

Mohandas Karamchand Gandhi had once written: 'I have come to the conclusion that immorality is often taught in the name of morality.... Civilisation seeks to increase bodily comforts, and it fails miserably even in doing so'.[17] While Jawaharlal Nehru could not contend with what he decreed as Gandhi's anti-modern ways, we may have paused to reflect on whether the legacy of Nehruvian high-handed policies combined with latter-day neo-liberal adaptations have created more ease or more misery. A visionary more than he was anti-science, there is a real need to recuperate Gandhi's fearless message from retrogressive, consumerist, statist and Hindu nationalist appropriations. We might remind ourselves of the freedom fighter's apocryphal response to the reporter's question when visiting Britain: 'What do you think of Western civilisation?'[17] Western or Eastern, Northern or Southern, British or Indian, the civil has been wrested out of the term. A century on, his answer still stands: 'I think it would be a good idea'.

Notes

1. Holloway, John. *Change the World without Taking Power*, 2002. Available on http://libcom.org/files/John%20Holloway-%20Change%20the%20world%20without%20taking%20power.pdf, p. 108, accessed 24 January 2019.
2. See Jain, Lochlann S. *Malignant: How Cancer Makes Us*. Berkeley: University of California Press, 2013.
3. As noted in Chapter 7, statistics here again are fiercely debated, differences largely dependent on factors to do with the method, scope, demographic and region. It is concluded that cancer is 90–95 per cent due to lifestyle and environment according to Anand, Preetha, Ajaikumar B. Kunnumakara, Chitra Sundaram, Kuzhuvelil B. Harikumar, Sheeja T. Tharakan, Oiki S. Lai, Bokyung Sung, and Bharat B. Aggarwal. 'Cancer Is a Preventable Disease That Requires Major Lifestyle Changes'. *Pharmaceutical Research* 25, no. 9 (2008): 2097–116. The figure is 70–90 per cent according to Whiteman, Honor. 'Most Cancer cases "Caused by Lifestyle, Environment—Not Bad Luck', 17 December 2015. Available on https://www.medicalnewstoday.com/articles/304230.php, accessed 24 January 2019. Other studies put lifestyle and the environment at a lower figure, for instance 42.7 per cent in Parkin, D.M., L. Boyd, and L.C. Walker. 'The Fraction of Cancer Attributable to Lifestyle and Environmental Factors in the UK in 2010'. *British Journal of Cancer* 105, no. Supp. 2 (2011): S77–S81.
4. See Jackson, Michael. *Existential Anthropology: Events, Exigencies and Effects, Methodology and History in Anthropology*. Oxford: Berghahn Books, 2005.
5. Jain, *Malignant*, p. 80.
6. Jain, *Malignant*, p. 40.
7. Cancer Research UK (2015) 'One in Two People in the UK will get Cancer, Experts Forecast', *Science Daily*, 3 February, https://www.sciencedaily.com/releases/2015/02/150203204348.htm, accessed 26 October 2019.
8. Gabrielle Hecht notes: 'Contaminants may be recognized and regulated in one place but not another; many industries distribute their hazards across international borders for precisely that reason.' Hecht, Gabrielle. *Being Nuclear: Africans and the Global Nuclear Trade*. Massachusetts: MIT Press, 2012, p. 42. See also Jean Comaroff on treatment for AIDS in the global south: 'the most devastating burden of suffering has shifted to parts of the world...where misery is endemic, life is cheap, and people are disposable'. Comaroff, Jean. 'Beyond Bare Life: AIDS and Biopolitics'. In *Theory from the South: Or, How Euro-America Is Evolving Toward Africa*, edited by Jean Comaroff and John L. Comaroff. Boulder: Paradigm Publishers, 2012, p. 201.

9. 'Pesticide Use "Threatens Human Rights", UN Advisers Claim'. *Chemistry World* 14 (2017): 11.

10. Pandey, Kiran and Rajit Sengupta. '9.6 million People Will Die of Cancer This Year'. *Down to Earth*, 17 September 018. Available on https://www.downtoearth.org.in/news/health/9-6-million-people-will-die-of-cancer-this-year-61646, accessed 24 January 2019.

11. Padmanabhan, V.T., R. Ramesh, V. Pugazhendi, K. Sahadevan, Raminder Kaur, Christopher Busby, M. Sabir, and Joseph J. Makkolil. 'Counterfeit/ Obsolete Equipment and Nuclear Safety Issues of VVER-1000 Reactors at Kudankulam, India', 2013, p. 11. Available on http://vixra.org/abs/1306.0062, accessed 24 January 2019.

12. See Digges, Charles. 'Rosatom-owned Company Accused of Selling Shoddy Equipment to Reactors at Home and Abroad, Pocketing Profits'. *Bellona*, 27 February 2012. Available on http://bellona.org/news/nuclear-issues/nuclear-russia/2012-02-rosatom-owned-company-accused-of-selling-shoddy-equipment-to-reactors-at-home-and-abroad-pocketing-profits; Roy Chaudhury, Dipanjan. 'India will Soon Get to Bid for Bangladesh Nuclear Power Plant Work'. *Economic Times*, 11 July 2018. Available on http://www.reuters.com/article/russia-nuclear-rosatom-idUSL5N0F90YK20130722; Available on https://www.moneycontrol.com/news/world/india-to-bid-for-bangladeshs-rooppur-nuclear-power-project-work-2696971.html, accessed 24 January 2019.

13. Güven, Ferit. *Decolonizing Democracy: Intersections of Philosophy and Postcolonial Theory*. London: Lexington Books, 2015.

14. Morrison, Toni. 'Introduction: Friday on the Potomac'. In *Race-ing Justice, En-Gendering Power: Essays on Anita Hill, Clarence Thomas, and the Construction of Social Reality*, edited by Toni Morrison. New York: Pantheon Books, 1992, p. xi.

15. Mbembe, Achille. 'There Is Only One World'. *The Con*, 10 July 2017. Available on http://www.theconmag.co.za/2017/07/19/there-is-only-one-world/, accessed 24 January 2019. Mitigating climate change has become another legitimating device for nuclear power. This is without paying heed to carbon emissions during the plant's construction, exorbitant costs when compared to the more immediate impact of renewables, nor the lingering question of radioactive waste. See Chapter 3.

16. McMurtry, John. *The Cancer Stage of Capitalism*. London: Pluto, 1998.

17. Gandhi, Mohandas. *Gandhi: 'Hind Swaraj' and Other Writings*, edited by Anthony J. Parel. Cambridge: Cambridge University Press, 1997, p. 37.

18. Prasad, Shambu C. 'Towards an Understanding of Gandhi's Views on Science'. *Economic and Political Weekly, Review of Science Studies* 36, no. 39 (2001): 3721–32.

Index

About the Author

Raminder Kaur is professor of anthropology and cultural studies in the School of Global Studies at the University of Sussex. Her other books include *Atomic Mumbai: Living with the Radiance of a Thousand Suns* (2013) and *Performative Politics and the Cultures of Hinduism* (2003). She is co-author of *Adventure Comics and Youth Cultures in India* (with Saif Eqbal, 2019), *Diaspora and Hybridity* (with Virinder Kalra and John Hutnyk, 2005); and co-editor of five books covering topics related with race, ethnicity, travel, migration, transnationalism, cinema, censorship, and arts and aesthetics. She is also a scriptwriter and has co-produced theatre and films (www.sohayavisions.com).